1995

D0893987

523.01   Wali, K.C.
C456w   Chandra.   CENTRAL

CHANDRA

CHANDRA

A Biography of S. Chandrasekhar

Kameshwar C. Wali

A Centennial Publication of
The University of Chicago Press

The University of Chicago Press
Chicago and London

Kameshwar C. Wali is professor of physics at Syracuse University, and specializes in high energy theory. Born in India, he has been a physicist in the United States since 1955 and first met S. Chandrasekhar in 1960.

NOV 4 1991

3 1116 01084 8023

523.01
C456w

Frontispiece: Photograph of S. Chandrasekhar by K. G. Komsekhar, 1984. Courtesy of K. G. Komsekhar.

The University of Chicago Press, Chicago 60637
The University of Chicago Press, Ltd., London
© 1991 by The University of Chicago
All rights reserved. Published 1991
Printed in the United States of America

00 99 98 97 96 95 94 93 92 91   5 4 3 2 1

Library of Congress Cataloging in Publication Data

Wali, K. C. (Kameshwar C.)
    Chandra : a biography of S. Chandrasekhar / by
Kameshwar C. Wali.
        p.   cm.
    1. Chandrasekhar, S. (Subrahmanyan), 1910–   .
    2. Astrophysicists—United States—Biography.   I. Title.
QB36.C46W35   1991
523.01′092—dc20   90-10845
[B]   .CIP
ISBN 0-226-87054-5 (alk. paper)
ISBN 0-226-87055-3 (pbk.; alk. paper)

∞ The paper used in this publication meets the minimum requirements of the American National Standard for Information Sciences—Permanence of Paper for Printed Library Materials, ANSI Z39.48-1984.

To My Parents

Contents

Illustrations   ix

Prologue: Tracking the Legend of Chandrasekhar   3

1 The Simple and True   13

2 Choosing the Unconventional: A Family Trait   34

3 Determined To Pursue Science   47
   Lahore and Madras, 1910–1930

4 Discoveries, Personal and Scientific   72
   Cambridge and Copenhagen, 1930–1933

5 Fellow of Trinity College   105
   Cambridge, 1933–1934

6 The Absurd Behavior of Stars: Eddington and
   the White Dwarfs   128
   Cambridge, 1934–1935

7 "I Must Push On in My Directions"   147
   Cambridge and Harvard, 1935–1936

8 Lalitha   168
   Madras, 1936

9 Scientist in the Midst of Political Turmoil   185
   Williams Bay, Wisconsin, 1937–1952

10 The Autocrat of the Editor's Desk: The *Astrophysical Journal*   206
    Chicago, 1952–1971

11 In the Lonely Byways of Science   229
    Chicago, 1972–1989

Epilogue: Conversations with Chandra   245

Notes   309

Appendix   327

Acknowledgments   329

Index   333

Illustrations

Subrahmanyan Chandrasekhar, 1984.   Frontispiece
1. *Man on the Ladder*.   Facing page 3

Following page 54
2. Ramanathan Chandrasekhar (1866–1910), Chandra's grandfather.
3. Parvati Chandrasekhar (1869–1916), Chandra's grandmother.
4. Sitalakshmi Ayyar (1891–1931), Chandra's mother.
5. C. Subrahmanyan Ayyar (1885–1960), Chandra's father.
6. Sir C. V. Raman (1888–1970), 1930 Nobel laureate in physics,
   Chandra's uncle.
7. Chandra with his older sisters and younger brother, Lahore, 1913.
8. Chandra at age six, 1916.
9. Chandra Vilas, Madras.
10. Chandra after receiving the Ph.D. degree, Trinity College,
    Cambridge, 1933.
11. Meetings in Russia, 1934.
12. Chandra as Trinity Fellow, Cambridge, 1934.
13. Sir Arthur Stanley Eddington, 1932.
14. Chandra with William H. McCrea, 1935.
15. Conference on White Dwarfs and Supernovae, Paris, 1939.

Following page 182
16. Chandra and Lalitha, Madras, 1936.
17. Farewell from family and friends, Madras railway station, 1936.
18. Meeting of physicists in Washington, D.C., 1938.
19. Chandra and Lalitha at the dedication of McDonald Observatory,
    Mount Locke, Texas, 1939.
20. Henry Norris Russell and Chandra, Chicago, 1941.
21. Chandra with George Greenstein, Williams Bay, c. 1945.
22. Chandra with Michael Lebovitz, Chicago, 1981.
23. Chandra with his brothers and sisters, Madras, 1961.

24. Lalitha, 1956.
25. Paul A. M. Dirac and Chandra, Williams Bay, 1958.
26. Chandra with Igor D. Novikov and Ya. B. Zel'dovich, Warsaw, 1962.
27. Royal Medal recipients, 1962.
28. Chandra receiving the National Medal of Science, 1967.
29. Prime Minister Indira Gandhi and Chandra, New Delhi, 1968.
30. Chandra receiving the Nobel Prize in physics, 1983.
31. Copley Medal winners, 1984.

CHANDRA

Fig. 1. *An Individual's View of the Individual (Man on the Ladder)*, by Piero Borello. Courtesy of the artist.

Prologue: Tracking the Legend of Chandrasekhar

As I stepped into Subrahmanyan Chandrasekhar's office for the first time, I was immediately intrigued by a photograph that faced Chandrasekhar from the wall opposite his desk. As I stood looking at it, Chandra[1] told me the following story: He had first seen the photograph in 1962 on the cover of a *New York Times Magazine* and had written to the *Times* for a copy. He was referred to the artist, Piero Borello. Chandra was somewhat surprised by Borello's reply; Borello said that certainly Chandra could have a copy, even the original, but only if he cared to explain why he wanted it. Chandra responded,

What impressed me about your picture was the extremely striking manner in which you visually portray one's inner feeling towards one's efforts at accomplishments: one is half-way up the ladder, but the few glimmerings of structure which one sees and to which one aspires are totally inaccessible, even if one were to climb to the top of the ladder. The realization of the absolute impossibility of achieving one's goals is only enhanced by the shadow giving one an even lowlier feeling of one's position.

Chandra's explanation could not have expressed more vividly what the picture portrays visually—the essence of the eternal human quest after knowledge; its triumphs, its frustrations, and its limitations, as well as one's own personal awe before the unknown.

For over five decades, Chandra has pursued research in physics, astrophysics, and mathematics, which has resulted in an immense increase in our understanding of the constitution and evolution of stars and of numerous other aspects of the physical universe. His work has brought him a measure of recognition attained by only a few. Yet, Chandra seems to embody the symbolic man in the Borello photograph, constantly aware of how much more there is to know. He remains unaffected by the outward symbols of success. The prizes, medals, awards, honorary degrees, and memberships in academic societies seem instead

3

to leave him with a vague sense of discomfort and an added burden of responsibility.

This image of a man climbing a ladder was to reappear many times during subsequent years in my discussions with other scientists, with his friends and relatives, and with Chandra himself. But when I began to work on this biography, my image of Chandra coincided with the general view among scientists: he has become a legendary figure—a preeminent astrophysicist, a great applied mathematician, and an extraordinary scholar. Chandra's monographs on stellar structure, stellar dynamics, radiative transfer, and ellipsoidal figures of equilibrium have become classics. His prodigious and diverse research, his phenomenal memory, his immaculate lecture style, and his astounding mathematical skills are all part of the Chandra legend. His discovery in the early 1930s of the upper limit (now called the Chandrasekhar limit) on the mass of a star that could evolve into a white dwarf has been hailed as one of the most important discoveries of this century. Altogether he presents a towering figure in science.

I first became familiar with the name Chandrasekhar in India during the early fifties, when I was studying for the M.Sc. degree in physics at Banaras Hindu University. This lovely campus in the historic city of Banaras on the banks of the river Ganges was then one of the few national universities, attracting students from all over India and a few from abroad. It was my first real exposure to the nation as a whole, since my undergraduate training had taken place in a small, regional community college. My special subject in physics was experimental spectroscopy, but I had no intention of continuing in that area beyond a master's degree. I was inclined toward theoretical physics, and I was intent on studying more mathematics as a preliminary.

It was there that I met V. L. N. Sarma,[2] a few years my senior, who was a doctoral research student in mathematics. His unassuming manner and his wide-ranging knowledge and interest in mathematics and physics drew us together; he became my mentor and close friend. He first told me about S. Chandrasekhar, the eminent Indian astrophysicist, whom he described as an "incredibly capable" applied mathematician whose contributions were far more outstanding than those of other Indian physicists whose names I knew: Meghnad Saha, S. N. Bose, and even C. V. Raman, who had won a Nobel Prize. In fact, in Sarma's opinion, Chandrasekhar would have been a Nobel laureate had he been in physics instead of astronomy and astrophysics, which, at that time, were not among the fields of study eligible for the prize. As it was, Chandrasekhar was a Fellow of the Royal Society (F.R.S.), and had won other equivalent high honors in his field.

Sarma and I did not discuss at length Chandra's scientific work or any details of his life. We knew, however, that Chandra had settled permanently in the United States. And we were well aware of the perception in India that for Indians to leave their motherland and settle elsewhere was a disrespectful act akin to treason; for devout Hindus it was a mortal sin. However, we felt Chandra was justified in remaining abroad for two reasons. First, the atmosphere in India was not conducive to continued, persistent scientific research. People who returned from abroad with foreign degrees and research accomplishments were soon engulfed either in the bureaucratic quagmire or in petty squabbles and personal rivalries. We were experiencing the stagnant atmosphere at Banaras Hindu University, which was no different from the general situation in the other universities around the country. A contentious faculty, constant disputes and legal battles, regional favoritism, and barriers put across the path of those individuals who were keen on continuing research all served to create an atmosphere of lethargy with little stimulation for creative work.

Second, Sarma and I, along with many others, absolved Chandra's permanent move abroad because we perceived it as a supreme sacrifice. We thought of his self-imposed exile as a patriotic act in the midst of India's struggle for independence. He had accepted exile to devote himself to science and thus to bring recognition and glory to India, demonstrating to the colonizers that Indians, given suitable opportunities, could excel in science, which was still the exclusive domain of the West. We considered his sacrifice equal to that of such stalwarts as Gandhi and Nehru, who had achieved distinction abroad, returned to India, and then abandoned lucrative careers to participate in the political struggle.

It was this legend of a glorious scientist-in-exile that I carried with me when I came to the United States to study theoretical physics at the University of Wisconsin in 1955. Soon friends (and often strangers on first introduction) asked me if I knew Chandra. We came from the same country, didn't we? Stories about Chandra and Chandra's own stories about other famous scientists he had met during his years in England were told and retold. My thesis advisor, Robert G. Sachs,[3] had known him well during the Second World War when they shared an office at the Aberdeen Ballistic Tests Proving Grounds in Maryland. Sachs had great respect for Chandra. "He has tremendous mathematical power," Sachs told me once. "He has a marvelous presence and is a masterful storyteller: he can repeat verbatim conversations he has had or heard twenty or so years ago."

I also came to know Carl Sagan,[4] then a graduate student in the Department of Astronomy of the University of Chicago, located at Yerkes Observatory, Williams Bay, Wisconsin. Sagan too had high regard for Chandra as an astrophysicist, but he said that as a person Chandra was extremely private and someone whom everyone held in great awe. Carl (and later some others in my interviews) described the two stairways which led from the basement of the observatory (where the students' offices were located) to the upper floors. One stairway led past Chandra's office; the other was on the opposite side. Students avoided the stairway past Chandra's office when he was known to be there. Sagan also told me the following story which made a lasting impression on me:

Chandra was giving a colloquium. Three walls of the lecture room had blackboards on them, all spotlessly clean when Chandra began his lecture. During the course of his lecture, he filled all the blackboards with equations, neatly written in his fine hand, the important ones boxed and numbered as though they had been written in a paper for publication. As his lecture came to an end, Chandra leaned against a table, facing the audience. When the chairman invited questions, someone in the audience said, "Professor Chandrasekhar, on blackboard . . . , let's see, . . . 8, line 11, I believe you've made an error in sign." Chandra was absolutely impassive, without comment, and did not even turn around to look at the equation in question. After a few moments of embarrassing silence, the chairman said, "Professor Chandrasekhar, do you have an answer to this question?" Chandra responded, "It was not a question; it was a statement, and it is mistaken," without turning around.

Stories like these, while intensifying both the aura around him and my desire to meet him, also erected a barrier against my ever coming to know him well. Moreover, his reputation for privacy and intimidation aside, I thought I stood even less of a chance as a fellow countryman because of my fear of being misunderstood. An experience I had soon after I came to the States made me particularly sensitive to this point. I had tried in vain to meet an Indian writer whom I greatly admired. On an irresistible impulse, I telephoned him. I would have accepted his refusal to see me with some disappointment, but what hurt me was his immediate question, "What can I do for you?" When I answered, "Nothing, I just wanted to see you," he refused to believe me. He kept on asking the same condescending question. I was afraid of similar embarrassment if I approached Chandra without good reason.

I finally met Chandra in 1960 when he came to Madison to give a colloquium on "The Equilibrium of Rotating Fluids." Dressed elegantly in a charcoal gray suit, white shirt, and a striped tie, he cut a handsome figure, though he appeared at the same time intense and stern. He delivered the lecture as if it were a theatrical piece; no hesitations, no errors, no pauses. It was a marvel to watch him explain the solutions of complex equations and their correspondence with the behavior of macroscopic rotating fluids. I was thrilled, but also nervous and uneasy, and so careful as to be awkward in what I said and did in his presence. I was thoroughly intimidated. The recollection of a minor incident shows how absurdly self-conscious I was.

After the colloquium, Sachs had invited my wife and me to an informal supper with Chandra. It was an early supper because Chandra had to return to Williams Bay that evening. Since he was a vegetarian, Jean Sachs had prepared cheese blintzes with strawberries and sour cream for the main course, which Chandra seemed to appreciate greatly. As the supper progressed, rather too quietly, I could only notice how elegantly Chandra ate; he cut the blintzes into pieces of just the right size without letting the cheese or the strawberries spread onto his plate. He left his plate so clean! And throughout supper, I felt he was watching how I was managing. I was certainly clumsy. When I poured the milk into my coffee cup, I let the milk drops run down the exterior contour of the pitcher. Consequently, when I put the pitcher down, milk soaked into the tablecloth. Immediately afterwards, to my great consternation, Chandra poured milk into his coffee cup, holding his spoon expertly under the spout to carefully collect the drops. Chandra, I was sure, was demonstrating manners to me. I felt mortified—no, crushed; the chance of getting to know him at all seemed to recede farther out of reach.

We later met briefly at some six or seven scientific meetings. Chandra showed no sign of having met me before (nor did I remind him). Each time was the first time. For five years (1964–69) we lived within one and a half blocks of each other in Hyde Park, the neighborhood surrounding the University of Chicago. Almost every Sunday our routines for purchasing the *New York Times* from the Medici coffeehouse and bookstore on 57th Street made us walk the same paths. Like the students at Yerkes, I used to be extremely careful to avoid him. If I sighted him from a distance, I quickly crossed to the other side of the street. During all this time I could never muster enough courage to take the necessary step—whatever that may have been—to get to know him.

The turning point came after I joined the Department of Physics at Syracuse University in 1969. In 1971, at the height of the Cambodian crisis, Chandra was invited to Syracuse to give a seminar, which had to be canceled because of a general student strike that closed down the university. Chandra's activities thus were limited to informal discussions with the Relativity Group in our department. My wife and I, however, had the pleasure and responsibility of entertaining him for dinner along with some other guests. In contrast with my first dinner encounter with Chandra, this was a marvelously pleasant evening. Chandra was in great form. He entertained us all with anecdotes and reminiscences. Such scientists as Lord Rutherford, Sir Arthur Eddington, John von Neumann, Paul Dirac, Enrico Fermi—legendary figures for our generation—came alive as Chandra narrated in vivid detail incidents, stories, and conversations he had had or had overheard. Rutherford and Eddington dominated the conversation at our dining table that evening as they had dominated the Cambridge scene in the early thirties—the Cavendish Laboratory meetings and seminars, dining at high table at Trinity College, and the after-dinner conversations. Chandra described for us, as I recall, the occasion of a Physics Society meeting in the Cavendish Laboratory in 1933, just two days after Dirac had won the Nobel Prize. When Dirac walked in, a few minutes late, there was naturally a large burst of applause which did not subside. When it had gone on for quite a while, Rutherford, who was in the chair, said, "That's enough. This is not the first time the Nobel Prize is coming around this way." Chandra also recalled how Rutherford had introduced young Werner Heisenberg when he visited Cambridge to give the Scott Lectures that same year. "Heisenberg was telling me," Rutherford said to the audience, "that alpha particles are important. . . . They could be constituents of heavier nuclei. I told him, 'young man, don't forget. I discovered them long before you were in your long pants.'"

Chandra also told us how Eddington became the leader of one of the two expeditions to observe the total solar eclipse of 19 May 1919 in order to measure the deflection that light suffered as it passed through the sun's gravitational field. Ultimately this would verify one of the predictions of Einstein's general theory of relativity. Eddington was chosen, not because, as was commonly believed, he was an ardent exponent of Einstein's theory and, at the time, one of the very few who understood it, but rather because he was a Quaker and a pacifist! His Cambridge colleagues did not like the idea of one of their distinguished company claiming deferment from the general conscription as a conscientious objector, so they gave him reason for a special deferment by appointing him as the leader of the expedition.[5]

That evening has stood out as something very special in my memory.

It was then that I realized how interesting it would be to know more about Chandra's life. Born in India, educated in England, he had lived all his professional life in the United States. He grew up and received his early education in India during an extremely important period of Indian history—1910–1930—a time when India was struggling for independence from the British. At the same time, British education was becoming internationally more influential in several fields, particularly science. While Mahatma Gandhi, Jawaharlal Nehru, Sardar Vallabhabhai Patel, Sarojini Naidu, and other political figures became household names, there were also men like Srinivasa Ramanujan, Jagadish Chandra Bose, Meghnad Saha, C. V. Raman, S. N. Bose, and Rabindranath Tagore who, through their accomplishments in science and art, had captured the imagination of young, college-educated Indians. Their recognition by the British and the West (Ramanujan, the first Indian to be elected to the Royal Society; Tagore and Raman, Nobel laureates) was a source of pride and inspiration for millions of Indians who struggled in poverty; Indians who had dreams but no means of transforming them into reality.

Chandra's years at Cambridge (1930–36) were years when great discoveries occurred in physics. They were also years of continuing British domination in science. When he arrived in the United States, American science was on the rise. Conditions in Europe and the impending war generated a large influx of brilliant scientist-refugees. Chandra's life is particularly unusual because in the 1930s it was extremely rare to find someone from India permanently settled in the States. He arrived with his new bride, Lalitha, in the heart of the Midwest—in the small town of Williams Bay, Wisconsin. What was life like for a young couple so far away from home during the war years when communication with India was so difficult? Sachs had told me a little about the humiliations Chandra faced at the Aberdeen Proving Grounds because of the color of his skin. How did Chandra face such problems and go on with his work? The many facets of Chandra's life—beyond his scientific accomplishments—would make an account of his life invaluable.

In 1975 I received the shocking news that Chandra had suffered a heart attack. I was afraid that so much would be lost if no one documented his exceptional life. Two more years passed before anything transpired, except my talking about it with interested friends and reading a few of Chandra's popular articles. The urge to know him better became stronger, but so did the fear of approaching him. I assumed, of course, that any attempt to probe into his personal life would be met with total dismissal. But Abhay Ashtekar,[6] who knew him better than

I did, thought otherwise. With his urging and encouragement, I approached Chandra during the General Relativity Conference in Waterloo, Ontario, in the summer of 1977, with the idea of writing one or two articles based on conversations with him. I was particularly interested in the controversy between Chandra and Eddington regarding the celebrated discovery of the Chandrasekhar limit, which has become a standard reference in the lexicon of physicists. Chandra's brief account in his Richtmyer Lecture[7] had aroused my curiosity, and I suspected that there was a great deal of drama behind the confrontation. Yet very few in the scientific community seemed to know the full story; indeed very few were even aware that the Chandrasekhar limit was bitterly contested by a preeminent astronomer.

To my surprise, Chandra was willing to talk with me, but he warned me that I was probably wasting my time. I began taping the first of many conversations with him on 17 December 1977. These conversations led me to discussions with some of his former students and associates, with other scientists, and with some individuals who knew him from other walks of life. Also, Chandra made available to me his extensive files of correspondence. With this surfeit of material, the original idea of writing one or two articles gave way to the idea of writing this book.

Chandra often tells his life story as follows:

I left India and went to England in 1930. I returned to India in 1936 and married a girl who had been waiting for six years, came to Chicago, and lived happily thereafter.

True. But the Chandra we know is the product of the complexities of three widely different countries: India, the land of his birth with its ancient culture and traditions, which undoubtedly influenced his early childhood and youth; England, the land of the colonial masters, where his scientific research mushroomed and matured; and finally America, his adopted homeland, where he continued his research and developed as a scholar, lecturer, and writer. This book is an attempt to understand the influences, both personal and historical, which affected his life; to trace the development of his personality and his scientific working style; and to explore that interstitial realm where scientific endeavor meets human subjectivity.

Lalitha, a student of physics herself, eventually gave up her own work in science to become an integral part of Chandra's life. In dedicating "A Scientific Autobiography"[8] to her, he said:

The full measure [of my indebtedness] cannot really be recorded: it is too deep and too all pervasive. Let me then record simply that Lalitha has been the motivating source and strength of my life. Her support has been constant, unwavering, and sustained. And it has been my mainstay during times of stress and discouragement. She has shared my life: selfless, devoted, and ever patient and waiting. And so I dedicate this autobiography, which is indeed my life, to her.

This book is not intended to be an appraisal of Chandra's scientific work, nor is it a scientific biography.[9] It is a biography of an individual whom I admired from a distance for many years. My associations with him since 1977 have only intensified my admiration for his accomplishments and unique personality. Therefore, the authenticity of my words here should be judged from a less routinely polemic point of view. Yet, fearing that the book would be perceived as merely laudatory, more like a memorial to a living person than a biography, I asked the people I interviewed to tell me something about Chandra's weaknesses or flaws. The typical response was like that of Victor Weisskopf, who said, "None. None. There is nothing to criticize. You can say of many people, either they are arrogant, or they are not nice to their colleagues. In science, you can say they are too superficial. Nothing like that is true for Chandra. He is the most positive human being I have met in my life as a scientist. He has that untouchable integrity which is so impressive."[10]

I am reminded of Peter Kapitsa and the episode which he narrated in his address to the Royal Society on 17 May 1966.[11] Kapitsa attended a small conference in 1930 in Cambridge to commemorate the centenary of the birth of James Clerk Maxwell, the first director of the Cavendish Laboratory. After the meeting, when Rutherford asked him how he liked the speeches of some of Maxwell's pupils, Kapitsa said that they were very interesting, but that all the speakers spoke only of the positive side of Maxwell's work and personality. He would like to have heard Maxwell presented as a living figure with all his human traits and faults which, of course, every man possesses, however great his genius. Rutherford apparently laughed and jokingly charged Kapitsa with the task of telling the future generations what he, Rutherford, was really like. Kapitsa went on to say:

And now when I try to fulfill his behest and I imagine Rutherford as I have to present him before you, I see that time has absorbed all his minor human imperfections and I can only see a great man with an astounding brain and great human qualities. How well I now understand Maxwell's pupils who spoke about him in Cambridge.

Many times, from 17 December 1977 until now, Chandra and I have conversed, sometimes for hours at a time. The man on the ladder still faces him in his office. There is also an original cartoon of Einstein by Vicky, the famous cartoonist for the *New Statesman*,[12] a picture of young Newton in a bathrobe, as well as photographs of Srinivasa Ramanujan and Edward A. Milne. A recent addition is the drawing of Eddington by Augustus John which was published with Chandra's Eddington Centenary Lectures.[13] As I have sat with him among the books, journals, files of correspondence, sketches, and thoughts, I have often caught a glimpse of the incomparable world which Res Jost[14] so aptly described on the occasion of awarding Chandra the Tamala Prize on 9 January 1984 in Zurich, Switzerland:

There is a secret society whose activities transcend all limits in space and time, and Dr. Chandrasekhar is one of its members. It is the ideal community of geniuses who weave and compose the fabric of our culture.

1

The Simple and True

The simple is the seal of the true.
And beauty is the splendor of truth.

With the words above, Chandra concluded his Nobel lecture on 8 December 1983 in the Royal Swedish Academy Hall in Stockholm. A thunderous applause broke the near absolute silence in the hall filled to its capacity with students, members of the Swedish Academy, invited guests, two other Nobel laureates of the year (William A. Fowler, co-winner in physics; Henry A. Taube in chemistry), and the relatives and friends of the laureates. Chandra, towards the end of his talk, was describing black holes in the astronomical universe, explaining the simplicity in the underlying physics and the beauty of their mathematical description within the framework of Einstein's theory. "They are," Chandra said, "the most perfect macroscopic objects there are in the universe."

The Nobel award brought an extremely private and somewhat shy individual into the limelight of public attention. For the newspaper journalists and the broadcast interviewers, neither the simplicity of the physics of black holes nor the mathematical beauty of their description was of major concern; the pronunciation of his full name—Subrahmanyan Chandrasekhar—seemed to present them with an astronomical difficulty in and of itself. And Chandra's innate gentility, civility, and patience were put to severe tests in answering some of their questions and also in responding to the public attention. For instance, a Hindu priest from a temple in Philadelphia wanted Chandra to perform a *puja* (worship) ceremony to celebrate his winning the award; and, on the occasion of a consulate dinner honoring him, a lady sitting next to him asked, "The work you are recognized for was apparently done fifty years ago. What have you been doing since?" When Chandra responded, "They also serve who only stand and wait," the lady said, "Oh! You have been waiting for the Nobel Prize all these years?" Chandra thought it wise to remain silent since it was apparent that the lady was not familiar with the poetry of John Milton.

Nobel Prize or otherwise, Chandrasekhar has not become a household name like that of Einstein. However, over the last fifty years, he has made many significant contributions to the fields of astrophysics, physics, and applied mathematics. His prolific contributions to widely diverse areas of these fields have made him a living legend, at least in scientific circles. He is the author of several books which have become classics in their fields, and he is the winner of all the coveted medals, awards, and prizes—the Gold Medal of the Royal Astronomical Society, the Royal Medal of the Royal Society (on the formal approval of Queen Elizabeth), the National Medal of Science (awarded by President Johnson), Padma Vibhusana (awarded by the President of India), and the Nobel Prize in physics followed by the Copley Medal of the Royal Society. As a teacher and a lecturer, he is a grand master who with elegance, grace, and scholarship literally charms his audience and keeps them spellbound. More than fifty students from all over the world have taken their Ph.D.'s under his guidance, and several of them have distinguished themselves as eminent scientists. He was also the sole editor of the *Astrophysical Journal* for almost twenty years and was chiefly responsible for making it the foremost journal of its kind in the world.

With such a background of achievement in science, Chandra presents two contrasting images to those around him. To those who have had no close association with him, his almost ascetic, highly disciplined, organized, and simplified life makes him seem completely unapproachable, someone to be respected from a safe distance. For instance, a student once told him, "Most people think you are an ogre." But those who have worked with him closely or made an effort to know him have a different experience altogether. "He is so intense in all his interests," says James Cronin, "that one gets the impression that he is averse to small talk. He is not. He is a man full of warmth and friendship with deep human concerns."[1]

Always impeccably dressed in well-tailored suits whose colors vary between dark charcoal and dark gray during the fall and winter, and between light gray and tan during the summer, he cuts a handsome, dignified figure. "There is a kind of fineness about him," says Marvin Goldberger, "both from a physical and from a philosophical point of view. He is one of the most elegant-looking people I've ever met." Goldberger recalls getting ready to go on a hike in Los Alamos when Chandra showed up in "his beautifully tailored summer suit," which was obviously inappropriate for the hike. "So I loaned him a pair of khaki jeans and a short-sleeved shirt. I had worn these clothes myself, but on him they were just absolutely transformed. There is something so unbelievably elegant about him as a person."[2] Victor Weisskopf first met Chandra when he was a Cambridge graduate student spending a year in the Niels Bohr Institute in Copenhagen in 1932. "The strange thing about Chandra," he says, "is that he has changed very little. He's got white hair, but apart from that he looks to me exactly like he looked at that time. [George] Placzek used to call him, 'young man with a dove's eyes,' because he had this naive almost animal kind of eyes, which he still has." In rhapsodic tones, Weisskopf continues, "right from the beginning, but even more later on, he became sort of the most pure example of the ideal scholar in physics . . . nothing of vanity, nothing of pushiness, nothing of job seeking, publicity seeking, or even recognition seeking. . . . His deep education, his humanistic kind of approach to these problems, his knowledge of world literature, and in particular English literature, are outstanding. I mean you'd hardly find [another] physicist or astronomer who is so deeply civilized."[3]

Indeed Chandra has a deep and abiding interest in literature and classical music. He cultivates them with almost the same degree of thoroughness and intensity as his science. "My interest in literature began in a serious way in Cambridge around 1932," Chandra recalls. "I used to devote most of the two to three weeks between terms to the study of literature. The real discovery for me at that time was the Russian authors. I read systematically, in Constance Garnett's translation, all the novels of Turgenev, Dostoevski's *Crime and Punishment*, *Brothers Karamazov*, and *Possessed*. Chekhov, I read of course all his stories and plays. Not all of Tolstoy's, but *Anna Karenina* certainly. Among English writers I started reading Virginia Woolf, T. S. Eliot, Thomas Hardy, John Galsworthy, and Bernard Shaw. Henrik Ibsen was also one of my favorite authors. I was introduced to him early when I was in India because my mother had translated *A Doll's House** into Tamil. Later I read *all* his plays and have seen performances of several. I am afraid I did not read as much as I would have liked to once I came to this country. At the beginning I devoted a month during the summer every year and then a

---

* *Et dukkehjem*, literally, "A Doll House."

month every other year, and then dropped it altogether for a number of years, since the time I could devote to literature became sparser and sparser. The only serious literary study I have accomplished since I came to the United States is that of Shakespeare's plays. I have read all of his plays at least once, and some, especially the tragedies, I have read three or four times."

Chandra and Lalitha live in a spacious apartment on the twelfth floor of an apartment building on Dorchester Avenue in Hyde Park. The large windows in the living room, dining room, and his study overlook the west side of Chicago in the distance and the buildings of the University of Chicago in the foreground. On a clear day, they have a magnificent view of the sunset. They live a comfortable but simple life organized around Chandra's work. There is no frivolous waste of time in getting through the necessary daily chores. Strict vegetarians by habit and not by any religious decree, their meals are simple, consisting of lightly spiced vegetables, noodles, cottage cheese, fruits, bread and butter, and various other dishes that Lalitha improvises. Periodically Lalitha turns out South Indian dishes, and Chandra prepares pancakes with the same care that he uses when he writes his equations.

Meticulousness dominates Chandra's life and work. "Meticulous, perhaps boringly," remarked a friend once. Even his rough calculations, his rough manuscripts are organized and written out so neatly and painstakingly that they could appear in print. Some "Chandra idiosyncrasies" have become legend, the stories of which are repeated, passed on at second and third hand. For instance, he turns the standard $8\frac{1}{2}'' \times 11''$ paper ninety degrees around to write long equations and to do the checking of complicated computations and algebraic manipulations. (His students invariably inherit this habit.) He is the only scientist in the world who uses Gothic symbols in his equations (even in his rough calculations), and one of the few who still uses a pen with ink—a real fountain pen. When one considers the enormous amount of work and the labor behind it, one wonders whether time stands still for him when he works. "He is younger than the youngest among us," says Basilis Xanthopoulos,[4] one of his young collaborators in current research. "He is an inspiration to an aging physicist like me," says Kip Thorne. "He continues to produce absolutely first-rate research at a pace that has slowed only modestly."[5] Similar sentiments are voiced by Alan Lightman, who says, "Usually one associates rigorous detail, painful calculations with a younger scientist; but here he is past seventy, doing hard calculations, which really inspires younger people—inspires me!"[6]

Along with research, teaching is an integral part of Chandra's life. Even during the war years when he commuted between Williams Bay, Wisconsin, and the Aberdeen Proving Grounds in Maryland, and later during the strenuous years of editing the *Astrophysical Journal,* he continued to teach and guide the research work of his students. He finds it extremely refreshing to be among the young, among the good and critically minded students. When people have on occasion asked him whether his collaboration with famous men, such as Enrico Fermi and John von Neumann, was the high point of his scientific career, he has responded, "It was interesting to work with those great men. But collaboration with my students was indispensable; in many ways it was crucial. I can easily imagine not having lost anything if I hadn't worked with Fermi or von Neumann; but I cannot say the same thing with respect to my students."

There is a famous story about how Chandra used to travel from Williams Bay to Chicago to teach a class of only two students. John T. Wilson, then the acting president of the University of Chicago, while introducing Chandra to the audience at the time of the second annual Nora and Edward Ryerson Lecture (22 April 1975), told it as follows:

In this day of "cost-effectiveness" and its frequent misapplication within the enterprise of higher learning, I cannot resist telling you that Chandra has furnished beleaguered provosts with an extraordinary example in defense of the educational traditions of this University. During the period of the mid-1940s and following, Chandra used to drive some hundred miles between Yerkes Observatory in Williams Bay and the University, week after week, to meet with a class of two students. Even at that time one might have raised a question of relative investment of time and energy, but I doubt that such a thought even entered his mind. When the Nobel Prize in physics was awarded in 1957, it went to the whole class, Messrs. Lee and Yang.[7]

More than fifty students have completed their graduate work under Chandra's supervision. Yavuz Nutku from Turkey, the thirty-ninth Ph.D. and one of the students with whom Chandra worked closely, perceives him as the "most unconventional scientist" he has ever known. "Forever learning, Chandra couldn't care one bit about the establishment. Everything he did was out of being curious in a productive way. He did it for one reason and one reason only—it would give him serenity and inner peace."[8]

Nutku was involved with Chandra in the study of post-Newtonian equations of hydrodynamics and other effects in general relativity. The problem that Nutku and Chandra were working on involved a great deal of hard work, horrendous calculations, a number of interesting details, but no striking or simple answers. During the long, arduous checking and rechecking, "I would often get a phone call at 7:30 in the morning," says Nutku, who had the habit of studying late into the night and could not easily wake up. "However, I transferred the phone to my bedroom and somehow acquired the very peculiar characteristic of picking up the phone in the middle of my sleep and carrying on the calculations. There were times when I would feel discouraged; I would not see the point of a program which appeared more like attempting to revive a dinosaur. On one such occasion, Chandra told me one of his favorite stories, which he was known to tell to most of his students." The story is from the Indian epic *Mahabharata*, and, as Chandra would tell it, purposely omitting the names of the characters in the story because of their unfamiliarity to his non-Indian students, it goes as follows:

There were five princes. When they were taking archery lessons from a famous master, one of the five princes became known as the greatest of them all. On one occasion, a visitor—a wandering minstrel—comes to the archery school and sees the five princes practicing. To him all of them appear extraordinarily good, nothing discriminates one from the other. When he encounters the master with his observation and asks him why one is picked as the greatest, the master leads him to the five princes. The master asks each prince to take aim at, but not shoot, the eye of a bird sitting on a tree. When they are ready, he asks each of them, "What do you see?" The first prince says he sees the bird's eye, the tree branches, flowers, and the sky beyond. The second prince narrows the list somewhat, but when it is the turn of the prince who is known to be the best archer of them all, he says, "Revered Master, it's strange. I don't see anything except the eye of the bird."

The moral—supreme concentration on the task at hand; allow no distraction or doubt. "Chandra would transmit an enthusiasm," says Nutku, "an enthusiasm, not in the ordinary sense that we will go and solve this or that difficult problem, but regarding how, in the end, after painstaking and lengthy calculations, things would fall into place. Miraculous cancellations would occur and simple results would emerge and so on."

Another one of Chandra's stories that many of his students heard was that of a milkman on his way to deliver milk in the early morning. His milk cart hits a rock on the road and topples over, spilling all the milk. The man of course gets upset, curses profusely, but proceeds on his way. A little later, a mother is taking her son to school. The boy stumbles over the rock, is hurt, and starts crying. The mother curses the rock and proceeds. This continues all morning—people stumbling, falling, and cursing. All this time, a blind beggar sitting at the side of the road wonders why all these people are cursing but doing nothing to remove the obstacle from their path. Finally, at noon, when there is a lull in the traffic, he gets up and removes the rock. To his surprise, he finds a bag of gold underneath it.

Indeed, during the course of his career, Chandra and his students have uncovered many a treasure hidden under stony problems stumbled against and neglected by others.

Chandra was quite formidable as a classroom teacher, however. He did not welcome interruptions, nor were they needed as he delivered masterful, well-organized, coherent sets of lectures replete with detailed steps. "Frivolous questions from people who did not appear to have studied the material thoroughly," says Carl Sagan, "were dealt with in the manner of a summary execution. On the other hand, questions that went deep were given serious attention and response. There was an electricity in those discussions (appropriately enough, as it was a course on electromagnetism that I was attending). I will never forget the excitement that Chandra exuded in discussing general relativity towards the end of that course."[9] Peter Vandervoort, a graduate student of Chandra's in the mid-1950s, recalls that his wife, a student of zoology, had a class in physics in the room next to where Chandra was teaching his course. "Right from the start," says Vandervoort, "my wife was impressed with Chandra's style, his appearance, and his lecturing. There was a kind of cadence, a rhythm and music to his lectures. She didn't like the physics course she was taking. She would sit in the back row, listen against the wall, totally entranced by the rhythm and the structure of the sound that came through the wall."[10]

As many would say, with respect to teaching and research, Chandra was quite unlike Fermi, whose equally masterful lectures were noted not for their rigor but for the physical insights that he brought in to cut through the maze of mathematical steps and arrive quickly at results. "Fermi did *not* have the slightest, I mean the very slightest, use for elegance," says Valentine Telegdi. "Fermi would use whatever means got him the answer. You know, if he would have to cut out the surface from paper and weigh it in order to get a numerical answer for integration, that is what he would do. When you were listening to Fermi, you felt the physics he was discussing could not be any other way. Everything appeared so clear and straightforward. Later, however, you found that was not the case. There were many gaps to fill, and what was obvious to Fermi was not obvious to all. You may be interested to know, there are two articles, one by Fermi and Chandrasekhar, the other by Chandrasekhar and Fermi. In the first article, there are essentially no formulas; only words. In the other article, there are very few words; only formulas."[11]

Generally speaking, students found it difficult to be relaxed with Chandra, again in sharp contrast with Fermi. "Fermi ate lunch with us practically every day," says T. D. Lee, who did his research under the supervision of both Chandra and Fermi. "We would eat in the Commons and have hot dogs. Now, it is unthinkable that Chandra would do that. With jacket and tie, Chandra was always proper. Fermi was always with his sleeves rolled up. Mrs. Fermi would invite us for square dancing and it was very casual. Chandra and Lalitha would invite us for lunch or tea. But it was always with style."[12] However, this reserved and formal behavior, when it came to scientific matters and discussions, did not in the least intimidate his students; it did not stampede them into agreeing with everything he said. "We had a lot of arguments," to quote T. D. Lee, "and a lot of fun. My office was in the basement, and he would come to my office practically every day. On that part it was very casual."

"There was a period of probation when you had to show that you were worthy of the master," Jeremiah Ostriker remarks. "During that period, he could be very forbidding and demanding. But once you passed that, he was really very gracious, friendly, and open to free discussions. I remember, this was very early on when I was just a green student, that I, like other beginning students, was somewhat terrified of him, but not Bimla Buti, who was already working with him. I came into the library, and there was this small woman pounding the table and saying, 'You are wrong, Chandra, you are wrong!' This banter back and forth between him and his students was not uncommon."[13] "My students, students with whom I have worked closely, are respectful in a way," Chandra says, "that is reminiscent of earlier times that we read of in books. At the same time they are not at all intimidated by what I say. They will react either positively or negatively, discuss and argue. If a person agrees with everything you say, then there is no point in the discussion."

Chandra's manner seems less intimidating to those who are not directly connected with his scientific work. When he was the editor of the *Astrophysical Journal*, Jeanette Burnett, Beverly Wheeler, and Jeanne Hopkins worked for him at the University of Chicago Press.[14] They found him quite easy to work with, despite his uncompromising standards and demands. They could tease him, joke with him, and on occasion they took trips with him to Williams Bay and back. When they would stop at wayside restaurants, Chandra would explain any physics questions they would ask. "When you'd ask him a question," Jeanne Hopkins said, "he focused his whole attention, his whole concentration span, on that question." Once she saw a student stop Chandra in the hall and ask him a question. "Chandra was just standing there," she said, "being very nice to him. He just doesn't seem to have the idea that he's better than anybody else. He's perfectly willing to take time and talk to you when he realizes that you're interested, and he makes you feel you are doing the most important work. He is one of those rare people. One in a million. When I'd see him, it just gave a lift to my whole day."[15]

Chandra also has that rare ability to inspire enthusiasm for hard work in others. Donna Elbert, who until 1979 was Chandra's technical scientific assistant, provides an exceptionally singular instance of the dedication that Chandra can inspire. Donna, whose father owned a local barber shop, was a nine-year-old girl when she first saw the young couple, Lalitha and Chandra, in the streets of Williams Bay. "How lonely they must be—so far away from home!" she thought. Later, after finishing high school, she went to Milwaukee with an interest in dress designing. "After working with a furrier for a short while, I did not see any future in the job," Donna recalls. "I decided to return to Williams Bay, find a job, and save some money to go to a design school. Upon learning that Chandra needed someone to assist him in computational and secretarial work, I applied for the job and got it."[16] Before long, impressed by the intensity of Chandra's personality and his hard work, Donna mastered his working methods, the special nuances in his manuscripts, the special care he took with everything he did. Then with Chandra's encouragement, she took courses in advanced mathematics and calculus offered by the University of Wisconsin summer school. "Chandra even drove me to Madison to register," Donna says. "Later he allowed me to arrange my work schedule so that I was able to complete a degree at the Art Institute School in Chicago."

Before long, Donna was able to assist Chandra with some of his complicated numerical work. "He has a woman assistant by the name of Donna Elbert," Marvin Goldberger said to me, "whom I used to accuse Chandra of chaining into a closet and making her carry out his horrendous computations. I also once accused him of inventing her as a name—that he actually did all these calculations himself." A woman of considerable skill in computations, Donna matched Chandra in patience, strength, and tenacity in handling and completing numerical details which otherwise would perhaps have been left alone. In 1978–79, when Chandra derived a long and complicated formula and arrived at an identity whose validity seemed difficult to establish by analytical methods, Donna checked it numerically, doing a laborious, long, and involved calculation on a small desk computer. "The fact that she was there and could do such work," Chandra says, "meant that I could carry on the work until it was aesthetically complete and not merely complete as far as the kind of information I wanted."

Thus as Donna played a vital role in Chandra's scientific work, she was a witness to the side of Chandra that is normally invisible to his students, associates, and colleagues. Goldberger, as many others, can say, "I've never heard him speak seriously in anger, be angry about anything. I've seen him outraged by what he considered not proper, but rarely did he ever raise his voice." Donna, on the other hand, has experienced Chandra during his occasional spells of impatience, anger, and stubbornness. "I could never understand how someone who is so rational in his work," she says, "can be so upset and irrational about nonscientific matters." Notwithstanding such occasional flurries, Donna was associated with Chandra for thirty years, from 1949 to 1979.

Regarding his scientific work, Chandra's most distinctive characteristic is perhaps his attitude towards science in general and to his own work in particular. In astronomy and astrophysics, mathematical theory plays a dual role. It serves to interpret observations, but more importantly, it provides a basis analogous to *experiments* in physics, chemistry, and biology. There are no controlled, reproducible experiments in astronomy. The laboratory is out in space. There is only one universe we know that came into existence some ten billions of years ago. Its beginning and early moments, its present large- and small-scale structures—the formation of galaxies, the evolution of stars, the fundamental constitution of matter—have to be inferred by theories based on our microcosmic understanding of terrestrial phenomena. The laws of physical science thus learned must then be extrapolated sometimes to vast scales of space and time and to extreme conditions of matter very different from those encountered on earth. To interpret and deduce what is observable from such extrapolations requires a mastery not only of diverse mathematical techniques but also of several branches of the physical sciences several orders of magnitude more complex than what is required, say, in any one special area of physics.

As a result, theoretical astrophysics covers a broad spectrum between two extremes. "The one extreme is to build, to throw together some kind of model of this week's important observational discovery," says William Press. "To be the theorist on the spot so that your phone can ring in the middle of the night, and someone can say, 'I've found such and such. What does it mean?' and you come up with something. A week later you submit your letter to the *Astrophysical Journal*."[17] The other end of the spectrum is the *pure* side that seeks model systems, seeks to understand them in rigorous, clean, mathematical detail. The model systems themselves may not correspond exactly to anything that is observed, but the mathematical analysis provides, say, a complete catalog of what to expect by way of observations. It is to this extreme, "pure" side of astrophysics that Chandra belongs. In Kip Thorne's words, "Chandra's research mode is very much one of sitting by himself with zero interaction for days struggling with equations, trying to make the mathematics fit into patterns, and then going to somebody like me and explaining what is going on, but explaining in mathematical terms rather than physical terms; mathematical in the sense of the symmetry properties of the equations he's working with, the algebraic relations between various terms in these equations, and where the equations had come from. He is looking for some kind of suggestions or ideas that make everything fit together better. You get the feeling that he knows he has the answer when the ultimate mathematical formula is simple, when everything has fallen into place in a nice, simple, coherent mathematical fold."

In this kind of pure mathematical astrophysics, Chandra's contributions, which have been almost unrivaled, involve a wide range of investigations. "There have been seven periods in my life," says Chandra. "They are briefly: 1) stellar structure, including the theory of white dwarfs (1929–39); 2) stellar dynamics, including the theory of Brownian motion (1938–43); 3) the theory of radiative transfer, the theory of the illumination and the polarization of sunlit sky, the theories of planetary and stellar atmospheres, and the quantum theory of the negative ion of hydrogen (1943–50); 4) hydrodynamic and hydromagnetic stability (1952–61); 5) the equilibrium and the stability of ellipsoidal figures of equilibrium (1961–68); 6) the general theory of relativity and relativistic astrophysics (1962–71); and 7) the mathematical theory of black holes (1974–83)."[18]

Indeed, in each of these periods, as Goldberger says, "he has produced an infinite series of papers followed by an infinitely thick book on the subject." These monographs and thick books have become classics which no serious student entering into research in astrophysics, and some areas of physics, can pass by; this is in spite of the existence of numerous later texts, treatises, and monographs.

William Press, Alan Lightman, and Saul Teukolsky were young graduate students at the California Institute of Technology when they first met Chandra in the early seventies. But long before, they had encountered S. Chandrasekhar in their studies of the Dover reprints of his books: *An Introduction to the Study of Stellar Structure, Principles of Stellar Dynamics,* and *Radiative Transfer.* "At the time," Press says, "I didn't realize he was still alive. You have the feeling that people who have done so much classic work aren't around anymore." The extraordinary amount of detail, formulas and equations running into the hundreds,* numerical tables and graphs, intricate references and cross-references, and the bibliographical notes at the end of each chapter which give short, but sufficiently complete, historical accounts of the development of the subject matter make these monographs unique.†

With each monograph, the culmination of a series of papers in a particular period, Chandra has essentially initiated a new field of research. In some sense, it seems as if he has annihilated it as well, in that he exhausts it and makes it appear as though there is very little for someone else to do. The field may be a tiny corner of astrophysics when Chandra undertakes it, but "he makes it a major field," says Press. "When things are worked out beautifully, then that defines a field; it hardly matters whether it was a field before. . . . *Stellar Structure* wasn't simply a book about stellar structure; it went a long way towards creating a new field." Therefore, the feeling that there is very little for someone else to do is quite transient. Soon one sees new applications, new directions previously unseen and undreamed of. Studying Chandra's work is like watching the course of a stream. A complete period concluded with a monograph looks as if the stream has stopped at a natural

*An article published in a German newspaper claims that if all the formulas, equations, and text that have appeared in Chandra's publications are put end to end, they cover the distance between the earth and the moon.

†John Sykes, currently lexicographer for the Oxford University Press, mimics the style of Chandra's papers and books in a very amusing parody: "S. Candlestickmaker, On the Imperturbability of Elevator Operators LVII, Institute for Studied Advances, Old Cardigan, Wales (communicated by John Sykes; received October 19, 1910)." This parody appears in an anthology compiled by R. L. Weber, *A Random Walk in Science* (London and Bristol: The Institute of Physics, 1973).

A note in the anthology says, "Professor John Sykes' famous spoof of Professor S. Chandrasekhar delighted the 'victim,' who arranged to have it printed in the format of *The Astrophysical Journal.* Some librarians bound it in series without noticing."

barrier. Soon, however, it creates a new path. There are only trickles at the beginning, but a new stream soon appears.

Often one comes across people who work in widely different areas— say, an engineer, a molecular physicist, or a specialist in colloid chemistry—all of whom can testify that one of Chandrasekhar's papers has given rise to new developments in their fields. "I thought that it might amuse you," wrote Peter Jakobsen to Chandra, "to know that the classical work that you did with Dr. Münch over thirty years ago on fluctuations in the brightness of the Milky Way has found a new application in this [soft X-ray background] field. I can assure you that it was quite an experience for us [Steve Kahn and myself] to discover after the fact that you had already solved the problem before we even were born." [19]

It is not surprising therefore that Chandra's work and his monographs evoke a feeling of respect and wonder. While they may seem intimidating and forbidding to the casual student, for the serious-minded scientist they leave an indelible impression of their enduring value in spite of the continual progress of astrophysics. They convince one of the innate values of science—the continuity, the interdependence, and the necessity of combining original research with scholarship.

In addition to their thoroughness, lucidity, and accuracy, the monographs have a highly personal, distinctive style. "Chandrasekhar is one of the most unusual examples of a scientist who has been able to inject his personal style into his work," says Lightman, who has made a study of aesthetic criteria in science. "He has an incomparable style," says Weisskopf. "Good English style is a lost art in physics, but he has it and this wonderful feeling for the essential, and a feeling for beauty." In the same vein, Lyman Spitzer says, "It's a rewarding aesthetic experience to listen to Chandra's lectures and study the development of theoretical structures at his hands. The pleasure I get is the same as I get when I go to an art gallery and admire paintings." [20]

Elegance and love for and attention to the language play as important a role as the scientific facts and the weaving of them into mathematical formulas. "I practice style in a very deliberate way," says Chandra. "I acquired my style from not only just reading, for instance, the essays of T. S. Eliot, Virginia Woolf, and Henry James, but also by paying attention to how they write—how they construct sentences and divide them into paragraphs. Do they make them short or long? For example, the idea of just using one sentence for a paragraph, or of a concluding sentence without subject or object, just a few words . . . 'so it is' . . . or some small phrase like that. I deliberately follow such devices. In fact there is one technique I started following when I was writing my book *Radiative Transfer,* and I have followed it since; that is, as you know, in music you repeat periodically the same phrase in exactly the same form. Very often in my books, when I have a key idea, and I have to restate it at a later stage, I don't leave it to chance. I go back and copy exactly what I had written before."

Indeed, in his monograph *Radiative Transfer,* for instance, he treats radiation transport in finite atmospheres in one chapter and in semi-infinite atmospheres in another chapter. When reading the two chapters, one notices identical English sentences describing identical equations wherever they occur. The description of the equations has the same similarity as the equations themselves in the two different physical situations. This exceptional care for elegance and beauty is most evident in his book *On Ellipsoidal Figures of Equilibrium.* He begins the book with a historical introduction of the problem which attracted the attention of the past masters. He points out the loose ends left in the investigations of Riemann, Dedekind, Jacobi, and Maclaurin and then, in subsequent chapters, goes on to describe the entire subject from his own perspective.* The problem comes alive again, bristling with new ideas, suggesting new directions of development. Even a casual reading of the book evokes in one that rare, magical feeling of oneness with the past and of the continuity of human effort.

Chandra's early education was at home and mostly on his own once he began to learn. He made up his mind quite early to be a scientist, in his teens, in fact. "On that account," he says, "I had, on my own, read and studied calculus and some higher physics. But once I registered for the B.A. (honors) degree in physics, I not only attended lectures in the mathematics department, but also I read books which were quite outside my course." He pursued this practice of independent studies throughout his student days. To know something purely for its own sake, and in a manner totally pleasing to himself, was an extremely important part of his life from the beginning. Chandra wrote in a letter to his father from Cambridge, a week before he was twenty-one, "He is wise who strikes an exact balance between learning for himself and attempting some research *consistent* with success." In another letter written after he had toiled over the two volumes of *Modern Algebra* by Bartel L. Van der Waerden while doing research in astrophysics, Chandra wrote,

*A large part of the original research in this book was done in collaboration with Norman R. Lebovitz. Chandra is concerned that, in spite of his best efforts, Lebovitz is not given enough credit for the part he played in the development of this subject.

"Perhaps you are surprised at how one could be so 'stupid' as to require nearly a couple of months of constant study to read a few chapters. I'm not going to say that I needed that time to master the subject—make it my own. I have now practically a book on 'Modern Algebra' worked out in my own way."

This attitude of striving to understand things in his own way, within his own framework, is reflected in his research. As Chandra says in the autobiographical account published with his Nobel lecture:

After the early preparatory years, my scientific work has followed a certain pattern motivated, principally, by a quest after perspectives. In practice, this quest has consisted in my choosing (after some trials and tribulations) a certain area which appears amenable to cultivation and compatible with my taste, abilities, and temperament. And when after some years of study, I feel that I have accumulated a sufficient body of knowledge and achieved a view of my own, I have the urge to present my point of view *ab initio,* in a coherent account with order, form, and structure.

The quest after perspectives is his principal goal; not the relative importance or unimportance of the subject, not how others perceive it, not whether it is going to bring fame and recognition. That being the case, once he is satisfied with the perspective gained in a particular area, he wraps it up in a book, packs up and leaves the area entirely, never to return, and is even reluctant to maintain a passing interest in that area. The affair is over. Another step on the ladder is there to be taken; new glimmerings of structure wait to be found. And so he seems to pass from one area to another with a complete sense of detachment and absence of emotion.* The hard work and the single-minded intensity invested in an area of pursuit for a number of years do not seem to trouble him in the least, for he is ready to start fresh in a new area. If necessary, he will attend classes, take notes, and study as if he were once again a student. Or he will teach a course in the subject he wants to learn.

It is the enlargement of his own vision that Chandra seeks, and this seems to urge him on to struggle and pursue the path of scientific creativity. It brings him into constant contact with younger people, which he finds enormously exhilarating. In this way he also escapes the most common fate of many great scientists who become captives of their own success and depart from the path of creativity. For some scientists of

---

*Somewhat like the mathematician David Hilbert. As Hermann Weyl says, "If he [Hilbert] was engrossed in integral equations, integral equations seemed everything; dropping a subject, he dropped it for good and turned to something else. It was in this characteristic way that he achieved universality." In Constance Reid, *Hilbert* (New York: Springer-Verlag, 1970), 245.

eminence, there will be the appeal of serving on national and international committees of various sorts, accepting public relations roles, acting the elder statesmen, counseling on matters with instant expertise (often far beyond their competence); there will be temptations to prestige and power in administrative circles, and to new ways and modes of life far different from those of a scientist. For others there is a different kind of subtle barrier that impedes their creativity—their desire to relive the past. Those who have made great discoveries and experienced the moment of glorious insight into some of nature's secrets often want to experience it again and again. To hack away at something less important, less fundamental, than their previous discovery appears demeaning to them. They perhaps fail to recognize that their sudden discovery was probably due to a fortuitous combination of circumstances—the sudden findings of new empirical facts, new techniques, the partial successes and failures of others. To expect all this to repeat itself is unrealistic, at best. For Chandra, there is no such deliberate homage to past glorious moments or momentous discoveries. For him, discoveries great or small happen through constant, sustained effort. No holy grail, but a personal vision, perspective, and understanding of an entire area are the only satisfactory motivating factors. He is at his happiest when problems take on their own momentum so that one problem generates another. It is as though he is obligated, compelled, to solve them, like the internal, irresistible urge of an artist to discover his own way of expressing himself. Chandra's work has an architectural quality to it—he is not simply putting in a nail here, a window there. Others may have contributed a great deal to its shape and dimensions, but the whole structure is put together in some uncannily durable and inspired way which is uniquely his own.

Chandra is often said to be a scholar in the classical tradition of Lord Rayleigh and Poincaré.* "Comparison with these two people is very sobering," says Chandra. There is, however, a great similarity between Rayleigh and Chandra. First of all, both of them have maintained an incredibly uniform record of publications; Rayleigh published an average of eight to ten papers a year for over fifty years, from 1868 till his death in 1918. Chandra, with minor fluctuations, has a similar record. Second, for both of them, systematic work, systematic chiseling at nature, was more important than being on the frontier or being initiators and

---

*As stated, for instance, in the citation of the Dannie Heineman Prize of the American Physical Society awarded to Chandra in 1974.

leaders in one particular area. As J. J. Thomson said in his memorial address on Rayleigh, "In science, there are two kinds of people. Those who write the first sentence of it, who may be considered as leaders, and those who write the last sentence. Rayleigh belongs to the second category." And so does Chandra. "There is a complementarity between a systematic way of working and being on the frontier," says Chandra. "To be systematic and to develop a subject to a certain completion requires that the subject is capable of being so treated. You cannot do that if the subject is something which is just beginning. [Max] Planck couldn't have written a book on quantum theory which was complete and final."

For Chandra, then, attaining a complete understanding of an area, grasping and internalizing it, is the essence of his scientific life. One's motivations for doing science may change with time. In the end, however, if one's motivations are not galvanized to pursue science for its own sake, then, according to Chandra, "one's scientific life has not matured properly." This is not such an uncommon attitude among scientists late in their careers, but it is extraordinary that in Chandra one finds it so early in his life and practiced so diligently throughout his lifetime. One is, therefore, naturally compelled to wonder why this is so. Could his unexpected encounter with Eddington be responsible? Could that be what a psychohistorian like Erik Erikson would designate as the "Event" in Chandra's life that formed a turning point? One cannot be sure, but it certainly had a profound effect on Chandra. While I treat this encounter in great detail later, let me encapsulate it here.

In 1930, when he was only nineteen years old, Chandra, after a brilliant undergraduate career at the Presidency College of Madras, India, went to Cambridge University to do research under the guidance of Ralph H. Fowler. On the long sea voyage from India to England, as a result of his musings and calculations, he arrived at a conclusion which, at first, seemed extremely puzzling. It undermined the then prevailing notion that the white dwarf, a densely packed star,* was the ultimate stage of each star in the course of its evolution. Combining Einstein's special theory of relativity and the new quantum mechanics, young Chandra had shown that if the mass of the star exceeded a certain critical mass, expressible in terms of fundamental atomic constants, *the star*

---

*White dwarfs have planetary dimensions but are as massive as the sun. In the early 1920s, three such objects were known, the companion of Sirius, known as Sirius *comes*, being the most illustrious. Its mass was estimated to be approximately the same as that of the sun, while its radius was approximately 20,000 kilometers, astonishingly small for such a great mass. It implied a density of 61,000 g/cm^3, or just about a ton per cubic inch.

would not become a white dwarf. It would continue to collapse under the extreme pressure of gravitational forces. After a few more years of hard work, he established the condition (the critical mass condition, the Chandrasekhar limit) on a rigorous footing and reported his findings at the January 1935 meeting of the Royal Astronomical Society of London. His findings raised challenging, fundamental questions: What happens to these more massive stars when they continue to collapse? Are there several terminal stages of stars other than that of white dwarfs?

The appreciation of the importance of this discovery by the astronomers should have been immediately forthcoming, but it was withheld because no sooner had Chandra presented his paper than Eddington began to ridicule the whole idea before the scientific community. He cast a cloud of disbelief over Chandra's work; he made it look as though Chandra understood neither relativity nor quantum mechanics, and as though there was a fundamental error in Chandra's derivations.

The whole affair was tantamount to a public humiliation. There was very little that Chandra could do. He was young and still a newcomer in the research arena who found himself pitted against an older, established, internationally renowned scientist. Eddington's authority, prestige, and fame prevailed. More than twenty years passed before the Chandrasekhar limit became an established fact and assumed its important role in astrophysical research.

This little-known episode provides a splendid example of the often unexpected and surprising ironies which may impede the recognition of what is exceptional in any field of human endeavor. Eddington was not only one of the twentieth-century pioneers in the study of stellar interiors, but he was also the man chiefly responsible for the early acceptance of Einstein's general theory of relativity. He already had a history of being able to recognize and accept new ideas. Einstein's famous paper on gravitation appeared during 1916 in the middle of the First World War, at a time when direct communication had been cut off between the scientists of Great Britain and Germany. Eddington, however, who received Einstein's paper from Willem de Sitter in neutral Holland, understood the significance of the new and revolutionary idea; he mastered both the physical content and the new mathematical machinery of Einstein's theory, and prepared a report on the theory of gravitation, which was published in 1918. This is believed to be the first complete account of Einstein's theory in the English language. His later books, *On the Theory of Relativity, Space, Time and Gravitation,* and *Mathematical Theory of Relativity,* were mainly responsible for the exposure of these ideas to the English-speaking world, and his name forever became associated with the very few who *really* understood Einstein's theory.*

Beyond that, he was also the leader of one of the two expeditions sponsored by the Royal Society of England in 1919 to observe the solar eclipse, measure the deflection of light as it passed by the sun, and thus test Einstein's theory. The expeditions were organized during the darkest days of the war. Einstein, still in Germany at the time, was, in political terms, an enemy scientist. Thus, Eddington's painstaking and bold confirmation of Einstein's predicted amount of light deflection was not only one of the most sensational events in the history of science, but also an event that exemplified the international character of science by refusing to recognize political and national boundaries.

It is not without irony, then, that this same Eddington failed to see the far-reaching consequences of a simple, straightforward application of Einstein's special theory of relativity which was at the heart of Chandra's discovery. Imaginative as he was, he did not foresee the exciting possibilities of terminal stages for stars other than the white dwarfs (although, as we shall see later, he himself came very close to being the first one to recognize the possible existence of black holes in the astronomical universe). Instead, he discarded the possibility as absurd. And it is beyond comprehension why, after prodding Chandra through months of labor involving hard numerical work, and in spite of being familiar with Chandra's work on an almost day-to-day basis, he chose, without giving the slightest prior hint, to publicly denounce Chandra's dramatically stated conclusions in a manner that ridiculed the young scientist.

From the point of view of the scientific world, the encounter was of minor consequence at the time. For physicists, it was a domestic matter among astronomers. Eddington's influence among physicists was on the wane as he had departed from the conventional rational approach to science into transempirical, metaphysical, and speculative ideas which were generally incomprehensible. He was no longer one to be taken seriously. For young Chandra, however, on the threshold of his research

*On New Year's Eve 1934, after dinner at high table, Rutherford, Eddington, and a few others (including Chandra) were in the combination room, conversing well past midnight. Eddington and Rutherford were smoking long clay pipes. During the course of the conversation, Eddington recalled his encounter with Ludwig Silberstein at a Royal Society discussion. Eddington said, "After the discussion, Silberstein came up to me and said, 'Well! Professor Eddington, you must be one of the three people in the world who understands relativity.'
I said, 'Oh, I don't know . . .'
To which Silberstein said, 'Professor Eddington, don't be modest.'
To which I replied, 'On the contrary! I am wondering who the third person is!'"

career, it was a different matter. Eddington still wielded great influence among astronomers. There was no quick way to prove the man wrong. As Eddington went on denouncing the idea of relativistic degeneracy and the critical mass condition in meeting after meeting, "It became a unanimous opinion among astronomers that Eddington was right and I was wrong," observes Chandra. "My few attempts at rebuttal made matters only worse. Some began to say I was a 'Don Quixote' out to kill Eddington and people like that."

Faced with such a situation, Chandra had to decide: should he continue to fight doggedly, meeting Eddington head on at every meeting? That is the course most would have taken. Instead, he withdrew gracefully; he stopped any further publishing of his work related to white dwarf studies; he summarized what he knew and where he stood in the form of his first book, *An Introduction to the Study of Stellar Structure* (1939); and he went on to do research in another area—stellar dynamics. After a few years of work in that area, and as he became interested in the problems of radiative transfer, it occurred to him that he should wind up his research in the form of a book, *Principles of Stellar Dynamics* (1943). Then, a few years later, he published the monograph *Radiative Transfer* (1950).

Thus one might attribute the origin of Chandra's distinctive pattern of work, the constant rejuvenation of himself by entering new fields, to this early dramatic controversy. Imagine for a moment that Eddington had acclaimed Chandra's discovery as fundamental and so announced it to the world (as he had done in the case of Einstein's theory); suppose Eddington had taken the concept of the black hole seriously. He would not only have been the first one to recognize the existence of black holes in the astronomical universe, but Eddington himself could have initiated work on gravitational collapse within the framework of general relativity some twenty years earlier! Consequently, he would have achieved much higher status than he now has within the scientific community. Along with his own increased stature, he would certainly have made his young protégé a luminary. Would such a widely acclaimed Chandra have followed the same path? Would he have become the Chandra of today? With such success when so young, with the burden of fame and recognition to bear, might he not have been pressured into a more standard pattern of being on the frontiers of research and striving after flashy discoveries, instead of pursuing his quest for personal perspective and satisfaction, beauty and completeness in his seemingly hermetic endeavors? Further, it would have been extremely likely that with Eddington's patronage, and his likely election to the Royal Society, he would have had no difficulty in finding a suitable position in India on his return from England in 1936.[21] If that had happened, the story of Chandra's life would have been very different.

Chandra's life stands out for its singular dedication to the pursuit of science, and for practicing the precepts of science and living up to its values to the closest possible limit in one's life. As Cronin says, "He is the closest form—closest Platonic form—of an ideal individual; a truly great scientist, one among a very few in this century." According to Ostriker, "He is his own worst critic. He sets his own standards higher and higher. He constantly reminds you of the man on the ladder in his favorite photograph in his office." With his extraordinary successes in astrophysics, physics, and applied mathematics, Chandra combines an extraordinary personality characterized by an intensity and fervor for completeness, elegance, and above everything else, gaining a personal perspective in his scientific work. He expressed this best during his three-minute speech after the gala banquet during the Nobel ceremonies in December 1983:

The award of a Nobel Prize carries with it so much distinction and the number of competing areas and discoveries are so many, that it must of necessity have a sobering effect on an individual who receives the prize. For who will not be sobered by the realization that among the past laureates there are some who have achieved a measure of insight into nature that is far beyond the attainment of most? But I *am* grateful for the award since it is possible that it may provide a measure of encouragement to those, who like myself, have been motivated in their scientific pursuits, principally, for achieving personal perspectives, while wandering, mostly, in the lonely byways of science. When I say personal perspectives, I have in mind the players in Virginia Woolf's *The Waves:*

There is a square; there is an oblong. The players take the square and place it upon the oblong. They place it very accurately; they make a perfect dwelling place. Very little is left outside. The structure is now visible; what was inchoate is here stated; we are not so various or so mean; we have made oblongs and stood them upon squares. This is our triumph; this is our consolation.

2

Choosing the Unconventional:
A Family Trait

An annual sum of 10,000 pounds [shall] be set aside for the revival and improvements of literature and the encouragement of the learned Natives of India and for the introduction and promotion of a knowledge of the sciences among the inhabitants of the British territories in India.

1813 Charter Act

Written accounts of family backgrounds are rare in India. However, Chandra's father, C. Subrahmanyan Ayyar, has pieced together a fascinating account of his father, Ramanathan Chandrasekhar (1866–1910), and his grandfather, Ramanathan (1837–1906).[1]

He has actually traced the family back to Mr. Ayyamuthian (circa early 1700s), who originally belonged to Vaigalathur, a village not too far from Tanjore. He went trading to the west coast of Malabar, returned with enough money to purchase some land in the Tanjore district of the Madras Presidency, and settled down in the village of Mangudi. He and the two succeeding generations lived off this land. Their education was in Tamil and Sanskrit, and that was limited mostly to the holy scriptures, the *puranas,* and other devotional literature.

Chandra's ancestors were Shaivite brahmans.[2] His great-grandfather, Ramanathan, had some fifteen acres of inherited land, the income from which, after paying taxes and expenses for cultivating, was barely adequate to provide food and shelter for his family. Landholders were a minority among brahmans, themselves a minority caste in every part of India.

Ramanathan, conversant only in Tamil, led the typical life of a village brahman of his class, supervising the cultivation of his lands and maintaining the family traditions of worship. Nevertheless, in a marked de-

34

parture from this conventional lifestyle, he sent his son, Ramanathan Chandrasekhar (referred to as R. C. hereafter, to avoid confusing the similar names), to get an English education. Brahmans in general, all over India, were the first to embrace western education and thereby take advantage of the new opportunities created by the British presence.

The Madras Presidency was then one of the most extensive territories under British rule. It comprised almost all of South India, excluding only the princely states of Hyderabad, Mysore, and Travancore. Thus it stretched north from the southernmost tip of the Indian peninsula (Cape Comorin) to halfway up the east coast of Bengal. A part of it on the west lay on the Indian Ocean, with the Bombay Presidency to the north. Another segment extended east from the border of the Bombay Presidency and Hyderabad state to the Bay of Bengal. Its population was a mélange of diverse cultures and languages (Tamil, Telugu, Kannada, Tulu, Malayalam, and several dialects).

The British conquest of modern India and Pakistan was completed by the year 1852.[3] The British East India Company was still in control then. By brilliant but ruthless tactics, James Ramsay Dalhousie, the last governor-general of India (1848–56) under the company's rule, annexed Punjab in the northwest and extracted enough territories from Burma in the east to make the British territories safe and secure, protected by the Himalayas to the north and northwest and the Burmese mountains and jungles to the east. Having made the frontiers safe, he began the modernization of India on a large scale. He introduced an all-India railway and telegraph system, built roads, and founded universities along western (i.e., British) models. He had thus put into effect with full force the reforms initiated by Lord William Bentinck, an earlier governor-general (1828–35).

Bentinck and Macaulay[4] correctly perceived that the country was in ruins politically, socially, economically, and physically. The Moghul empire of the seventeenth and eighteenth centuries had crumbled, and local wars had ravaged every area, leading to widespread social and economic breakdown. India was exhausted; it was without inspiration; and it was hopelessly divided between castes, creeds, and races. Its rich past heritage and its ancient religion were only a decaying facade. But those among the British who considered themselves wise detected a vast store of energy behind the facade. And the best way to tap it, they thought, was to bring the "West" to India. Bentinck and Macaulay offered the following resolution as a panacea for the ills of Indian society:

. . . that the great *objects* of British government ought to be the promotion of European literature and science; the available funds should be henceforth employed in imparting to the Native population knowledge of English literature and science through the medium of the English language. (The Resolution of 7 March 1835)

English replaced Persian (the court language of the Moghuls) as the official state language and was used in the higher courts of law, while local languages (called vernaculars) replaced Persian in the lower courts. Bentinck and Macaulay together put the education provision of the 1813 Charter Act (which had left vague the question of oriental versus occidental learning) into full use in the service of western education in India.

The aim of Macaulay, the chief architect of English education in India, is described in the following quotation:

. . . our aim should be to create a class of persons who would be Indian in blood and colour, but English in tastes, in opinions, in morals and intellect.

Macaulay's education drive, however, touched only a very small segment of the people, and it was not a result of any deep understanding of their needs. The new policies had more to do with changes in the political and social thinking in England toward the end of the eighteenth century. It became clear then that the East India Company's rule extended over a large territory, and the British parliament began to consider the future of their possessions. Idealists like Edmund Burke did not think it appropriate that the administration of such a vast territory should be left to a mercantile company whose administrators were totally unaccountable to the British parliament.[5] Besides, there were other mercantile concerns who wanted to see an end to the company's trade monopoly and missionaries who wanted access to the subcontinent.

The company had banned missionaries from India, because it was afraid of an uprising or revolt against itself and the British since the missionaries were certain to interfere with local traditions and customs. The evangelizers, of course, thought otherwise. To them, there were millions of souls to be saved by the light of the gospel, which would dissolve the mists of superstition and cruelty enshrouding the "dark nation" of India. Strange as it may appear, they found allies among the British radicals whose gospel was rationalism and humanism. The radicals fought equally hard for direct British intervention in India to change it, modernize it, and abolish "barbaric and cruel" customs such as *suttee* and *thuggee*.[6] As the combined voices of evangelists and radicals got stronger, they began to be heard in the British parliament. Bentinck and Macaulay responded to these voices and proceeded to undertake the westernization of India with zeal. However, before this campaign could take root, become a reality, be accepted as inevitable, and be lived with as with other prior foreign intrusions, India made one last effort to resist it.

The resistance took a bloody turn at the Sepoy Mutiny of 1857, the immediate cause of which is attributed to the cartridges for the new Enfield rifles. Before loading these rifles, the *sepoys* (Indian soldiers employed by the British) had to bite the tips of the cartridges, which were smeared with cows' and pigs' fat. For Hindus and Muslims alike, this was distasteful, unclean, and definitely against their religion. They began to wonder: Was this meant to defile them and force them to Christianity? Were the missionaries behind it?

When the *sepoys* refused to obey orders, they met with brutal punishment, dishonor, and were discharged from the British services. These discharged, and hence idled, soldiers joined ranks with the orthodox majority, which was already full of smoldering discontent. The suppression of *suttee* and *thuggee* was seen by them as an alien intrusion in the "sacred" practices of their traditional religion. Their combined ranks were joined by landlords, who were dispossessed because of large-scale land settlements,[7] and by those who had lost their ancestral land holdings. Their predecessors had acquired such holdings because they were learned or had special religious merit. The successors had no such claim to scholarship or religion, and hence the awards were taken away. In addition, there were also those from the princely states that were annexed by Dalhousie on the pretext of misrule or the lack of a natural heir, although Hindu religion and tradition sanctioned fully the legitimacy of an adopted heir.[8] All these elements, the fears of the orthodox and the resentments of the dispossessed, outweighed the benefits the reforms brought to the emerging, small westernized class.

The result was fourteen months of bloodshed and a full-scale Anglo-Indian war, which, for a short while, threatened to lead to the expulsion of the British and the reestablishment of the Moghul and Maratha regimes. Ultimately, however, the British prevailed, because there was, as yet, no united Indian leadership. Most of India (especially the Deccan and the south) remained passive while fierce, localized battles were fought in northern and central India. The Sikhs of Punjab were solidly on the British side, due largely to their dislike of the reestablishment of a Moghul regime in Delhi. For the westernized class, there was nothing useful or progressive in the uprisings, only a revival of the old defunct, feudal, and factional regimes of the rajahs and nawabs. Hence its members remained passive. Thus, a last convulsive movement to resist the British ended. At the same time, however, it marked the end of the East India Company's rule,[9] which had lasted exactly one hundred years, and the advent of the era of India as a full-fledged British colony.

Soon the bitterness of the bloodshed faded away to make way for new reforms, but with due caution paid to the lessons that the uprising had taught: leave the socioreligious customs and traditions alone; teach not "western moral values" but western technology; build public works; industrialize; and improve communications. These reforms and measures, while appearing altruistic, in fact provided raw materials and wealth that meant riches and employment in England. The Imperial heyday had begun. In India, a sudden upsurge of English education ensued since the British were there to stay, their superiority acknowledged by both the orthodox and the new westernized classes. "For the former it was the superiority of power, for the latter the magic of the new knowledge."[10] It would take decades before the magic of that new knowledge would give rise to nationalism and the demand that Indians benefit from the acclaimed western values of individual freedom, human dignity, and self-rule.

English education did lead to the creation of a cadre of "anglicized" civil servants who became the backbone of British rule in India. At the same time an English education exposed many Indians, like Chandra's grandfather, to European thought, literature, philosophy, and science. Some Indians, both in the service of the British and outside of it, became intensely aware of their own heritage and the contributions of their great men and women of the past. The grip of age-old traditions was also quite strong. Children read and heard from their elders the stories from the *puranas* and from the epics, *Ramayana* and *Mahabharata*. They listened to devotional music and songs. The ancient Hindu ideals of self-sacrifice, learning, strict celibacy until marriage, honesty, and respect for elders and teachers were constantly drilled into the minds of the young. To know and master western knowledge was acceptable; but western culture was another matter. Certain of its aspects—indulgence in alcohol, eating of meat, unrestrained free association between the sexes—were contrary to Hindu precepts. These had to be abandoned. Needless to say, the complex situation produced tensions and conflicts as families and individuals within families varied greatly in their capacity to synthesize and blend the new with the old. Constant and rapid change was the order of the day.

The fact that the brahmans were among the first to adapt to such change and take advantage of western education is commonly attributed to their strong heritage of learning and scholarship. For centuries, brahmans were regarded as the sole guardians of the Hindu religion because they had the exclusive right to study Sanskrit, the sacerdotal language, the language of the holy scriptures. For brahmans and non-brahmans alike, knowledge of the scriptures was essential for everyday life. This created a mutual dependency between brahmans and non-brahmans; the former had to maintain the tradition of learning while depending economically on the labor of the latter. Brahmans guarded the sacred, the gates of heaven, while the nonbrahmans labored, cultivated land, traded, did menial tasks, and provided the daily bread.

For villagers, it was not easy at that time (nor is it so easy even now) to acquire a western education. Sometimes for higher elementary and definitely for high school studies, one had to leave home and go to a larger town. Generally there were no boarding or lodging facilities attached to the schools. Rooming houses were unheard of in most of the towns. One had to seek a relative, a family friend, or a friend of a family friend to provide food and shelter. It was an imposition which carried with it responsibilities and sometimes strenuous household chores for the young student. The burden of feeding an extra mouth and sharing living space which was often very cramped was at times too much for the host family. Customs, traditions, social pressure, and a sense of false prestige made it imperative to accept such responsibilities with an outward show of grace and equanimity. But inward resentments were unavoidable, leading to domestic conflicts, and often making the student and his family the target of subtle insults or even overt humiliation. It often required, needless to say, great strength of purpose and character to pursue studies under such circumstances.

Fortunately Chandra's grandfather, R. C., was spared most tribulations of this kind; for his great-grandmother had planned and established a small household for his high school education in the nearby town of Kumbakonam, which had an English school. Kumbakonam was the largest town in the Tanjore delta region, situated only a few miles away from Mangudi, R. C.'s home village. The town was known (and continues to be known) for its temples and sacred ponds. Myth has it that the Goddess Ganga (the river Ganges) herself springs up once every twelve years to fill one of the biggest and most sacred ponds in the town. Ganga does this by coming directly from Banaras, the sacred city in the north, thousands of miles away. The underground journey takes her twelve years; once every twelve years, Kumbakonam becomes one of the most popular pilgrimage sites in South India.

Ramanathan brought rice and other necessities to his son at regular intervals from the home village. Thus helped and supported, R. C., a tall, handsome young man with a robust physique, graduated from high school in 1881. He then continued his education, first at a two-year college in Trichinopoly, and later at Christian College in Madras. Had he completed his B.A. degree as expected, he probably would have ended up in a high government post; that was, after all, the principal aim of western education for a great majority of Indians. R. C., however, followed a different path. He took his new education seriously.

He read English literature and philosophy extensively, studied mathematics and physics, and in general pursued what interested him rather than what was required of him. "Exposed to the ideas of freedom of thought, expression, and free inquiry, he became an agnostic," says Chandra's father, C. S. Ayyar, in his "Family History." "He came under the influence of Colonel Ingersoll by reading the latter's Free Thought Lectures; R. C. also read the works of Charles Bradlaugh and Herbert Spencer. He spent a great deal of time in discrediting Mme Blavatsky, who was the leader-founder of the Theosophical Society. As a result of these extracurricular activities, he did not obtain his degree because he failed in Tamil, which was one of the required subjects. After that he was unable to continue his education as a full-time student."

Although R. C. had departed from traditional ways in pursuing western education, tradition had prevailed in other aspects of his life. While still a high school student, he had married Parvati;[11] now after four years the time had come to take on family responsibilities. He opened a lower secondary school in a town between Kumbakonam and Tiruvanakoil, his wife's village. Within a short time, however, the venture failed and they moved back to Trichinopoly, where he was appointed to teach high school courses at his old college. He was paid a salary of ten rupees a month, which was far too inadequate to live on, even in those days. Furthermore, during these years of hardship, R. C.'s maternal uncle was living with R. C.'s family while attending school. With some financial help from his brother-in-law, R. C. managed to live modestly with his wife in a two-room tenement.

It was during such strained family circumstances, and in the midst of wildly turbulent times in India, that Chandra's father, C. Subrahmanyan Ayyar, was born in 1885 in the village of Tiruvanakoil near his mother's family home. A second child, C. Venkataraman (later known as C. V. Raman), was born in 1888.[12] They were the oldest children in a family that would eventually include five sons and three daughters.

In spite of his growing family, heavy teaching responsibilities, and a meager income, R. C. continued his studies. Fortunately, the university rules had changed and made it possible for him to take the examinations in parts, one subject at a time. Taking advantage of this flexibility, he finally obtained his B.A. degree in 1891, only the third person from his area to go so far in higher education.

It was while he was teaching at Trichinopoly that R. C. became involved in what is known locally as the "St. Joseph's College Commotion." A minor incident in local history, but one that reveals much about R. C.'s personality and maybe sheds light on some of his reasons for going into education rather than government service. In a wider perspective, it also reveals the gradual beginning of the national movement in South India, and the nonviolent passive resistance that was to become the way of struggle for national independence.

Trichinopoly,[13] an ancient city situated on the banks of the river Kaveri, has a famous temple at its center built on a great bare rock, like the Rock of Gibraltar, rising 273 feet above the city. The temple's annual festival was a big event of the city; like other Indian temple festivals, it was highlighted by a procession of the deity in the temple chariot. In such processions, the chariot is drawn by men who push and struggle for the chance to put their hands on the thick rope used to pull it through the streets to the accompanying music of drums, horns, and other instruments. As the procession moves through the city, it stops in front of the temples of other friendly deities, so that the deities may greet each other. It is a joyful occasion; the musicians compete with each other, they perform special routines, and all the people celebrate. Every year the rock temple chariot stopped before the Ganesa temple in the city. Ganapati or Ganesa is the elephant-headed God of Learning, the son of Shiva. The Ganesa temple, however, happened to be in front of "Clive House," home of some Jesuit missionaries who taught at St. Joseph's College.

One year in the early 1890s the missionaries complained to the district authorities about the traditional affair, which took place between ten and eleven in the evening. They did not want the chariot to stop before the Ganesa temple, because the gathering and the loud music disturbed their sleep. They got an ordinance to that effect, and the temple committee responsible for the festival meekly acquiesced to the order. When the news got around to the students of R. C.'s college, they were outraged at what was an obvious, flagrant violation of the religious rights of Hindus. Under R. C.'s leadership, they decided to stop the chariot as usual in front of the Ganesa temple. R. C. masterminded an ingenious plot. All was quiet and normal in the streets as the procession neared the temple. At just the right moment, students in large numbers started coming into the street from all directions. The chariot had to come to a complete stop. Then in a peaceful and orderly way, each student went up to the chariot with a coconut to be offered to the deity. The chariot was stopped in front of the Ganesa temple, and Clive House, for more than an hour, whereas normally it would have stopped for only ten or fifteen minutes. The crowd was quiet, peaceful, and orderly, giving the police no chance to intervene. R. C. stood under a nearby street-lamp, like an innocent bystander, as though he was surprised by the whole operation.

It is quite likely that this incident may have prevented him from entering government service. As it was, the Jesuits had their revenge by not allowing one student,[14] who was on one of their scholarships, to appear for his B.A. examination on the excuse that he left their "hostel" premises without prior permission.

In 1892, R. C. was chosen to be an assistant to the principal of Mrs. Ankitham Venkata Narasingrao College in the city of Vizagapatam.[15] In culture and language Vizagapatam was almost a foreign land. Yet, leaving his eldest son (Chandra's father) to take care of the aging Ramanathan, he proceeded to the distant place with his wife, an infant daughter, and Raman, then just four years old. It proved to be an arduous journey, taking twenty-one days. They first had to go out of their way to Guntakal in order to take the train to Bezwada (Vijayawada); then they traveled by construction trains and canal boats to Rajahmundry. Finally they rode for four days in carts pulled by bullocks to reach their destination. At that time the railroads were just being built. The very next year, Chandra's father could join them in one day.

The journey marked the beginning of R. C.'s real career as an educator, and the starting salary of eighty rupees a month at least offered a little better financial security than his previous salary. R. C. soon established a reputation as a very good teacher and administrator. Within a few years of his appointment, he became a professor of mathematics and at the same time held the position of vice-principal of the college. He played an important role in the improvement of elementary education in the Madras Presidency, especially in mathematics, by providing a well-coordinated syllabus in arithmetic, algebra, and geometry. Even more remarkable was the fact that he never stopped his own education, even though his personal life was plagued by severe financial difficulties, the responsibilities of a large family, and his own health problems. As vice-principal he was making approximately 125–175 rupees per month, still not enough to support a family of eight children, aging parents, and other relatives in need. He was frequently in debt. In 1892 he was seriously ill; a normal medical procedure (tapping for hydrocele) led to serious complications, due to medical incompetence, which then required major surgery.

He continued to study on his own for master's degrees in mathematics, physics, and English literature. Studying for advanced degrees was the only way for someone in R. C.'s circumstances to keep up with new developments in his fields of interest. Academic degrees also determined the levels of courses one could teach, and he wanted to teach advanced courses and to write advanced textbooks. However, it is not easy actually to get such degrees while shouldering the dual responsibilities of teaching and administration, and he did not succeed in getting them. But R. C. accomplished something that proved to be far more valuable and important for his children and for his grandchildren: he acquired a great many books on a variety of subjects, resulting in a fine home library. This ready availability of books was certainly one of the most significant factors in the childhood development of his sons and grandsons. Raman's early interest in physics and Chandra's early start in mathematics were mainly due to this fortunate situation.

R. C. was unconventional in other ways. As already mentioned, he was an agnostic most of his life, not an easy thing to be in a society dominated by tradition, orthodoxy, mechanical and faithful observances of rituals, and superstition. Although in his later life he came under the influence of Annie Besant[16] and did become a theosophist, he maintained an open mind. In no way did he subject himself or his children to the prevailing religious customs and discipline. He was far ahead of his time in letting Raman choose his own bride. He was devoted in his youth to many types of sports, body-building exercises, and Indian martial arts, quite out of line with the usual preoccupations of brahmans. In his history, Chandra's father writes,

R. C. played both football and cricket. He organized outdoor sports like long-jump, high-jump, pole-jump, walking on skates, etc. In his younger days, he was never satisfied with less than four sets of tennis. And, at the Vizag club, he used to play billiards when he could not move about very much.

"I have gathered that my grandfather was a very strong man," Chandra says, "and that he kept himself fit by constant exercise." He was, from all accounts, known to be a gentle and considerate person.

R. C.'s life, remarkable in so many ways, came to an end at the relatively early age of forty-six. It may have been his own doing. Loving vigorous exercise, he was prone to excess and liked to demonstrate his powers. One occasion of such showmanship[17] led to the breakdown of his health soon after he took his new position in Vizagapatam. In 1907 he underwent surgery for a double hernia. This surgery, though minor and commonplace now, was a major operation in those days. Complications ensued, and he was left in a weakened, almost disabled state. When he had an attack of pneumonia in 1910, he died, just seven months before Chandra's birth.

In spite of his short life, he left a profoundly new heritage for his family. Higher education came to be accepted as the common pattern rather than an exception. Many of his descendants have distinguished, successful careers in the arts and sciences, in industry, and in government service. In just one generation, a typical brahman family, living off the land and leading a traditional village life, had been inducted into the modern world.[18] This pattern was repeated in many families, reflecting the emergence of a new middle class under British rule.

C. S. Ayyar's and Raman's primary education was mostly under the tutelage of their father. They learned Tamil, arithmetic, and English at home till they were seven years old, a tradition C. S. Ayyar would continue with his children. For their secondary and higher secondary education, they attended the college in Vizagapatam where their father was the vice-principal and sometimes their teacher. R. C.'s zest for learning and his varied interests, as we have seen, were phenomenal. He did his best to pass them on to his children. He was not satisfied to teach them just what was required in English, algebra, geometry, trigonometry, and physiography. He urged them to read on their own. "He read with me," recalls Chandra's father in his history, "Scott's *Lay of the Last Minstrel,* Shakespeare's *Coriolanus* and portions of Milton's *Paradise Lost.*" R. C. shared with his children his love for Indian classical music and his fine collection of books. The children also experienced his free thinking and the questioning mind that he brought to religion. Thus, C. S. Ayyar and

Raman had a strong, though unusual, springboard from which they launched themselves into successful careers.[19]

C. S. Ayyar was a brilliant student. After completing two years of college in Vizagapatam, he attended Presidency College in Madras, where he earned a bachelor's degree and then studied civil engineering for two years. He took the Superior Accounts Examination[20] of the Public Works Department, secured first rank, and entered British government service as a deputy accountant-general. Soon he found himself serving in various capacities in the audit and accounts offices of most of the major British railway companies in India, rising to the position of a chief auditor. His work took him all over India.

Raman's career was even more dramatic. He graduated from high school at eleven; completed college at fifteen; got his M.A. at seventeen; and took the All-India Competitive Examination, which chose civil servants for the Finance Department. He received the highest score and, at the age of eighteen, he became an assistant accountant-general.

Although he pursued a career in the government railway service, C. S. Ayyar remained keenly interested in music and literature. He played the violin well, wrote books on Karnatic (South Indian) music, traveled abroad, and even wrote fiction.[21] However, he followed these pursuits only in his spare time, and more fully only after his retirement from government service in 1940. His family responsibilities and the advancement of his career, which brought financial security and better opportunities for his children, took precedence over dubious adventures in the arts or sciences.

In contrast, Raman always wanted to pursue original research in physics. Though his father persuaded him to take the competitive examination and accept a position in government service, he continued doing research in his spare time. Ten years later, he gave up his lucrative position as a finance officer for full-time research, against the advice of his family, friends, and colleagues.

Raman eventually became a celebrity in India, almost as much of a household name as Nehru or Gandhi, and one of the most colorful personalities in the world of science. He made original research not only respectable but a way to world fame and immortality, certainly something no position in government service could offer. For many young people, he became a model for their aspirations to overcome parental pressures and pursue their own independent courses.

The two brothers entered government service in 1907 and were posted to Calcutta. At that time, C. S. Ayyar had already been married

for five years to Sitalakshmi, who had been chosen for him in the traditional manner when he was seventeen years old and just out of high
school. His eleven-year-old bride was the second daughter and fifth
child of Rao Bahadur N. Balakrishna Aiyar.[22] Raman, on the other hand,
chose his own bride, Lokasundari Ammal, when he was eighteen, and
married her just before he was to depart for Calcutta. The two families
lived together in Scott's Lane, off Bowbazzar Street.[23]

In 1909, C. S. Ayyar was transferred to Lahore to serve as the assistant auditor general for the Northwest Railways. Lahore was then the
capital of the Punjab province in British India, the headquarters of
Northwest Railways, and the railway junction for British railway lines
from Karachi, Peshawar, and Calcutta. This historic city[24] on the banks
of Ravi, one of the five rivers of the Punjab,[25] had in 1910 a population
of approximately 200,000, predominantly Muslim. In the partition of
India in 1947, Lahore became a part of Pakistan. In language, culture,
and distance, no place in India could have been farther away from the
Ayyars' native village.

Chandra was born in Lahore on 19 October 1910.

3

Determined to Pursue Science
Lahore and Madras, 1910–1930

From a purely scientific point of view, the most crucial incident was my meeting with Sommerfeld when he visited Madras in 1928.

Chandra

Chandra was the first son of C. S. Ayyar and Sitalakshmi. He had two older sisters, Rajalakshmi (Rajam) and Balaparvathi (Bala). Chandra's birth, after that of two daughters, probably caused great rejoicing and happiness because in Hindu tradition only a son preserves the family lineage. As an adult, he assumes his father's responsibilities, inherits property, and cares for the family. Also, only a son can perform certain annual rituals (*shraddha*) that ensure perpetual bliss for deceased ancestors.

Later the Ayyars had three more sons (Vishwanathan [Vishwam], Balakrishnan, and Ramanathan [Ramnath]) and four more daughters (Sarada, Vidya, Savitri, and Sundari). Chandra, as the firstborn son, inherited his paternal grandfather's name: Chandrasekhar. But soon his elders called him "Ayya," and his younger brothers and sisters called him "Anna."

When Chandra was six, the family moved from Lahore to Lucknow in Uttar Pradesh, a province of northern India. Two years later, C. S. Ayyar became deputy accountant-general in Madras. There he established a home where the family stayed while he traveled, because he was frequently transferred from place to place throughout his career.

The Ayyars' house[1] in Lahore was outside the city walls, not far from a large public garden known as the Lawrence Gardens. Chandra has few memories of his childhood, but he does remember frequent visits to the gardens and the Anarkali bazaar,[2] which is even now a well-known and popular shopping place in Lahore, teeming with products from all over

47

the world. He also distinctly remembers the beginning of the First World War in 1914, which coincided with the birth of his younger brother Balakrishnan.

Besides his sisters and brothers, he had many other companions as he was growing up in Lahore. These were the children of family friends from Madras who also worked in the government service at Lahore. These Tamil-speaking families formed a close-knit community. They visited each other often, celebrated their native festivals, and maintained a distinct cultural identity. Since there was no real mixing with the pre-dominantly Muslim community or the Hindi or Punjabi-speaking mi-nority, Chandra learned only Tamil as a child.

A healthy and handsome child, Chandra was "unbearably mischie-vous." At least that is what his youngest uncle, Ramaswamy,[3] used to hear from Chandra's mother. Chandra used to pick on his oldest sister, Rajam, by teasing her and quarreling with her over toys. "He used to take the lion's share of everything," recalls his sister Bala. "He would break his things first and take my elder sister's. He would keep them for himself, because he would claim they were his. They were given to him."[4]

Chandra does not remember these incidents, but he does remember the day his primary education began, when he was five years old. On this auspicious day, the first *Vijayadasami* day, which falls in Octo-ber and is an important festival for the Hindus, he sat by his father's side and wrote three letters of the Tamil alphabet on the sand spread before him. Beginning with that special festival day, which is celebrated in nearly all brahman families,* Chandra started studying; he learned Tamil from his mother, English and arithmetic from his father.

Chandra's parents began all their children's education at home. This practice was common among middle-class families, since the public or municipal schools were poorly run. Usually they were one-room schools with inadequately trained teachers; they also mixed several social classes and castes and, therefore, were not popular among those that were privileged by high caste or wealth. Few, if any, private schools existed, and those few were reserved for the children of the ruling British or of the well-to-do princes, maharajahs, or their surrogates. Schools run by

---

* *Vijayadasami* is celebrated slightly differently in different parts of India, but the essential aspects of the ceremony are more or less the same. Offerings are made to the God of Learning (Ganapathi) or the Goddess of Learning (Saraswathi) at a *puja* ceremony with relatives and family friends, followed by a sumptuous dinner or distribution of sweets. The boy or girl and the father sit side by side; the assembled people, one by one, write greetings, good wishes, Gods' names, on the sand spread before them. The ceremony ends with the child writing the first few letters of the alphabet on the sand.

Christian missions concentrated their efforts among the poor and the lower castes, who were easier targets for conversion to Christianity.

Above all, parents in the middle or upper-middle class found teaching their children a pleasant diversion from their dull clerical or bureaucratic work. Besides, English normally was not taught in the primary schools, and parents who had an English education were eager to begin their children's English lessons as early as possible. English was a prerequisite for a job in the British service. Most educated parents dreamed that their sons* would be ICS (Indian Civil Service) officers one day, then "district magistrates," or "collectors," or "assistant commissioners" of a province, the highest position to which an Indian could aspire. There were other All-India Competitive Examinations which selected candidates for other government services like the audit, police, and military. A few aspired to high positions in these services, while others sought law degrees and dreamed of going to England for further studies so they could return as barristers and practice law in the high courts of India.

Chandra's home education was quite disciplined. Recalling those early years, he says, "My father used to teach me in the mornings before he went to his office, and then after he went to the office, my mother would teach me Tamil. In the afternoon she would supervise both the English lessons and the Tamil lessons we had to do." Chandra's parents did not have to worry about his studying or about exercising discipline. He enjoyed learning English,[5] and arithmetic caught his fancy very early. "I remember very well," he says, "that my father used to assign lessons and exercises. I used to do far more and very often went far ahead of the assignments. I found that I could study the books on arithmetic on my own. So when my father came home, I had done one chapter (or more) ahead of what he wanted." At first, Chandra's father was amazed, but he and others soon realized that they had an exceptionally bright child in their midst.

There were, of course, precedents in earlier generations of his family—notably his uncle C. V. Raman—who exhibited precocious brilliance in school and college. So Chandra was allowed to follow his own course; when they moved to Madras in 1918, his father hired private

* At the time, higher education for a daughter was considered detrimental to her chances of marrying into a good family. Early marriages prevented a significant number of women from undertaking higher education. Still, a few grades of English education for a girl was considered an asset—a qualification to marry an educated husband who, if he was lucky, would have a job in the British Raj.

tutors to teach him.* Not until 1921, when he was eleven, did Chandra go to a regular school. He was accepted straightaway into the third form, skipping two years of normal high school.

The Hindu High School in Triplicane, where Chandra was a student, was considered the best school in Madras. Situated on a street named "Big," the school was an imposing red building kept cool by large shady trees. Chandra and his two brothers, Vishwam and Balakrishnan,† and his uncle Ramudu all attended this school in successive years. In the beginning, Chandra found formal education neither easy nor pleasant. Education at home had opened their minds, as Balakrishnan notes, "to a free play of their own without being cramped by a load of educational lumber on the head during the early years." Chandra, who had studied only what he liked (mainly English and arithmetic), was suddenly required to study history, geography, and general science and to take periodic examinations. Thus his first year in school was a disappointing beginning. Only when he discovered that the fourth form curriculum included algebra and geometry did he get excited. Without waiting for classes to begin, he began studying these subjects during summer vacation. "I remember getting the books of my higher class," says Chandra, "and reading them ahead of the classes. I remember reading Pierpoint's texts on geometry; I went right through the first two books before I got into my fourth form. When I got to the fourth form, I knew all the geometry and all the algebra they were going to teach, and in fact more— permutations and combinations, solving cubic equations, and so on. Similarly in my [next three] summer vacations, I started studying conic sections, coordinate geometry, calculus, and differential equations." With this kind of drive and motivation Chandra did extremely well in high school and became a freshman at the Presidency College in Madras when he was only fifteen years old. He was regarded as a prodigy, especially in mathematics, both at home and at school, sometimes to the consternation of his younger brothers who came after him in the same school, as they naturally became targets for comparison.

These early years of learning were happy years for Chandra. Though

---

*Tutors came to the house and taught each of the brothers individually. Chandra's father did not consider the schools to be good enough. It was certainly a tremendous advantage to have the privilege of individual instruction so one could learn at one's own pace. However, Chandra's father sent his daughters to regular schools. Private tutors were expensive, and in the matter of education, boys always came first; girls took second place.

†Vishwanathan (1911–79) was the General Manager of the Tata Steel and Iron Works at Jamshedpur, Bengal; Balakrishnan is professor emeritus of pediatrics at Jawaharlal Nehru Medical Center at Pondichery.

the family was growing—a new child every two years or so—C. S. Ayyar's income provided a comfortable life. Household help spared the children from the responsibility of regular chores. When the family first moved to Madras they lived in a spacious rented house named Brindavan in a pleasant residential area called Purashawakam.

C. S. Ayyar's assignments frequently took him away from home, sometimes for long periods. But because he was in the railway services, he and the family received free railway travel or reduced fares, so they got together more frequently than they could have otherwise. The children traveled to all parts of India—a privilege very few Indians could afford. Thus life changed markedly for C. S. Ayyar and his family. Grandfather R. C.'s efforts had paved the way for a new urban life for his children and grandchildren.

However, education and urban life could not change completely centuries of tradition. Chandra's father, a highly cultivated individual, widely read and traveled, still accepted certain customs and practices. And when it came to family matters, he was traditional and authoritarian, demanding unquestioned obedience from everyone, very much a father of his generation. Reserved and undemonstrative, he remained aloof from his children; they, in turn, could not share their innermost thoughts or feelings with him.

Tall for an Indian and strong and handsome, C. S. Ayyar was an imposing personality. At home he wore the typical South Indian dress: a shirt over a *veshti* (*lunghi*). For work, he donned a shirt, necktie, jacket, and trousers, but he always wore his turban, the hallmark of the *Madrasis*. Turban-wearing Indians dominated all branches of British government service as well as the fields of science and education. Raman's turban became as famous and as familiar to the public as his discovery and his Nobel Prize.

If the jacket, tie, and trousers signified westernization, the turban signified a reluctance to totally abandon the old ways. In this, Chandra's father exemplified the prevailing conflicts of the Indian society ruled by the West. One gets a very good introduction to these conflicts and, at the same time, to the conflicting elements in C. S. Ayyar's own complex personality by reading his book called *Life's Shadows*. Written in the form of short stories, the book depicts his thoughts on what makes an ideal father, an ideal son, an ideal daughter, and so on. He was indeed a curious blend of the old and the new.

"C. S. Ayyar is a dynamic personality," wrote G. H. Ranade, a noted Hindustani musicologist interested in South Indian (Karnatic) music. He spent Christmas week of 1941 as a houseguest of the Ayyars while attending the annual conference of the Music Academy of Madras and recorded his impressions of C. S. Ayyar and his family:[6]

. . . I reached C. S. Ayyar's place late in the evening, when it was almost getting dark. After greeting me, he was quick in action and thought alike. In one breath, he pointed out to me the room which I was to occupy, containing also the library of his music-books and instruments, gave orders for tea and bath for the guest (myself), cut pieces of tape for my mosquito curtain, inquired casually about any incidents of my journey and did many other similar things for me. It was certainly an odd time for a bath, but according to him, one must have a hot bath after a long journey, and I soon saw and felt the wisdom of it!

Mr. Ranade, being a Maharashtriyan from Poona near Bombay, had some misapprehensions about the south:

I soon realized that Sri C. S. Ayyar, perhaps because of his wide travels and long stay in almost all the major provinces of India, is more Indian than most of us, and he might easily pass for the model of the cultured citizen of the India of the morrow. In fact his house is a small democracy, in which he plays the role of a friend, philosopher and guide to his children. He, however, sometimes insists on maintaining the dignity of the head of the family, would rather roar than speak, run than wait and thinks that the other man's clock always goes too slow! The next moment, however, he would cool down and his people love and venerate him all the more for the simplicity and the straightforwardness. This nature of his reflects equally upon his intellectual accomplishments and pursuits. But here he has sometime to come in conflict with people, who have not known him long enough, or who have their own axes to grind. The latter sort take him to be either impetuous or imperious.

In contrast, Chandra's mother had married young and received only a few years of formal elementary education. When she entered her husband's large family, she found herself dominated by a very strong-willed mother-in-law and a grumbling and peevish aunt, who had the misfortune of being a widow, a virgin widow at the age of thirteen. The trials and tribulations of almost every Indian girl—married quite young and transferred to an alien family, required to behave in certain set ways, scrutinized constantly by older and younger relatives—cannot be told in a few pages. What is important here is that Sitalakshmi managed to survive it all. And on her own, with some help from her husband, she continued her education while bearing ten children. She learned English well enough to adapt Ibsen's A Doll's House into Tamil, translated a long story by Tolstoy for a Tamil journal, and contributed several articles to newspapers. She championed her daughters' wish for an education[7] and wanted them to marry educated husbands so they would become part of educated families.

Since C. S. Ayyar traveled frequently, Sitalakshmi became the family guardian, keeping them together, teaching them, and attending to their needs. She imposed no strict religious discipline; yet she infused her children with the cultural heritage of Hinduism and the ideals it stood for. According to Balakrishnan, "Even before I had learned the Tamil alphabet, I came under the influence of that perennial source of inspiration of the youngest boy or girl in our land—Valmiki. Mother used to read aloud every morning from Pandit Natesa Sastri's literal translation of the *Ramayana* after having duly bathed and purified herself for the ceremony."[8] All the brothers and sisters gathered around her and listened. She influenced them, especially Chandra, who, as the eldest son, surely held a special place in her heart.

The five older children were all close in age, and they became good companions. All highly motivated, good students, they did not need outside friends, though they visited the children of family friends occasionally. The entire family occasionally rented a horse-drawn carriage and went on evening outings to the beach on the Poonamallee High Road. On such occasions, Chandra demanded and got to sit in the front seat next to the coachman. Neither this nor the rivalry between himself and Rajam nor his marked brilliance in school seemed to create turmoil between him and his brothers and sisters, except that the girls had an easy upper hand; they teased their brothers no end by reciting from Mother Goose, "What are little boys, little boys made of? Snaps and snails, and puppy-dog tails; . . . What are little girls made of, made of? Sugar and spice, and all that is nice . . ."

Chandra was sensitive, warm, considerate, and caring of others. His performance in school was so brilliant that the others did not compete, become envious, or hold grudges. "He was thought of as a genius by all the people," says Balakrishnan, "and shown marked deference by all the members of the family. His interest in mathematics was evident from the beginning as he used to do an incredibly large number of problems every day. In tests, he always scored more than one hundred percent by answering more questions than was necessary [that is, if one was required to answer, say, any eight out of ten questions, he would answer all ten], and he would be always right."[9] Teachers were naturally awed and showed great respect. One, for instance, asked him one day to name his favorite vegetable. When Chandra said that it was a ladyfinger (okra), the teacher admonished all the students to start eating that vegetable every day. It must be good for the brain. Look at Chandra!

Chandra's happy days as a boy ended in 1921–22. First, he noticed the discrimination his sisters experienced at home. The girls did not have private tutors; instead they were sent to public schools because their education was not considered as important as the boys'. Then, in 1921, Rajam married A. S. Ganesan (who later became a physicist). The next year, it was Bala's turn. Both girls, in their early teens, had to stop going to school soon after marriage, since that was the prevailing custom. The girls, too young to appreciate their new station in life, were extremely unhappy about the restrictions imposed upon them. Chandra, aware of their unhappiness, was struck by the unfairness. He saw no reason for the termination of their education. Bala's bridegroom was not, in her mother's opinion, a good choice; he had failed his university entrance examination, and Chandra's mother, who valued education highly, disapproved, and vehemently protested. But the marriage took place since C. S. Ayyar, the head of the family, had decided. As Chandra recalls, "Here was Bala who was told who she would marry, and she knew perfectly well that her mother was not satisfied with the choice; then she was stopped from going to school. It shattered her and it shattered me. As a young boy, I realized that something that was making these girls unhappy was being done. And somehow or other it made me very angry. My mother's unhappiness also affected me deeply. I sensed the injustice done to my sisters and my inability to do anything. . . . Looking back all those years, [I know that] my personal warmth for my father was somehow diminished. . . . I may not have so recognized it at the time, but I do remember how totally distraught I was that this kind of unfairness was going on."

Disillusioned, Chandra found ways to channel his energies into algebra and geometry. He discovered books on conic sections and calculus in his grandfather's library, withdrew from his family, and concentrated on his studies, learning on his own.

In 1924, Chandra's father built his own house (No. 46, Edward Elliot's Road) in Mylapore, a prestigious suburb of Madras.[10] Named Chandra Vilas, it remained the family home throughout C. S. Ayyar's life.* Chandra was in his last year of high school and vividly remembers the new house going up, beginning with its ground breaking, then the foundation and the first bricks being laid. He kept the roll of the workers, whom he remembers as "charming carefree men and women with

---

*Chandra Vilas is still in the family. Chandra's youngest brother, Dr. S. Ramnath, has made it his retirement home. Ramnath was the chief resident officer of the Vikram Sarabhai Space Center at Trivendrum.

2. Ramanathan Chandrasekhar (1866–1910), Chandra's grandfather.

3. Parvati Chandrasekhar (1869–1916), Chandra's grandmother.

4. Sitalakshmi *née* (Divan Bahadur) Balakrishnan (1891–1931), Chandra's mother, 1914.

5. C. Subrahmanyan Ayyar (1885–1960), Chandra's father, 1917.

6. Sir C. V. Raman (1888–1970), 1930 Nobel laureate in physics, Chandra's uncle. Photograph courtesy of S. Radhakrishnan.

7. Chandra with his older sisters and younger brother, Lahore, 1913. *Left to right:* Rajam, Chandra (age three), Vishwam, and Bala.

8. Chandra at age six, 1916.

9. Chandra Vilas, the Ayyars' home in Madras where Chandra lived before going to England.

10. Chandra after receiving the Ph.D. degree, Trinity College, Cambridge, 19 December 1933.

11. Meetings in Russia with astronomers and astrophysicists, 1934. (a) *Below left to right:* Viktor A. Ambartsumian, Nikolai Kozyrev, S. Chandrasekhar, Evgenii Ya. Perepelkin, and Dmitri I. Eropkin, at the Pulkovo Observatory in Leningrad. (b) *Opposite left to right:* Mme Inna N. Leman-Balanovskaya, Evgenii Ya. Perepelkin, S. Chandrasekhar, Boris P. Gerasimovich, O. A. Mel'nikov, and Innokentii A. Balanovskii, at the Pulkovo Observatory in Leningrad. (c) *Next page:* At the Shternberg Institute, Moscow. *Left to right:* A. A. Kancheev, the director of the institute, . . . , S. Chandrasekhar, and Vorontsov-Velyaminov. Inscription on back of photograph is reproduced below it.

a

With gratitude for a
brilliant lecture 47 years
later — and hoping to
hear more!
Moscow, 'Astron Inst
of Univers!
R.Bell/(Zeldovich/. IIC
Ull and/Shandarin/
(Polnarev/
(Novikov) (V.N.Lukash/
N.Tutukov

Vladimir G. KURT   Leonid P. Grishchuk
Dhompancety

12. Chandra as Fellow of Trinity College, Cambridge, 1934.

13. Sir Arthur Stanley Eddington (1882–1944), 1932. Photograph courtesy of The Bettmann Archive, UPI/Bettmann Newsphotos.

14. Chandra with William H. McCrea, London, December 1935.

15. Conference on White Dwarfs and Supernovae, Paris, August 1939. The photograph shows all conference participants. *Left to right:* (*front*) Frederick J. M. Stratton, Cecilia Helena Payne-Gaposschkin, Henry Norris Russell, Amos J. Shaler, Arthur S. Eddington, Sergei Gaposschkin; (*back*) Carlyle S. Beals, Bengt Edlén, Pol F. Swings, Gerard P. Kuiper, Bengt G. D. Strömgren, S. Chandrasekhar, Walter Baade; Knut Lundmark is standing between Chandrasekhar and Eddington.

joyous faces." The family moved into Chandra Vilas the summer of 1924, and Chandra's remaining years in India were spent there. It is a magnificent, two-story, brick-and-concrete house which is quite spacious, with nine rooms including a kitchen and dining area, a back porch, a middle living space, and a front verandah. Chandra occupied the front room (the "Bay Room") on the second floor. He could sit by the window and look over the coconut grove adjoining the house. There were only a few scattered houses nearby, one of which belonged to Sarvapalli Radhakrishnan, the noted Indian philosopher and educator who later became the President of India (1962–67). Another nearby house, Sri Vilas, belonged to Mrs. Savitri Doraiswamy, wife of Captain Doraiswamy, a medical officer in the Indian armed services. Lalitha, Chandra's future wife, was one of their daughters.

Chandra's first two years in college (1925–27) proceeded smoothly. His private studies in mathematics put him far ahead of his classmates, yet that alone did not decide the class grades. Grades were (and to a large extent still are) a strong, integral part of Indian education and also the measure of a student's brilliance. Chandra invariably received the highest grade in class without sacrificing his independent interests.

He threw himself into his studies, so much so that his father became concerned about his health. Vidya recalls that their father often used to admonish him to stop working so hard and stop reading all the time. "Go to the beach and get some fresh air," he used to insist. On such occasions, Chandra used to disappear from his father's sight. He would go to a neighbor's house, read there for hours, come home, and tell his father that he had gone to the beach.

In college Chandra studied physics, chemistry, English, and Sanskrit. He found himself drawn to physics rather than chemistry, and he also seemed to have an innate enthusiasm for English. He read extensively, often aloud to his brothers. Balakrishnan remembers some of Chandra's favorites: Oliver Goldsmith's *The City Night,* Robert Louis Stevenson's *An Autumn Effect,* Edward V. Lucas's *A Philosopher That Failed,* the description of Egdon Heath in *The Return of the Native,* and the ending of *Tess of the d'Urbervilles* both by Thomas Hardy, Ruskin's *Sesame and Lilies,* and Shakespeare's *Merchant of Venice.* Although he enjoyed Sanskrit, Chandra totally neglected its study until he realized belatedly that he was required to pass an examination in Sanskrit. Since there was not enough time to study on his own, he sought the help of a Sanskrit *pundit* and took a crash course.

After completing his intermediate two years at Presidency College with distinction in physics, chemistry, and mathematics, Chandra's next step was to work toward a B.A. honors degree. He wanted to take mathematics honors; he had not only excelled in his mathematics studies, but he had also come under the spell of the legendary Srinivasa Ramanujan. Chandra was not quite ten years old when, on a certain day in April 1920, his mother told him that an item in the newspaper reported that a famous Indian mathematician, Ramanujan by name, had died the preceding day; she told him further that Ramanujan had gone to England some years earlier, had collaborated with some famous English mathematicians, and had returned to India only very recently. He was well known internationally as a great mathematician. Two years later Chandra heard the name again, this time from his uncle Raman during one of his visits to Madras. "I remember a conversation between Raman and my father," Chandra says. "Raman was telling my father that he [Raman] had been nominated for a Fellowship of the Royal Society by Gilbert Walker and Lord Rayleigh. Then Raman said that Ramanujan was the first Indian to be elected to the Royal Society in 1918, and J. C. Bose[11] was the next one in 1921."

Raman had spoken in praise of Ramanujan. The earlier comments of his mother and those of Raman made an unforgettable impression on Chandra. Later, in college, he had gone to the library to find Ramanujan's articles. A few more years would pass before Chandra understood the extent of Ramanujan's genius and achievement. But what he had heard and read spurred his interest in mathematics, making him want to pursue mathematics honors.

Unfortunately C. S. Ayyar had different ideas. He saw no future in mathematics and wanted Chandra to take physics. After completing the B.A. honors degree, Chandra was to go to England for the Indian Civil Service (ICS) examination and become an ICS officer. By common consent that was certainly the right thing for such a brilliant young man to do; anything else was a waste of talents and opportunities.

Chandra did not mind taking physics, but he had no intention whatsoever of entering government service. He wanted to do basic research in either mathematics or physics. No ICS officer could compare with a Ramanujan or a Newton. His uncle had already set a precedent in the family by devoting himself to pure research. "The two scientists' names I knew were Ramanujan and Raman, and to some extent they were my role models," recalls Chandra. And, fortunately for him, his mother gave him strong support. "You should do what you like. Don't listen to him;

don't be intimidated," she told Chandra when his father became insistent about the civil service, a better job, and a better life. Chandra remembers his mother's support and says, "If I consider my grandfather, who was a professor of mathematics and enormously interested in educating his children, I can understand that his interest in educating the children was by and large motivated by his desire that they get into positions of greater upward course. My father took that same view. But it is curious that my mother did not take that view. One would have thought that my father would have been much more open to intellectual achievement per se, but to him well-being in life played a bigger role. Whereas with my mother it did not."

At least the ICS examination was not an immediate concern; Chandra would not even be able to take it for three years. In the meantime, he would register for B.A. honors in physics to please his father. Once he made that decision, Chandra, during the summer of 1927, read Arnold Sommerfeld's *Atomic Structure and Spectral Lines,* a classic treatise on what we now call the "old quantum theory" of the atom. The book was advanced and difficult for someone who had just completed his first two years of college; yet Chandra was enormously pleased because he was able to work his way through it. Thus, essentially on his own, he gained a formidable background in both physics and mathematics.

When school started, Chandra officially became a physics honors student, but he attended lectures in the mathematics department. He studied the prescribed physics books on his own and took all the required tests. Since the mathematics courses were more difficult, he wanted to attend those lectures. It is to the credit of his professors[12] that they made it possible for him to follow his own course. As Chandra says, "My college professors allowed me to do what was normally forbidden. I was given complete freedom. They were highly sympathetic with the idea that a young man was devoted to science. They helped me every way they could. For example, in order to be able to take books from the university library you had to be a graduate student. I was given access to the library in my first year in honors as a special case."

However, Chandra was not all simply books and learning. "He was no book worm," says V. S. Jayaraman,[13] a classmate of Chandra in high school and college. "He participated in sports, played tennis, was known as an extremely good debater, and an all-round scholar. I remember him as a young man of strong convictions. He once said in a discussion that God was a hypothesis invented by man to regulate man's conduct and establish order and behavior between man and man." Chandra himself recalls a field and track competition in which he once participated. "I remember running in a race," he says. "As I was halfway around the track, well behind the others, I heard a loud applause. As I passed the finish line and continued to run, since I had one more lap to complete the race, I heard some people laugh and some cheer me. I ran the last lap all by myself, although I knew the race was over a long time before. I still won a medal though, a medal for sportsmanship."

Life at Chandra Vilas was quite comfortable and pleasant for the most part. The family had a full-time cook, two other servants, and a man to take care of the garden. All his brothers and sisters were scholarly and ambitious, each preoccupied with his or her own work. "We all got up early in the morning (by 6 AM)," says Vidya. "After morning coffee, we all retired to our respective rooms for our studies. Around 10 AM, we would start getting ready to go to school. After having a bath, we would have our morning meal cum lunch and off to school. The dinner brought us all together again."

This dormitory-style life had its lighter and fun-filled moments. Early one morning, for instance, the coffee was a bit delayed (probably because their mother was preparing something special). While waiting, Chandra gathered all his brothers and sisters on the back verandah and turned the group into a choir. As Vidya recalls,

We all started singing—a two-line Tamil nursery rhyme—about a fox that tried to get the grapes on a vine beyond its reach—with constant repetition of the same two lines. Chandra was "conducting" the chorus as it were, leading it in a crackling, drumming tune of his own. We all became exuberant, shouting at the tops of our voices. Gopal Aiyya, a neighbor and family friend, came dashing over, alarmed at what he thought was screaming and crying for help. "What is going on? I thought some thieves had come," he said looking at us all. Chandra came forward and said very seriously, "Music, Sir, it is music."

Another morning that Balakrishnan and Vidya remember very well was when Chandra initiated a game in which one had to include the phrase, "I doubt it," or "I don't doubt," or "without any doubt" in every sentence one uttered. If one failed to construct a meaningful sentence this way, he or she lost, and the game continued with the rest. When Chandra persisted for a long time after everyone else had lost out and it became a bit exasperating to the rest, Chandra's mother intervened and said, "Stop it now; if you continue, I'll pour this soup on you." Chandra's instantaneous retort, "I doubt it," brought the house down with laughter, including their mother. Chandra was the obvious hero in the family. As family friends used to say, in jest or half-seriously, "He [Chandra] is holding court and all his brothers and sisters are in attendance."

During the years of Chandra's high school and college career, his father was away most of the time. He was transferred to Calcutta, Lahore, Bombay, and even to Vizagapatam, where he had lived as a boy, for a short-term assignment. C. S. Ayyar's visits home during this period were not particularly pleasant for the young people in Chandra Vilas. He tended to be overbearing and self-righteous. Since his children already were good in school, he worried about their futures, about what they should do and what they should not do, and about their health. He lectured and preached that too much work and study and not enough play was unhealthy, that they needed more sleep, that everyone should be in bed by 10 PM, and so on. He imposed this rule rigorously even when examinations were imminent and intense preparations were necessary. Chandra remembers how his uncle Ramudu had to study under the light of a partially shaded kerosene lantern and how he himself had a hard time before his tests, when he had a lot of physics to make up because of his extracurricular studies in mathematics.*

Chandra, though young, showed considerable maturity and understanding when dealing with his father. Instead of letting conflicts lead to confrontation, he adapted himself to the situation. He was determined to pursue pure science, a goal he had set for himself as a high school student. However, he remained loyal to his father. He always behaved properly, reassuring C. S. Ayyar that he understood him and respected his advice. In 1928, for instance, after completing the first of his three-year honors course in physics, he wrote the following in a letter to his father (9 April 1928):

. . . I have only to feel gratitude for what all you have done for me. I am nowadays coming to like mathematical physics more than pure mathematics. If it had not been for your kind, yet persistent advice for my taking physics, I would have taken mathematics.

The benefit I will gain by taking physics will certainly be far greater than what I could ever acquire with mathematics and it was only your advice that

---

*All this rings an extremely familiar bell in my own personal life. How devious I had to be to avoid the attentions of my father during those hours of study late into the night or very early in the morning! How patient one had to be while listening over and over again to the stock platitudes regarding frugality, good health habits, value of time, etc., etc.

made me take a proper turn at a critical time. This only shows to me and makes me feel gratitude to you for knowing me much better than I do myself.

But in reality Chandra took physics because it did not alter his determination to do science. He spoke more freely to his brother Balakrishnan in a letter from England on 12 May 1931:

. . . A mind at 17 can certainly be worked into any particular frame, but the heart, the inner craving to satisfy certain ideals, if it is deep-seated, can *never* be shaken and *should not be shaken.* I wonder if you know this—when I was in the fifth form and later, I used to go to the beach and pray—to mold my life like that of an Einstein or a Riemann. However shameful and miserable my progress has been—that it is—my personal aspirations were there. *When I was asked to take physics in preference to mathematics, I did so because I knew that my inner aspirations were left untouched by the choice.* . . .

The year 1928 became an extraordinary time for Chandra. In February and March of that year, his uncle Raman, along with K. S. Krishnan, made a fundamental discovery in the molecular scattering of light. Later known as the Raman effect, the discovery provided a very useful tool in unraveling the molecular structure of substances using ordinary visible light. In March, when Raman was on his way to Bangalore to announce his discovery, he visited his brother's family. Young Chandra watched as Raman demonstrated the spectrum to enthusiastic and admiring visitors, heard him explain the effect, and heard others rank it with the Compton effect,[14] for which Arthur Holly Compton had received a Nobel Prize in 1927. To someone in the group who said that if he (Raman) only had discovered the new effect the previous year, he would have shared the Nobel Prize with Compton, Raman quickly responded, "No, I don't think so. It is much better this way—I don't have to share it with anybody. I will get all of it."

Soon thereafter Chandra went to Calcutta to spend the summer months working in the very laboratory where the discovery was made. Raman wanted to give his young nephew a start in experimental work; he asked him to assist one of his senior research scientists, who was working on X-ray diffraction by liquids. Within the first week, however, Chandra broke the apparatus and that brought a quick end to his future as an experimentalist. "I was never enthusiastic in the first place to do experimental work," Chandra says. "I went into it simply because Raman asked me to do it. Therefore, I was glad in a way that it ended in a failure. Yet in many ways my spending the summer in Calcutta in 1928 was, during that time and retrospectively, a more exciting experience because the atmosphere in the Indian Association for the Cultivation of

Science during those months was one of a group of people who knew that they were involved in a great discovery. It was an extremely exciting time to have been associated with such a group, particularly for a young man of my interests." The young man knew enough theoretical physics to participate in the excitement.[15] He could even explain the significance of the new effect to experimental physicists in the laboratory. He was staying with Raman, who, of course, was in a constant state of euphoria. They went for walks together and took occasional ferry rides down the Hooghly River; Raman talked about his discovery, telling Chandra about the worldwide acclaim it was receiving. He told him that S. N. Bose[16] had visited him and, after viewing the spectrum, had said, "Professor Raman, you have made a great discovery. It will be called the Raman effect, and you will get the Nobel Prize."

Chandra also came to know extremely well K. S. Krishnan, the principal collaborator and a junior colleague of Raman. Although twelve years apart in age, the two struck up a friendship that lasted through Krishnan's lifetime.

Thus inspired, Chandra returned to Madras to begin his second year of undergraduate studies. Soon he received a letter from Krishnan informing him that Arnold Sommerfeld* was going to be in India in the fall on a lecture tour and that Presidency College in Madras was on his itinerary. For Chandra this was most exciting news—a rare opportunity to hear Sommerfeld, especially because he had already worked through Sommerfeld's book. Perhaps he would be able to meet Sommerfeld, discuss physics with him, and impress him with his knowledge of physics in general and the quantum theory of the atom in particular.

Sommerfeld came to Madras in the fall of 1928 and lectured before the science students of Presidency College. Chandra knew from Krishnan where Sommerfeld was going to stay in Madras. As Chandra recalls:

I went to visit him in the hotel and told him I was interested in physics and would like to talk to him. He asked me to see him the following day, and so I went. He asked me how much I had studied. I told him I had read his *Atomic Structure and Spectral Lines,* an English translation. He promptly told me that the whole of physics had been transformed after the book had been written and

---

*Arnold Sommerfeld (1868–1951) was a German physicist noted for his original research and his teaching. He may indeed be called the father of theoretical physics in Germany. "What I especially admire about you," Einstein wrote to Sommerfeld in 1922, "is the way, at a stamp of your foot, a great number of talented young theorists spring from the ground." A list of physicists who studied under him includes such names as Peter Debye, Paul Ewald, Gregor Wentzel, Wolfgang Pauli, Werner Heisenberg, Otto Laporte, Albrecht Unsöld, Walter Heitler, Hans Bethe, Rudolf Peierls, and Herbert Fröhlich.

referred to the discovery of wave mechanics by Schrödinger, and the new developments due to Heisenberg, Dirac, Pauli, and others. I must have appeared somewhat crestfallen. So he asked me, what else did I know? I told him I had studied some statistical mechanics. He said, "Well, there have been changes in statistical mechanics too," and he gave me the galley proofs of his paper on the electron theory of metals, which had not yet been published.

Someone with less intellectual fervor, determination, and passion for study would have become terribly discouraged by the encounter. Instead, Chandra immediately launched into a serious study of the new developments in atomic theory that already had stunned Europe but had not yet made their way to India. Sommerfeld had studied the behavior of electrons inside a metal, applying the newly discovered quantum statistics of Fermi and Dirac. Chandra had enough mathematical preparation to understand the new statistics, a discussion of which was contained in Sommerfeld's paper. He began immediately looking for research problems where he could apply his newfound knowledge and, amazingly, within a few months he had written a paper entitled "The Compton Scattering and the New Statistics," which he thought contained significant enough results to merit publication in a prestigious foreign journal, such as the *Proceedings of the Royal Society*. But this journal only published articles by its Fellows or articles "communicated" by one of them.

As luck would have it, Chandra thought of someone who might be willing to help him: Ralph Howard Fowler, whose paper in the *Monthly Notices of the Royal Astronomical Society* he had come across as he was browsing through the newly arrived journals in the university library soon after his encounter with Sommerfeld. Fowler had entertained another application of the new Fermi-Dirac statistics in an entirely different area—astrophysics, specifically the theory of collapsed configurations of stars, namely, the white dwarfs. Within weeks of the discovery of the new statistics, Fowler had extended its application from electrons in atoms in the laboratory to electrons in stellar matter. Fowler's paper was subsequently to have a great impact on Chandra's life and career; at the time, however, Fowler, a Fellow of the Royal Society, was someone who knew Fermi-Dirac statistics and who might be interested in Chandra's paper. Therefore, he sent Fowler his paper in January 1929. After a few months of anxious waiting, Chandra heard from Fowler. Both Fowler and Neville F. Mott had read his paper, made some suggestions to improve it, and indicated that if Chandra accepted these the paper would be published. Chandra had no difficulty in accepting the suggested changes because they were chiefly matters of style, section headings, and so on. The paper was published in the *Proceedings of the Royal Society* later that year.[17]

Chandra had given a preliminary account of the paper during the January 1929 meeting of the Indian Science Congress[18] held in Madras. It was a historic meeting, the first one held since the discovery of the Raman effect in the spring of 1928. The honor of presiding over it naturally belonged to Raman, and his presence at the meeting had attracted scientists from all over India as well as a number of foreign delegates. Immediately after Chandra's presentation in the physics session, Parameswaran (one of Chandra's physics professors at the Presidency College) had stood up and pointed out that the author of the paper was just a student in his second year of the B.A. honors course, and that he had written the paper without any guidance or advice from anyone. It was a dramatic moment for young Chandra when the audience greeted the remarks with thunderous applause. His scientific career had begun.

The discovery of the Raman effect and the chance to work closely with the people responsible for that discovery, as well as the Sommerfeld encounter that led him to do original research in physics, reinforced Chandra's decision to pursue pure science.

The following summer, the summer of 1929, Chandra returned to Calcutta. The excitement of Raman's discovery was still running high. Many new visitors were investigating various aspects of the new effect which was now, without hesitation, called "the Raman effect." Peter Pringsheim had sanctioned the use of the term in an article in *Naturwissenschaften*. Raman reprinted the article and circulated it. Chandra's own research activity was proceeding at a hectic pace. While waiting for Fowler's response, he had studied the effect of a magnetic field on Compton scattering and prepared a short paper for publication in the *Indian Journal of Physics*. However, when he got the encouraging response from Fowler, he included his new results in the original paper, making it longer and more complete. He had corresponded with Jesse W. Dumond in America, who had independently thought of the idea that the new statistics would have an effect on the shape of the shifted line in Compton scattering. Two other papers of Chandra's were going to be published in the *Philosophical Magazine*.[19] He had other notes and articles in progress. The most remarkable aspect of this flurry of activity was that Chandra was only an eighteen-year-old undergraduate student.

His achievements did not go unnoticed, especially since he was working where most of the noteworthy scientists of India were working at that time.

Chandra's final year in college was as eventful as the previous ones. First, Werner Heisenberg of Germany came through Madras on a lecture tour during October. As the secretary of the student science association, Chandra was in charge of Heisenberg's visit to the college. Further, Krishnan, who was also in Madras at the time, entrusted Chandra with the responsibility of showing Heisenberg around Madras. In a rented car, Chandra spent a whole day taking him around to the well-known temple sites at Kanchipuram and Mahabalipuram. That night they drove along the seashore of Madras. It was an exhilarating experience for young Chandra to be in the exclusive company of such a famous man, who, still in his twenties, was one of the pioneers of the new quantum mechanics and the discoverer of the uncertainty principle. "I spent the whole of Monday with him till he left for Colombo by Boat Mail," Chandra wrote to his father on 16 October 1929. "I discussed with him my papers also. In one day by merely talking to him I could learn a world of physics."

A few months later, 2–8 January 1930, Chandra attended the Indian Science Congress Association meeting held in Allahabad. The host and the president of the physics section of the Congress was Meghnad Saha, the eminent Indian astrophysicist,[20] whose theory of ionization a decade earlier had unlocked the door to the interpretation of stellar spectra in terms of laboratory spectra of atoms of terrestrial elements, providing information about the state of stellar atmospheres, their chemical composition, the density distribution of various elements, and then about that most important physical parameter—the temperature.

Chandra had learned all of this from Eddington's book *The Internal Constitution of Stars*[21] and was aware of the high esteem Eddington had accorded to Saha and of Saha's election to the Royal Society in 1927. But Chandra was not aware that Saha was acquainted with his own work; so when he met Saha at the Congress and introduced himself, he was pleasantly surprised by Saha's compliment on his paper in the *Proceedings of the Royal Society*. Saha said that it was very suggestive and that one of his students was working on extending Chandra's ideas. He introduced Chandra to his students, who also seemed to know about his work, and he invited Chandra to his home for lunch with a small group of research workers all older than Chandra. The small lunch turned later into a dinner invitation with such distinguished senior Indian sci-

entists as J. C. Ghosh, D. M. Bose, and J. N. Mukherjee. Saha persuaded
Chandra to extend his stay in Allahabad so that he and his students
could discuss more with him. Chandra, so young, did not expect to be
treated almost as an equal by an internationally renowned scientist of
Saha's stature.

Chandra was not allowed to revel too long in his newfound stature in
the world of science. He was soon made painfully aware that he was an
Indian, a colonial subject to the British Raj. On his return trip to Madras, he first went to Delhi to pay a brief visit to his father. When he
boarded the express train at Delhi bound for Madras, he found himself
sharing a first-class compartment with an Englishman and his English
lady companion. "Since my father was in the railways, I had a free first-
class ticket," Chandra recalls. "I heard the lady say to the man, quite
openly so that I could hear, that it was so disgusting to share the compartment with an Indian. Then she said that it was some consolation
that the Indian was in European dress [Chandra was wearing a suit].
When I heard that, I was really very annoyed, so when the train started
moving, I went to the washroom and came out changed into *veshti* and
shirt, in typical South Indian style. When she saw me changed, she was
totally and absolutely horrified and furious. She said to her companion
that she couldn't possibly stay in that compartment. And when the train
came to the next stop, she called the station master and told him so. She
wanted either me to be moved or them moved to a compartment where
there were no Indians. As there were no other unoccupied first-class
compartments, the station master asked me whether I would move to a
second-class compartment which was practically empty; there was only
one other person in it. I said, 'No.' Why should I? I had a first-class
ticket. He offered that second-class compartment to the couple, willing
to move the one person already there to first class. She refused. He left
saying that there was nothing he could do. But when the train started
moving, she pulled the emergency chain and made the train stop. The
station master was upset and again told her there was nothing he could
do. The lady was adamant and was ready to pull the chain again, but her
companion prevented her from doing so. After a while they moved to
another compartment—second class or something, I don't know."

This kind of occurrence was not uncommon. Educated Indians in
government services acquiesced for the most part, but students were de-
fiant. "We were all emotionally involved in the Indian National Congress movement," says Chandra. "We all wanted to be a part of it,
although most of us did not go to the extreme of abandoning our stud-

ies." Chandra had intentionally done poorly[22] in the history paper of his final high school examination to protest the fact that there was too much British history required and not enough Indian history. In his first year of the honors course, he had joined the crowds with other students to welcome Jawaharlal Nehru to Madras and had attended his lecture on the seashore, although forewarned not to do so by the principal of the college, P. F. Fyson, an Englishman. When Fyson called all the students that were absent that day to levy a fine, he was surprised to find Chandra among them and said, "You too? I must fine you too?"

When Chandra returned to Madras, it was time to prepare for the final examinations in March; it was time to study the required subjects, do the required experiments, and write the required laboratory reports. Chandra's teachers had been exceptionally generous to him. His physics professors not only allowed him to forgo physics lectures so he could attend mathematics lectures, they had exempted him from the frequent tests that the others had to take. While his classmates[23] labored on the "practicals" in the laboratory, Chandra would be reading Sommerfeld or Compton. Parameswaran, professor in charge of the laboratory, would come along and notice that he was not doing the experiments he was supposed to do. "I see you are reading," he would say. "Do the experiment when you can, over the weekend perhaps?"

The other students could not help but notice what they considered to be favoritism. What kind of "apple-polishing" is going on here, they wondered aloud. Somewhat jealous, "half-playfully, but half-seriously, they thought that I was getting away with doing all sorts of things," Chandra recalls.

They also did not understand why Chandra put such an enormous effort into extra work[24] that did not matter at all when it came time for tests and finals. "Why do you do it?" they would ask, and Chandra would sometimes respond by telling them the following story:

An old gardener was planting some mango tree seedlings in the woods. A King and his courtiers, seeing the old man's efforts, said laughingly to themselves, "Look at the old man planting mango trees. How young he thinks he is!" So they came near him and the King said,

"Old man, I see you are planting some mango trees."

"Yes," the old man said.

"How old are you?" asked the King.

"I am past eighty-five, going on ninety," responded the old man.

The King and the courtiers laughed and said, "Do you expect to live till the mango trees grow and bear fruit?"

"No," said the old man, "of course not."

"Why do you do it then?" the King asked.

"Well," said the old man, "the mangoes I eat are from the trees that my father and grandfather planted. I hope my children and grandchildren will eat from these trees."

The King and the courtiers stopped laughing. The King was so pleased that he gave the old man five gold pieces.

And after the King and his retinue left and were far away, the old man chuckled and said, "And they thought I couldn't get anything for planting these trees."

Like the old man in the story, Chandra did not have to wait too long for his extraordinary efforts to bear fruit. Within days of his return from attending the Science Congress in Allahabad, he was called into the principal's office. Principal Fyson told Chandra (after warning him that the matter was strictly confidential) that he was going to be offered a Government of India scholarship to pursue his research in England. The scholarship was special, more or less created for him. It was not going to be advertised in the usual way, inviting applications from all qualified candidates. An exception was even going to be made regarding the strict rule that any scholarship for a specific year had to be included in the budget of the previous year. All this must have delighted Chandra, but he did not show his emotions when he reported this news to his father. He gave credit to Parameswaran for this surprising development.

Indeed, with papers published in the *Proceedings of the Royal Society* and in the *Philosophical Magazine,* Parameswaran did not need to try very hard to make a strong case for Chandra with his English colleagues—Fyson, the principal; Earlam Smith, former professor of chemistry who had become Director of Public Instruction; and M. A. Candeth, former professor of history who had become the Deputy Director of Public Instruction. Candeth and Smith knew Chandra since he had taken their courses in his junior and senior intermediate years of college. Hence, the scholarship took shape promptly. On 12 February 1930, Chandra was asked to meet with Candeth and Smith; on 15 February 1930 he had an interview with Mr. Subbaroyan, the Education Minister of the Madras State Government. The scholarship was more or less assured, the only stipulation being that Chandra had to agree to serve either in the Madras state service or at the Presidency College after his return. If he chose Presidency College, he could expect a special chair in theoretical physics. Chandra saw no difficulty in agreeing to such a stipulation. He was told, of course, that the scholarship depended upon his successful completion of the honors course and his securing a first-class grade. That did not present much of a problem to

Chandra, but when Candeth asked whether one could expect him to be elected to the Royal Society in four years, Chandra felt a bit disconcerted. "It is, however, rather difficult to satisfy those who start with a good premium of good opinion," he wrote to his father on 12 February 1930. "I told him that if I could secure it in 1940, I should consider myself lucky.* I explained to him how P. A. M. Dirac and G. P. Thomson both of them are not Fellows yet."

During the next few months, Chandra devoted all his energies to studying for the final examinations. As expected, he came through them successfully; his grades set a new record. And on 22 May 1930, he received official notification that he had been awarded the Government of India scholarship. He could proceed to make travel arrangements and execute a bond that stipulated that he would serve for five years in the Madras government service after the completion of his postgraduate studies "at such a salary as may be offered to him by the said Government of Madras." Further, Chandra had to agree to return to India when the Madras government or the India office in London directed him to do so if they thought his studies were completed. In case he did not comply with these requirements, he was to repay the government the entire amount of the scholarship including whatever bonus he was awarded plus his passage money.

Thus for Chandra the opportunity to go abroad for advanced studies, ordinarily so difficult a matter, was given to him in a remarkably simple way. Even so, the decision to leave his home was not an easy one, for he faced an extremely difficult personal conflict.

His mother had been seriously ill since late in the summer of 1928, just before Chandra's encounter with Sommerfeld. "I remember it all very well," Chandra says. "I was standing outside Chandra Vilas talking to a friend. I came home late and saw mother lying down in considerable pain. I was rather upset that I hadn't come back earlier. Next morning we went and saw the doctor, I think his name was Dr. Keshava Pai, a well-known doctor in Madras at that time."

First the illness was diagnosed as pleurisy; later as intestinal tuberculosis. Chandra cared for her, accompanied her to the doctor and the hospital, and brought her medicines. Her suffering and her inability to retain any food caused him a great deal of anguish.

By 1929 she was very ill. During the fall of that year her problems became so serious that they tried to pump out the accumulated fluids

---

*He was elected to the Royal Society in 1944.

from her stomach. Tragically, the pump broke in the middle of the operation and, to Chandra's extreme annoyance, there was no other machine available in the entire city of Madras; the treatment had to be left incomplete. After two years and after every kind of treatment, although she had ups and downs, it had become clear that she was not going to get well again. If Chandra went to England, he might never see his mother again.

Traditions and pressure from friends and relatives mounted against his going abroad and leaving his mother in such a condition. Had not his uncle Raman, who because of his own poor health had never left India, still accomplished a great deal? In fact, Raman had made a great virtue of never going abroad. He said that he was indebted enormously to the British civil surgeon who had recommended against his going to England, where because of his poor health and vegetarian diet, he was sure to get sick. Raman had stayed in India and developed his native genius independently, uncorrupted by western influences. But more than anyone else's opinion, it was Chandra's own feelings about leaving his mother that made his decision so difficult. He almost chose to refuse the scholarship and remain in India.

But Sitalakshmi herself intervened. She told Chandra, "You must go. You must pursue your own ideals to the utmost." To people who advised her against sending her eldest son so far away, she said, "You can have Lakhsmi *kataksam* or Saraswati *kataksam* (Lakhsmi, the Goddess of Wealth, can shower grace on you, or Saraswati, the Goddess of Knowledge, can shower knowledge on you). But to have both showering on you is only one in a million, and I just ask for that for Ayya. He is born for the world, not for me. I can't come in his way because he is born of me. He is born for the world. That is the single gift that a mother can give."

With such words of self-denial, she persuaded the reluctant Chandra to proceed with his preparations to go abroad. She would get well. She would even accompany him to Bombay to see him board the ship and to bid him farewell. And indeed, she did have a brief period of recovery before Chandra's departure.

One could find less noble and more earthly reasons for Sitalakshmi's extraordinary attitude. There was, first of all, the example of Ramanujan, who had returned to India from England with world fame. At first, Ramanujan's mother had been dead set against his going to Cambridge at G. H. Hardy's invitation.[25] It required no less than the intervention of the family goddess to make her change her mind. The goddess, as the story went, appeared in a dream and told her not to stand in the way of her son's fame.

Sitalakshmi, who shared none of Ramanujan's mother's religious concerns about her son going abroad, no doubt wanted Chandra to return from England with a glory like that of Ramanujan; she also wanted him to surpass his uncle Raman in scientific achievement. Raman, genius that he was, never cared too much about being nice to people. In her early married life, he was unabashedly rude to her, making snide remarks about her appearance, about the extent of her education and her accomplishments, and comparing her unfavorably with his own wife, Lokasundari, who was an accomplished *vina* player. Chandra's mother insisted that Chandra should under no circumstances take Raman's help in achieving his goals. And indeed the scholarship and the recognition he received at such a young age was all Chandra's own doing, without the slightest intervention of his famous uncle.

Encouraged by his mother's recovery and her insistence, Chandra made his plans for going to England. He booked passage on the *Lloyd Triestino*, an Italian liner scheduled to depart from Bombay on 31 July 1930. He signed the necessary contract, executed the bond, and informed the Director of Public Instruction of Madras that he intended to use the scholarship for doctoral research studies at the University of Cambridge. He wanted to study under the supervision of Fowler, so he asked that admission to Trinity College, Cambridge, be secured for him.

This request went to the office of the High Commissioner for India in London in June 1930. On 24 July, a week before his planned departure, Chandra received a cable informing him that it was not possible to get admission to Cambridge and that he should think of joining University College, London.

Chandra was determined to work under Fowler at Cambridge. He did not know anyone that he wanted to work with at University College, London. He thought that the High Commissioner's office had made an error in seeking admission for him just as a postgraduate student instead of as a Ph.D. research student. He wrote to the High Commissioner's office, informing them of the possibility of this error and asking them to look into the matter again. Hoping for the best, he proceeded as planned and left Madras for Bombay on 22 July 1930.

The customary fanfare at the Madras railway station brought together relatives and friends. His teachers, Appa Rao and Parameswaran, came to bid Chandra farewell and wish him success. It was an exciting moment for young Chandra, not yet twenty; but it was also a moment tinged with sadness because of his mother's illness.

His mother had planned to go to Bombay to see him off, but a day or two before his departure from Madras, she had a slight temperature and Chandra's father forbade her to make the journey. She seemed to have had a premonition that she would never see Chandra again. She insisted that he be garlanded as he got on the train. According to Chandra's sister, their mother told Chandra, "I won't be there to receive you with a garland; so I want to see you go with one." There was also the sad situation of his paternal grandaunt who wanted so much to be a part of the farewell gathering at the station. C. S. Ayyar forbade her to do so because she was a widow. (The presence of widows was thought to bring bad luck—a superstition that Chandra would have no part of.) She had lived with Chandra's family all her life and was very fond of Chandra, who always felt that "she was so much sinned against." As he left Madras, he thought of her left alone in the house when everyone else came to the station.

4

Discoveries, Personal and Scientific
Cambridge and Copenhagen, 1930–1933

. . . if things come from home, it may be some relief to the dull bread and insipid butter and the neverending boiled potatoes and cauliflower for dinner.

Chandra (letter to his father, 14 November 1930)

Chandra arrived in the humid swelter of Bombay a week prior to his scheduled departure for Cambridge, England. Since his father was posted in Bombay at the time, there was a place for him to stay. Chandra was nineteen years old. He hadn't the slightest idea at the time that he was embarking on a journey which would take him away from India for the rest of his life. He expected to return to India after the completion of his doctoral degree in three years and then take a good position at Madras or some other university.

The week went by quickly. He took care of all the last minute travel matters—picking up the tickets, traveler's checks, cabin labels, and so on—and visited family friends in the Bombay area. Chandra was invited to give a talk at the Royal Institute of Science during that week. The self-assurance and confidence with which he discussed his scientific work was in sharp contrast with what was expected of a young student. The crew-cut hairstyle he maintained at the time only augmented his boyish appearance. As one of the very few who was fortunate to go abroad for higher studies, he was envied, but given a royal farewell.

The *Lloyd Triestino* left the Bombay docks near the Gateway of India on the afternoon of 31 July 1930. There were the usual huge crowds come to bid farewell to their friends and relatives. Chandra's father, his sister Sarada, and Ramanathan[1] and his family stayed on board the ship until the very last minute.

Chandra left behind a loving and caring family—parents, brothers and sisters, numerous aunts, uncles, and cousins—and a large group of friends. He was leaving India as a celebrity. Everyone expected great things from him. He would return with the fame and recognition of a Ramanujan or Bose. He would make new discoveries, like his uncle, Sir C. V. Raman. But Chandra would soon find himself alone in an alien culture, far away from home for the first time in his life with no friends to share his concerns, worries, and anxieties. He had to adjust to the cold English climate, learn to survive on English food, and carry on with his studies and research. His beginning research in India, which had accrued a measure of fame and recognition because of his youth and his relationship to Raman, was of little consequence in Cambridge, where some of the foremost scientists of the time dominated the scene.

After leaving Bombay, the ship entered the Arabian Sea and encountered severe weather which lasted three or four days. It sailed at half its usual speed, arriving in Aden, now part of Yemen, two days behind schedule. Chandra became violently seasick during the first two days; his discomfort, augmented by the sudden loneliness so far from home on a rough sea, made him think of his mother. He recalled how much she had wanted to come to Bombay to bid him farewell. Now he could actually feel her suffering the attacks of vomiting which accompanied her illness. Would she recover? Would he ever see her again? He felt there was practically no chance of his returning to India before completing his Ph.D. studies. Three years and six thousand miles was a long interval of time and space.

He also thought of his father. During the week before his departure, he had often sensed his father's concerns and fears for his young son going so far away to a foreign country with its corrupting influences. Chandra's father wanted to counsel him but had hesitated to say anything directly, perhaps because during the nineteen years of his life, Chandra had given him no cause for concern. His behavior had been exemplary in all respects. Chandra, however, took it upon himself to reassure his father. From aboard the ship, on 7 August, he wrote, "I know that during our stay in Bombay you wanted to advise me at length on the necessity for discreetness of conduct etc., but you somehow failed to give *full* expression to it. I however fully understand; it will be [my] earnest endeavor during my stay there to keep straight and be true to myself and, more than all, be worthy of the very fine cultural atmosphere which has surrounded me all along from my childhood."

Chandra remained faithful to these words. The customary festivities on the ship—the gala nights of dance and music, inexpensive liquor, rich food, and carousing—were not for Chandra. As the rough weather receded and the voyage from Aden onwards became smooth with sunny, clear skies, he rented a deck chair and sat immersed in a book or occasionally participated in a deck game. He was frequently rebuked by his fellow Indian travelers who were enjoying their maiden voyage to the West and the novel experiences it had to offer. Why was Chandra sitting and reading all the time? Why not enjoy this once-in-a-lifetime chance for fun and frolic? Chandra was accustomed to such rebukes, having endured them for most of his college years. To those who would listen, he told the story of the old man planting mango trees. But most of the time, he kept on reading.

There were a dozen or so other students from various parts of India on board. His cabinmate was a former classmate, C. A. Ramakrishnan, who was going to England for the Indian Civil Service examination. The Indians were all seated together at adjoining tables in a corner area of the dining hall, ostensibly for the convenience of serving their special dietary needs, though in reality it was an arrangement made to ensure that the whites on board would not be "contaminated" through proximity to the browns. "I remember a German couple," Chandra says, "who had a ten- or twelve-year-old daughter. The girl, shy at first, became quite friendly with me when I spoke to her in German. She would come on the deck to see me. I soon sensed, however, that her parents did not approve of this. She stopped coming to see me." The little girl was color blind, but not her parents!

When the other Indian students left him alone, Chandra's preoccupation with his studies attracted the attention of a missionary from Kerala. Father Saldhana, rebuffed by the Hindu Indians when he tried to perform his missionary duties, found in Chandra a silent, polite listener who was not outrightly abrasive like the others. He used Chandra's courtesy towards him as an opportunity to unload on Chandra, with all his missionary zeal, the superiority of the Lord Jesus Christ. According to him, the Hindu Gods were primitive, evil, and corrupt. Chandra recalls that Father Saldhana slipped a book or two into his cabin. When his cabinmate happened to browse through them, he became so angry at their crudeness, inaccuracies, and falsehoods that he threw the books across the cabin.* But Chandra did not mind listening to the mission-

*While Chandra was describing this experience, I was reminded of an amusing story which illustrates the flavor of such conversion attempts in action. A missionary had gathered a large crowd of villagers together at a crossroad. Standing up on a makeshift platform, he was telling them the story of a hawk in hot pursuit of a helpless dove. The dove flew to Brahma, the Hindu God of Creation, and pleaded for its life. Brahma said he was helpless. He was the creary. "Of course, I never left him in doubt that I did not share his views," Chandra says. "I mean, he knew I listened to him more out of politeness than anything else. I think still he appreciated that, particularly when compared with the rudeness he met from others. He was a missionary, but he was also gentle, friendly, and anxious to please. Why be rude to him? I let him have his opinions." In fact, Chandra has a fond memory of this tall, dark priest in white clothes, standing on the dock and waving a touching farewell when they parted at Venice.

The food on the ship did not present much of a problem to Chandra. Very quickly he adapted the menu to his own needs. Since he had been brought up in a strict vegetarian tradition, fish, eggs, and meat of any kind were prohibited. He was aware of the myth that in the cold countries of Europe one had to eat meat in order to maintain one's body temperature and remain alive, but Chandra had no intention of changing his eating habits. He would not touch meat. If absolutely necessary, he might include eggs in his diet, which he was advised to do. He described his first experience with these dietary maneuvers on shipboard in some detail in a letter to his father (12 August):

The food on the steamer is very good. In the morning we get tea and biscuits. At 8:30 AM we have breakfast. I take porridge—it is just like wheat diet in Nestle's milk—and some corn flakes, again in milk, some bread and butter, boiled potatoes eaten simply after being salted, coffee and fruits. We get *Varuvals* and *Pooris* at times!

At 1 o'clock we get lunch. There I take rice with the coconut powder which I got prepared in Madras, boiled potatoes again, some vegetables which are simply boiled. Cabbages, turnips, beans, ladies fingers and the like. Some sweet cakes, fruits and toast.

At 4:30 goes the tea-bell. Here I take some toast, biscuits and tea.

At 7 PM we have the Dinner. I take rice as in lunch, some vegetables and soup *if it is of vegetable make,* potatoes made into a mash (pepper and salt we have to add ourselves), vegetables and some tea and fruits.

I have not found the necessity to take eggs yet. I suppose it will be different in Cambridge.

As good weather returned and with it a smooth passage, Chandra began to think of his physics. He had with him a paper he had com-

---

ator, but he had no control over the nature of his creation. After all to kill and eat was the hawk's nature. The dove then flew—flapping its wings as hard as it could—to Vishnu, the Sustainer of Life. Vishnu said he was helpless too; it was all a part of nature. Helpless also was Lord Shiva, to whom the dove flew next in utter desperation. The hawk was practically on its neck when the Lord Jesus appeared. *"What did the Lord Jesus do?"* asked the missionary. Before he could answer his own obviously rhetorical question, a passing schoolteacher raised his hand and in a most innocent tone said, "Jesus killed the hawk and ate the dove."

pleted just before his departure from Madras. In it he had developed Fowler's theory of white dwarfs further. Fowler, by appealing to the then-new quantum statistics of Fermi and Dirac, had demystified a long-standing puzzle regarding the high densities encountered in white dwarfs. Chandra, by combining Fowler's ideas with Eddington's poly-tropic[2] considerations for a star, had been able to obtain a more detailed picture of a white dwarf star. One of his conclusions was that the density at the center of such a star was six times the average density (i.e., one ton per cubic inch!). It suddenly occurred to Chandra to ask the question: If the central density is so high, will the relativistic effects[3] be important?

The point was that, according to Fermi-Dirac statistics, no two electrons could be in the same quantum state. Hence, starting with low or zero momenta, if one assigned appropriate, different momentum states to each electron, one might find, because of the high density, electrons in momentum states corresponding to energies comparable to or higher than the rest mass of the electron. In that case, the special relativistic effect, namely, the increase in the mass of the electron depending on its velocity, must be taken into account. Chandra quickly found that this was indeed the case. He had three books with him which contained all the theoretical tools he needed to incorporate the changes in Fowler's theory due to relativity.* Chandra worked out the calculations, expecting to find a neat, relativistic generalization of Fowler's theory. But, much to his surprise, he encountered something totally different.

If his calculations were correct, there was a limit to the mass of a star that would evolve into a white dwarf. If a star's mass was greater than this limiting mass, it could not evolve into a white dwarf star, as Fowler's theory would then imply an unphysical negative radius for the star. The expression for the limit on the mass, Chandra discovered, was in terms of fundamental atomic constants and the average molecular weight of the stellar material. Hence it could be calculated. It was a startling result: A limiting mass for a star in terms of fundamental atomic constants! Chandra was delighted, intrigued, and at the same time confused. What happens to stars whose masses are greater than this limiting mass? He noted down the results of his calculations. He would write a paper about these findings and discuss it with Fowler at the first opportunity, along with the results in his other paper. He knew that a great

---

*The three books he had on board with him were Eddington's *The Internal Constitution of the Stars*, Compton's *X-Rays and Electrons*, and Sommerfeld's *Atomic Structure and Spectral Lines*.

deal of work would be necessary to understand the result fully and to establish it on firmer ground by getting rid of some of the assumptions he had made in order to simplify his calculations. In the meantime, his long journey was coming to an end. The ship docked at Venice, and after a night there, he continued on to London by train.

Chandra arrived in London on the afternoon of 19 August 1930. The Indian Students' Union at 112 Gower Street provided temporary accommodations for arriving Indian students. London did not make a strong first impression on Chandra. He was already familiar with its sights and sounds from books and from hearsay. Its Victorian buildings and red double-decker buses only reminded him of Bombay. However, what Chandra remembers well is seeing Dirac's book *Principles of Quantum Mechanics* displayed at the H. K. Lewis and Company Bookstore on the corner of Gower Street near the hostel where he stayed. The first thing he did on the morning after his arrival was to buy Dirac's book.

The next order of business was to straighten out the matter of his admission to Cambridge. He had arrived in England without being formally admitted to any college. Since he was a government scholar, the responsibility of securing admission for him rested with the offices of the Director of Public Instruction in Madras and the High Commissioner for India in London. Between them they had confused the situation by seeking Chandra's admission for a master's degree at Cambridge, which they could not secure because all the colleges were full. Chandra wanted to be a research student, working for a doctoral degree under the supervision of Fowler. For this status to be granted, the board of research of Cambridge University had to give its permission. But the High Commissioner's office had not communicated with the board, nor had it provided the necessary documents supporting Chandra's qualifications for research, nor had anyone communicated with Fowler regarding the matter.

Chandra went to the High Commissioner's office and met with the secretary there. He was totally unprepared for the bureaucratic red tape he encountered. The secretary was completely unsympathetic and even obnoxious. The entire experience became very humiliating. "What do we know about your qualifications?" the secretary asked Chandra. "We don't know how to judge them and recommend you to the research board." To Chandra's response, "But, Sir, I got a first-class first. The next one trailed me by 280 marks," the secretary shrugged and retorted, "But who can say what the criteria were?" Chandra pleaded; he talked about his published papers and the attention which other scientists had already given to his work. But it all fell on deaf ears. "Young man," the secretary said, "that is all no good. We want an authoritative and strong statement from Fowler regarding your merits. Anyway, why don't you just study at the University College, London?"

This last remark infuriated Chandra. He could not make the secretary understand that he had not come to England simply for the glamor of a foreign degree. He knew exactly what he wanted—to do research under Fowler at Cambridge or to go back home. He rushed to Cambridge and arranged to meet with Mr. Whitehead, the head of the advisory board for Indian students at Cambridge. He was not much help either. He was an ICS officer in no position to judge Chandra's research qualifications, and he was unwilling to make any special effort. The colleges were full and that was that.

Things looked gloomy, even desperate. To make matters worse, Fowler was on vacation in Ireland. Chandra had written to him from Madras as soon as he had learned about the trouble with his admission; he wrote again from Cambridge. There was nothing to do except wait for Fowler's response. Chandra rented temporary accommodations, bed and breakfast for one pound a week, in a house on Jesus Lane in Cambridge and started writing his paper on the startling results of his shipboard calculations. On the advice of Dr. Vaidyanathaswamy (a mathematics professor at Presidency College), he had called on S. Chowla,[4] another Indian student, during his first visit to Cambridge. Chowla had come to Cambridge a year earlier to study under the well-known mathematician J. E. Littlewood. Like Chandra, he had come to Cambridge without being formally admitted, but when Littlewood accepted him as a student, the matter was settled. Chowla assured Chandra that the same thing would happen in his case. Once Fowler returned and said that he was willing to accept Chandra as a research student, everything would be all right.

Indeed, a week or so later, Chandra learned that Fowler had written directly to Mr. Priestley, the president of the research board, and to Mr. Winstanley, the senior tutor at Trinity College. As a result, Mr. Priestley gave Chandra permission to register as a research student. Mr. Winstanley would get him admitted to Trinity provided the High Commissioner's office requested him to do so. But "Mr. Secretary" to the High Commissioner continued to be a bureaucratic stumbling block. Fowler's letter was not strong enough for him, although it said that he knew Chandra's work, that it was fairly good, especially considering how young he was, that studies at Cambridge should do him great good, and that he was quite willing to supervise Chandra's work. Chandra was baffled. How could it be any stronger? He said to the secretary, "You asked me to get the endorsement of the research board and a letter from Fowler. I did that. Now you say that the letter is not strong enough. Fowler has so many brilliant young men working under him, he has no reason to write any stronger letter. But you who are paid to help me, why don't you write a strong letter and get me admitted?" Chandra thought of reminding the secretary of a remark by the Right Honorable Srinivasa Sastri,[5] who, after he had spoken nearly two hours against the Rowlatt Act[6] and still found the government unconvinced, said in great despair and anger, "My Lords, you wanted us to bring the tigress's milk, which we did. Now you want the *male* tiger's milk. How is that possible?" The secretary finally agreed to write a note, but not a strong recommendation. Further, he was audacious enough to say that there was not a ghost of a chance that Chandra would be admitted.

Chandra returned to Cambridge disappointed and distraught. "I wish I had not come at all and had refused the scholarship," he wrote to his father on 3 September. "One can well imagine the feelings of one who has come six thousand miles from those near and dear to him, and to find the one compensation, that of a distinctly favorable atmosphere in which to work, breaking down, and carrying down with it, Hope." Fortunately for Chandra, this sense of everything falling through lasted only one day; on the evening of 4 September, Mr. Winstanley told him that he had been admitted to Trinity on the strength of a personal letter from Fowler. All he had to do was register his signature in the red notebook at the office and proceed to make boarding and lodging arrangements through the Bursar's office. Chandra was so overjoyed that he sent a telegram to his father at once and that very night sat down to write a long letter to him, recounting in detail his misadventures with bureaucracy.

In the end, however, Chandra realized how fortunate he had been in gaining admission. He recalled how Sommerfeld's paper had introduced him to the new statistics; the new statistics had drawn his attention to Fowler's paper; Fowler's paper had introduced him to stellar structure and to Eddington's book; and soon thereafter, he had been able to extend Fowler's work [leading eventually to his own discovery of the celebrated critical mass condition—the "Chandrasekhar limit"—on stellar masses that could become white dwarfs]. As he wrote to his father, "I have got admission purely due to the accident that I happened to know Fowler for the last two years. Why I should have written then to Fowler,

God alone knows. I suppose that was because Fowler was to help me two years later!"

Once the question of admission was settled, Chandra lost no time in finding more permanent "digs" in Cambridge. "My lodgings are the cheapest a student can get in Cambridge without at the same time inconveniencing oneself too much," he wrote his father on 23 September. His quarters were in fact quite comfortable. He had two small, connected rooms—a bedroom and a living room-cum-study—on the second floor of a house very near Trinity College (1 Park Street). The bedroom was furnished with a wardrobe and its own washbasin. In the study were a table, a chair, a bookshelf, a fireplace, and near the window, a corner table at which Chandra used to write his letters. Chowla, who had become Chandra's friend during the anxious weeks of waiting for admission, had lived in the same house. Now he had moved to college rooms. The landlady, Miss Smith, was middle-aged, single, and "awfully kind, generous, quiet, and businesslike." The house, which was very quiet and orderly, suited Chandra extremely well. Chowla had told him about other noisy and messy houses with husbands, children, and chatty landladies, who would certainly have been a nuisance to Chandra's scholarly and spartan way of life.

The beginning of Michaelmas term was on 7 October, two weeks away. Cambridge is dreary during term breaks, with few people around and nothing exciting happening. Chandra launched on a hectic schedule of his own work, which had actually begun during the gloomy days of his struggle for admission. "I have been getting deeper into my studies," he wrote his father, "and have made good progress in quantum mechanics for the first time, in as much as I have written a paper on "Position as a Differential Operator in Quantum Mechanics."[7] He had also completed another paper on the density of white dwarfs during the same time, and he started working on relativistic Fermi-Dirac statistics, which was to result in another paper in a matter of weeks. Hence, when Fowler returned from his vacation just before the reopening of the university, Chandra was ready to prove to Fowler that his confidence in him had not been misplaced. The two met on 2 October 1930 in Fowler's rooms in Trinity College. Famous though he was for his research, Fowler was still a lecturer with no office of his own. He used to meet his students in the library of the Cavendish Laboratory or in his rooms at Trinity.

Fowler greeted Chandra by saying, "So you had quite an adventure getting here." Chandra took an instant liking to him. Strong and healthy, exuberant and happy after his vacation, Fowler appeared to enjoy life. Chandra gave him two papers he had brought along and discussed some details and his conclusions in them. One paper, which Chandra had completed in India, clearly extended Fowler's own work. "Splendid," Fowler said as he browsed through the paper. Chandra would hear that phrase, a characteristic mannerism of Fowler's, quite often in the future. Fowler was not so sure about the other paper, which gave an account of the startling discovery Chandra had made during his journey. Fowler said he would send it to Edward Arthur Milne, who was more familiar with such matters. On the whole, Fowler was extremely pleased with his new young student who had exhibited so much independence and initiative. He advised Chandra to attend lectures by Dirac on quantum mechanics, Jeffreys on operational calculus, Pars on generalized dynamics, Littlewood on function theory, and his own lectures on statistical mechanics. Chandra became dizzy with so many famous names and new and fascinating subjects to study, plus the excitement of meeting Fowler in person. He almost stumbled down the stairs as he backed out of Fowler's rooms. Chandra remembers well the alarmed expression on Fowler's face at the top of the stairs as he said out loud into the stairwell, "Steady, steady!"

Chandra was inspired by this first meeting to work hard and bury the loneliness that he had slowly begun to feel in his studies and research. "I do not intend making any friends here as that merely wastes one's time," he wrote to his father on 2 October. "I have to realize *more fully* that I have come down 6,000 miles, not to fool away my time, but by utilizing opportunities in the proper way, to at least compensate for the anxiety which my coming is bound to cause in others." He set a rigid schedule of hard work for himself. From the time he got up at seven in the morning until late in the evening, except for short breaks for meals, there was nothing but work. However, he did miss his home and family. "There is nobody to talk to," he wrote to his brother Balakrishnan on 16 October. "[M]athematics has become my morphia. Dull and stupid, it helps me to forget everything else. I take a walk in the evening, but Cambridge is so small that you come back to the same place in half-an-hour. One feels like the ragged (in thought) rascal running round and round."

The opening of the university, with its lectures and seminars, was of some help in overcoming his intense loneliness, but, at the same time, it aroused within him the sudden feeling that he was essentially a nobody in these new surroundings. In India, he was somebody special; he had published papers; he was *known*. He had access to and was known by some of India's well-known scientists. All that, however, did not amount to much in Cambridge, where some of the world's foremost scientists were at work. As Chandra says, "It was a shattering experience, coming to Cambridge, suddenly finding myself with people like Dirac, Fowler, and Eddington, and living in a society that was altogether disconnected from me."

But Chandra was inwardly driven in his pursuit of science. He said to himself, "Advances in science are not made by one discovery or another, or writing and publishing a paper, but by earnest study and a lot of work." Hence, although there was essentially unlimited freedom for a research student at Cambridge, with no requirements to complete course work or to pass examinations, Chandra followed the advice of Fowler and attended lectures in mathematics and physics. During his first year in mathematics, he studied differential equations, complex analysis, group theory, and modern algebra. And in physics, he attended lectures by Dirac on quantum mechanics, by Eddington on relativity, and by Fowler on statistical mechanics.

Since Chandra had studied little or no modern mathematics as an undergraduate, he had to work hard to keep up with the lectures in mathematics. By contrast, physics was comparatively easy. He had studied statistical mechanics and relativity on his own; Fowler's and Eddington's lectures were, therefore, more like a review for him. So were Dirac's lectures during the first term; Dirac essentially copied onto the blackboard in his neat, small handwriting from his book which had just been published. Chandra had already studied the book thoroughly. Still, he continued to attend the lectures (actually he attended the same ones three times during the course of his student days in Cambridge), because he enjoyed Dirac's virtuosity in presenting his ideas from his own point of view with utmost clarity and economy. To a careful listener to his lectures, it was, as Chandra would explain much later in a radio interview, like listening to great music. It is not unusual for one to listen more than once to a Beethoven symphony and discover something new every time.

During the second term, Dirac's lectures rose to an advanced level. The attendance at his lectures, which had begun with "a high tide" of forty students at the beginning of the first term, had dwindled down to eight students by the end of that first term; only four of them (including Chandra) continued to attend. "Dirac's lectures are masterly," Chandra wrote to Balakrishnan (21 January 1931). "Now he has come to that region of quantum mechanics where the demarcation between science and metaphysics is hazy (even as most people's notions of it are), and his lectures are profoundly philosophical and deep—the like of it one can perhaps hear only in a few other places—Göttingen (Hilbert), Copenhagen (Bohr), Zurich (Weyl), Leipzig (Heisenberg), and Berlin (Einstein, Laue, Planck). I am attending also Eddington's lectures on Relativity. He is in a way a contrast to Dirac. Dirac is always serious—'a martyr to meditation' (Joad). But Eddington is always funny and humorous."

In spite of the fact that he was officially a student of Fowler, Chandra could not see him often. He recalls how he used to stand outside the library and wait sometimes for two hours or more, trying to catch Fowler's attention. He would see other people like Dirac or Sir Harrie Massey go in to talk to Fowler, but then Fowler would just pass by Chandra, smile courteously, and walk away. "When this happened for several days in a week and for several weeks, then I gave it up," says Chandra. Of course, Chandra would never have thought of approaching him directly. He was, at the time, too much in awe of people like Fowler, Rutherford, J. J. Thomson, and Eddington because their scientific contributions and worldwide reputations set them apart from all the scientists he had known up to that point. His own contributions thus far, interesting and important though they were, had not seemed to attract a great deal of attention. And Chandra was not one to push himself forward. The initiative had to come from those he admired.

In the same way, although Chandra attended Eddington's lectures regularly, he made no attempt to meet with Eddington for almost his entire first year. He would see Eddington on the streets of Cambridge, "smoking his pipe, looking very content and happy," and say to himself, "wouldn't it be marvelous to be in a state like his? He is so well known; he has got the Royal Medal, the Gold Medal of the Astronomical Society, and so on." That kind of recognition and achievement seemed far beyond his reach.

By contrast, Chandra's first impressions of Dirac were rather amusing. "I saw Dirac yesterday," he wrote to his father on 10 October 1930. "A lean, meek, shy young 'fellow' (FRS) who goes slyly along the streets. He walks quite close to the walls (as if like a thief!) and is not at all healthy. (A contrast to Mr. Fowler—a strong, 'big,' healthy, middle-aged man, quite happy, full of joy of life)." And though he shrank away from Fowler and Eddington, Chandra came to know Dirac quite well in the latter part of his first year. Dirac became his official advisor when Fowler left Cambridge on sabbatical at the beginning of the second term. Ten years his senior, Dirac was a well-known person in the scientific world with his outstanding contributions to the new quantum mechanics, quantum field theory, and the theory of the electron. However, Cambridge had not found a vacant chair for him yet,[8] so he was a lecturer and had rooms in St. John's College.

"He was very human, extremely cordial to me in a personal way," Chandra recalls. "Even though he was not very much interested in what I was doing, he used to have me for tea in his rooms in St. John's about once a month. He also came to my rooms for tea and, on some Sundays, used to drive me out to fields outside Cambridge where we used to go for long walks on the Roman Road."

"I remember an amusing thing about Dirac on these walks," says Chandra. "There were periodic dips on the Roman Road. When we came to each of them, Dirac would try the experiment of going down with as little exertion as possible and use the energy to find out how far he could climb on the other side. He always used to say, 'converting potential energy into kinetic energy' as he went down, and ' kinetic energy into potential energy' as he went up."

Beyond that there was little or no conversation between them on these long walks. Since both were disinclined to idle talk and were reticent, they went together but followed their own lines of thought silently. On one occasion, however, Chandra recalls Dirac asking him why he was doing astrophysics, remarking that if he (Dirac) ever became interested in astronomy, he would do cosmology. Indeed in his later years, Dirac did turn his attention to cosmology and suggested a model of the universe to accommodate his Large Number Hypothesis,[9] including the possible variation of physical constants[10] with time.

As Dirac was not directly interested in astrophysics, he could not be of much help to Chandra in his research. He did, however, offer critical advice and communicated one of Chandra's papers for publication. Chandra continued to do research on his own and to submit papers for publication while he attended the lectures. His research efforts won him election to the "Sheep Shanks Exhibition" and an associated award of forty pounds. It was a special honor bestowed every year to one candidate for proficiency in astrophysics. Eddington then sent Chandra a congratulatory note and an invitation to meet with him on 23 May 1931. At the same time, Chandra had begun a correspondence with

Milne, to whom Fowler had sent Chandra's paper on critical mass. While Chandra had not received any response to that paper, he found Milne receptive to his subsequent work on relativistic ionization and on stellar atmospheres. Milne had done some pioneering work in these areas and Chandra's contributions supplemented Milne's work.

Milne's encouragement as well as his critical comments were undoubtedly of great help to Chandra in those early days at Cambridge. Within six months, they had established a strong rapport and Milne was suggesting collaboration and joint publications. Chandra was quite pleased with this, as he noted in a letter to his father on 30 January 1931. "Prof. Milne is exceedingly good to me. I allow myself the vanity of quoting from one of his many letters—for vanity it is—'You have worked out the relativistic degenerate star most beautifully—I wish other people understood my analysis as completely as you do.'" Milne, however, did not appreciate the importance of Chandra's discovery of the critical mass.

It was also during the first year that Chandra was introduced to the meetings of the Royal Astronomical Society, London. Meetings were held every second Friday of the month, and Fowler wanted Chandra to attend the 14 November 1930 meeting. He would "formally introduce"[11] Chandra to the society and also introduce him to Milne, who was reading a paper that day. "The paper is supposed to cause a lot of sensation," Chandra wrote to his father (14 November) from London, "as Professor Milne and others (including me) believe that he has completely *destroyed* Eddington's view of the interior of stars. Milne's results are in a sense a generalization of my own on the density of dwarf stars. (The paper I wrote here.) Mine is one of the limiting cases of Milne's formulae. I think he will refer to my papers." Chandra, therefore, expected a heated debate between Milne and Eddington, and perhaps he too would be drawn into the arena. He became nervous just thinking about his debut into this historic society.

As it turned out, the meeting was a bit of a disappointment for Chandra. Milne had no time to go into the details of his work. Neither did he mention Chandra's contribution. There was no time for discussion after Milne's talk. Nevertheless, these meetings were an important part of Chandra's career in Cambridge. He contributed papers regularly to the meetings and was asked several times to present his work. Chandra says, "I have always felt that the RAS played a fairly important role for me in my early years because, between 1933 and 1936, I must have read papers to the RAS six or seven times, which is more than the normal share. People there were extremely cordial. They were very hospitable to my papers, and all my papers were published promptly."

While Fowler, Dirac, Eddington, and Milne were formidable scientists to have as mentors, they were not the same as friends. But Chandra had determined, in his exuberance for his studies, not to waste valuable time in making friends. He allowed only a couple of exceptions.

First, there was Chowla, who was well suited as Chandra's friend. He was an equally hard-working and studious young man, working on number theory. "He sometimes comes to my room and we study together," Chandra wrote to his father on 7 April 1931. "He minds his work and I mine. (Life is so lonely here, that it is some consolation to have somebody with us). I sometimes go to his rooms and study. But we don't talk much except during walks. For instance, once we studied nearly four hours without a single word passing between us. Yet, the feeling that somebody is with us makes it easier."

The second exception was Harold Gray. When Chandra had gone to see Chowla that first day in Cambridge, Gray was there and it was he who had suggested to Chandra that he go and see Winstanley. Gray and his fiancée, Frieda Marjorie Picot (Freye), belonged to a Methodist group (Wesleyans) which worked for world peace, and they were extremely active in pacifist campaigns. They disapproved of Britain's colonialism and imperialism and were profoundly sympathetic with Gandhi's non-violent struggle for India's independence. Gray was a physicist, a Fellow of Trinity, and since 1928 a member of the Cavendish Laboratory, which hosted researchers like J. J. Thomson, Ernest Rutherford, Peter Kapitsa, and P. M. S. Blackett. When they would get together, Gray would bring news about the work at Cavendish and stories about Rutherford and the others. "One of the most memorable instances of my friendship with Gray," Chandra recalls, "was my running into him one day when he told me with great excitement, "Chadwick has found the neutron!" Gray provided Chandra with an important link to pure physics.[12]

Freye, Harold's fiancée, was a Girtonian,[13] an "out student" at Newnham College, studying for her Cambridge Tripos in Theology. She was totally blind when Chandra came to know her. He did not know nor did he inquire whether she was blind since birth. "She was an extremely nice person," Chandra recalls. "I don't think anyone would call her a beautiful girl. She was not. But she had a manner and an elegance which were extraordinary." Harold and Freye became good friends of Chandra.

They introduced him to some of Cambridge's pleasures: punting on the Cam, Sunday picnics on the riverbank, and afternoon teas. Chandra could go to their place unannounced, as one did in India, without any hesitation whenever he felt like it. "In those lonely days during the first year, this kind of friendship meant a great deal to me. Their friendship was enormously influential in making my life a little better," says Chandra. Besides, Freye persuaded Chandra that science was not everything; she started him on a serious and systematic study of literature and became his unofficial tutor. Under her influence, Chandra read I. A. Richardson's *Literary Criticism* and became a devout fan of Virginia Woolf and the Russian masters such as Tolstoy and Dostoevski. The Sunday picnics turned inevitably into study and discussion hours. Still, such excursions into leisure and friendly companionship were probably too few and far between in a strict schedule of work.

Chandra's severity with himself continued in matters of diet as well. He could get along with corn flakes with milk, bread and butter, and tea for breakfast; fruits, bread and butter for lunch; and tomato soup, some potatoes and vegetables for dinner during the first two or three months. But soon it became a drudgery. He wrote his father (26 November 1930): "My food here is quite dull. There is no zest, no 'looking forward' to meals as in India. I take food because it is 9 AM, or 7:30 PM. The clock reminds one of food, not appetite. How can one relish potatoes, bread and butter, corn flakes, all of them having no taste if it is not sweet?"

At the same time, he had to reassure people at home not to be anxious about his health, not to worry that he might be risking starvation. He was doing well with vegetarian food, which now included eggs, since his father was constantly urging: "You must take eggs to keep up your nourishment in that country." Surprisingly, Chandra's father was more flexible than Chandra in his attitude towards food, at least for his son. He once wrote, "There is no necessity to *absolutely stick to* vegetarian food and eggs only—if you are unable to maintain physical efficiency in the cold climate. Live in Rome as Romans do, is a wise proverb. . . . Occasionally fowl or meat may be taken if you feel it appetizing."

But Chandra was determined to remain a vegetarian. "There may not be any special virtue—I don't maintain there is—in being vegetarian," he wrote in reply on 4 November 1931, "and in not doing what the Romans do when living in Rome, but I have so far found *no* difficulty in being vegetarian, or any necessity to change over and do 'what the Romans do.' I *am* going to be a vegetarian. Because I just want to be an exception and tell bold-facedly and *honestly* that it is not only possible to be a vegetarian in England for a stretch of three years, but that I have actually been one. Also the statement that considerations of physical fitness require one to be nonvegetarian is just false. In any case I have no iota of belief in that statement."

Chandra's father also used to urge Chandra to make occasional trips to London where he could get good Indian-style vegetarian meals in restaurants or at the home of his former classmate Ramakrishnan who lived in London and cooked excellent curries, spicy rice (*pongal*), and other favorite dishes from Madras. But for Chandra such trips were too expensive. Day return tickets cost fourteen shillings (except on Thursdays after 1 PM, when tickets were half-price). "I should prefer to buy a good book than to go to London just for a meal," he told his father. The best solution he said was "not to think about it [food]; take the same old dull things, repeating, 'there is nothing good or bad, but thinking makes it so.'" He did, however, look forward to occasional parcels of food from home containing *papadams* and chutney powders. His landlady would toast or fry the *papadams* for him, and a little chutney powder added to the boiled potatoes gave a great deal of zest to his meals.

Financially, his scholarship of 340 pounds a year was adequate to pay tuition, board, and lodging. It afforded a comfortable, but by no means luxurious, life. Money had to be saved for books, clothes, and hopefully a short tour of the continent before his return to India. Chandra was determined not to burden his father or anybody else for additional support. He lived frugally and saved. For example, on his trip from India when he had to wait in Venice until the next morning to take the train to England, he had spent the entire night sitting in the train station instead of finding accommodations in a hotel. "Why should I give 3 pounds for one night's sleep? I had good sleep sitting."

Writing letters home provided a much needed, and probably the only, diversion from his routine. During his years in Cambridge, he wrote twice a week to his father and at least once a month to his younger brother Balakrishnan. Others in the family heard from him too, but not so frequently. Nephews and nieces received picture postcards, children's books, advice on what books to read, and so on. When letters arrived, Chandra's father would gather everyone around in Chandra Vilas and read aloud "Ayya's letter from England." He would then mark the date received, his corrections (spelling lapses, grammatical mistakes), and put the letters away, neatly stitched in a file ordered according to the dates received. Thanks to him and Balakrishnan, Chandra's letters of those years have survived to the present. The letters to his father reveal in depth Chandra's life, his concerns and worries, his work, study, and leisure routines, his excursions and walks, the scientists around him, financial details (how he spent and saved), and his health as indicated by his weight ("Did you weigh yourself?" was a frequent query of his father). To Balakrishnan he wrote more about matters of mind. Together the letters provide a remarkable record of a loyal, duty-bound son and a loving, caring brother who, amid all his preoccupation with research, lectures, and studies, had time to worry about every family detail—"How is Chandra Vilas? . . . Has the new house been completed? . . . How is Sundari's [his youngest sister] English? . . . Have you found a new servant to comb the hair of the girls?"

When Balakrishnan wanted advice regarding his future course of studies and career, Chandra wrote him a fifteen-page letter. His brother wanted to be a writer, to pursue a scholarly career. His father, however, wanted him to become a doctor. Balakrishnan felt helpless; he deplored the fact that he had no proven capabilities like his elder brother so that he could fight back and insist on his own choice of career. A few excerpts from his long letter of 12 May 1931 reveal Chandra's concern and sensitivity to his brother's problem:

You mean implicitly that my capacities were proved by my earlier career. If you think so—or for that matter, if anybody thinks so—I can say that it is all wrong. Do you think, that just because one read a little calculus and conics when his equals did not, read a little statistical mechanics which others of his age did not, happened to publish a few papers, Do you think that he is better than his friends, or Do you imagine he has in any case *intrinsic* merit? You are woefully mistaken if you think so. . . .

I wish I could divulge to you the sorrows of my heart, and tell you how I feel at times that my heart will break by the oppression of my ignorance. You are mistaken and so are others, if they think that I have proved anything at all. My progress I only know too well is positively shameful. . . .

What is essential, however, is to have the ideal of gaining knowledge and to work steadfastly towards the ideal. One should not care to worry about what happens. One must lay sound foundations, one must have enough enthusiasm, one must have a passion, one must be filled by the joy of study. That is enough. Age does *not* matter. It is *never too late* to begin. . . .

. . . if your heart yearns after a scholarly career, then the gates of knowledge are as much—if not more—open to you as to anybody else who happened—just happened—to get a better start—according to the onlooker's impressions. . . . In your choice of Medicine or Arts, the essential point to discriminate is as to whether by this choice you do not damage your inner feelings and aspirations. . . . If your aspirations are deep-seated, do not on any account damage them or molest them.

In advising Balakrishnan on whether to take medicine or arts courses, Chandra goes on to say:

. . . it is important to recognize that choosing one of two different subjects in the arts course is essentially—fundamentally—different from choosing either the "arts" or the "technical course." For instance, to choose Physics or Mathematics is *not* the same as choosing between Medicine and Physics, or Engineering and Mathematics. Physics and Mathematics will make one pursue the *same* ideal. Physics and Medicine will *not* enable one to pursue the same ideal. Hence choosing Physics or Mathematics, is choosing *one* of two paths towards the *same* ideal. Choosing Physics or Medicine is choosing one of two essentially dissimilar ideals.

His mother's health was constantly on his mind. In every letter he inquired after her health and ended his letters with "Namaskarams to mother." On his twentieth birthday he had written to her (in conformity with "protocol"),

I bow down to you (though I am at this great distance) for your kind blessings on my birthday. I complete twenty years today and I hope that in years to come, with the strength of your blessings and love, I will be deserving of that life which has been given to me.

As her health went through rapid changes, sometimes better and sometimes worse, and the disease took its course, he always hoped that she would recover. He used to urge her to include eggs in her diet, to build her lost strength, offered suggestions to his father concerning new treatments, and waited every week anxiously to hear about her. The news came suddenly in the form of a telegram on 21 May 1931:

MOTHER PASSED AWAY THURSDAY 2 PM BEAR PATIENTLY.

As one can imagine, Chandra was distraught. He sent the return message:

TERRIBLE TO REALIZE IT TRUE. WHAT CONSOLATION IS THERE AND WHAT CONSOLATION CAN I OFFER WHEN I AM SO DISCONSOLATE. WE MUST ALL BEAR PATIENTLY.

He wrote to his father the next day:

It is so difficult to bear all this. Life seems so tyrannous. Difficult as it is for all of us, I know it must be most difficult for you. What can I say, being so young and immature . . .

Mother had such a difficult time in her illness during her last few months. She ought to have lived and seen her children grow to age and all become settled. It seems so unjust and cruel to take a loving mother's life away when there are young children needing her help and presence so much. I hardly know how we are going to bear this. It must be borne, since it has happened and the only thing we can perhaps do is to shake our fist against Him who has shown his "sport with us."

He wanted to know how his mother spent her last moments. He had written a letter in Tamil just for her. Did it reach her before her death? Did someone read it to her?

Oh! My heavens, little did I dream that when I left her at the station that it was my last farewell to my mother. It is so hard for me at this distance. I had always hoped that she would recover. It is all over, all over.

His father's letters, written during the week before his mother's death, arrived a few days later, so Chandra had to live in reverse the ordeal his family had already passed through without him. The letters described his mother's worsening condition, the intense pain she was suffering. Her death was imminent as the right lung had completely collapsed and the disease was affecting her brain. She had stopped taking any nourishment. Yes, Chandra's letter for her had arrived before her death and it had been read to her. She lay in a semiconscious state for over a week prior to her death, but hearing Chandra's letter, she had brightened momentarily. Once during the week, in a delirious and semiconscious state, she had recounted her past life, mentioning each child in the order of their birth. She had expressed her hopes and aspirations for each of them: Vishwam to go into the forestry service. Balakrishnan—a doctor and a poet. Sarada married to an ICS officer. Vidya, a professor of music. Savitri and Sundari, to be well married. Ramnath, an engineer. Good grade jobs for her sons-in-law.

Balakrishnan also wrote that week: "Far away from the rest, alone in a strange land, the news of your great loss would have been a great shock to you. Unlike here, you have no one to share your grief with you. I hope you will bear it as best as you can. . . . the delay was slow and certain. Many a time I spoke to her; you filled all her thoughts and the mention of your name was sufficient for her to forget all her ills as it invariably produced a bright face and cheery smile. She would then tell me how you had promised to return with name and fame, how you would prove a worthy rival to your uncle."

It was so sad to read all this. Chandra could not function. Alone, with no one to share his grief, he would go over to the riverbank, sit, and weep. Letters from home, with the long time gaps between their having been written and his receiving them, mostly reopened the wounds rather than soothed. "Bear patiently," he told himself. He had to. On 28 May he wrote to his father, "Time helps to heal wounds ever so sore they may be. That appears to me the tyrannous aspect of time. However I have consoled myself sufficiently to begin the daily work."

Ironically, his first meeting with Eddington was on 23 May, the day after he received the telegram announcing his mother's death. He had kept the appointment, received congratulations from Eddington, and discussed his work, all the while feeling empty inside.

Work was the only panacea for loneliness and grief. He was working on stellar coefficients of absorption with Milne and was planning to spend the summer in Oxford. But after the news of his mother's death, Chandra felt the need for a change from the drudgery of Cambridge and the past eleven months of ceaseless study and research. He thought that a few months on the continent might provide the necessary diversion. Because he had studied German back home, he decided to travel to Germany, where he would have a chance to put his knowledge into practice. Also, since the German universities did not close till the end of July, he would have a chance to see them in session.

Chandra chose to go to Göttingen, where the Institut für Theoretische Physik was located. Max Born was its director at the time. Chandra wrote to him, asking for permission to work at the institute, and was happy to receive a prompt reply from Born, who said, "It happened accidentally that your friendly letter was received by me just after I had read your beautiful works. It will give me great joy indeed if you will come to us and share in our work."

So Chandra left Cambridge on 1 July 1931 and crossed the channel from Dover to Ostende, a port in Belgium. From there, a train journey took him to Göttingen via Köln and Kassel. A two-hour stop in Köln gave Chandra the opportunity to visit its famous cathedral and to enjoy a view of the Rhine from the bridge next to the cathedral. When Chandra arrived in Göttingen, one of Born's students helped Chandra get settled in a pension (57 Stegenmuhlenweg). It turned out to be quite a comfortable place and, more importantly, as he wrote his father on 9 July, "I get very good food, better food as a matter of fact. The chief reason being that the Germans take considerably less meat than the English and considerably more vegetables and consequently I get better cooked vegetables than in Cambridge. In Cambridge, the food was positively disgusting! The house also has brilliant co-inmates, and all for only 165 marks per month (165 marks is slightly less than 165 shillings)."

The "brilliant co-inmates" included some prominent mathematicians and physicists of the time. There was, for instance, the Russian mathematician L. Schnirrelmann, the German physicist Lothar W. Nordheim, and the Russian physicist Georg Rummer, who was a student of Born, and Edward Teller from Hungary. In addition, there were short-term visitors like Léon Brillouin and Heisenberg. Heisenberg was a professor at Leipzig, but he used to come to Göttingen to see Born. Ludwig Biermann, a noted theoretical astronomer, was another frequent visitor from the nearby observatory.

Schnirrelmann and Chandra soon became good friends; they went on long walks together on Sundays in the Kaiser Wilhelm Wald outside Göttingen. They went on an excursion together to Kassel, where the Kaiser Wilhelm palace is situated. The palace is famous for its art gallery and an artificially constructed waterfall in its gardens. "The art gallery there contains a remarkably large collection of original Rubens, Rembrandt, Van Dyke, and Tiscen," wrote Chandra to his father on 26 July. He also witnessed the magnificent "water-display" from the fall (which was turned on and off periodically for the tourists).

Although Edward Teller was not working in astrophysics, he liked to talk with Chandra about astrophysics and Chandra's work. That led to a friendship which was renewed in Cambridge in 1933–34 when Teller came there as a visitor. Their friendship continued in the United States till the 1950s. Thereafter, the contacts between the two became infrequent owing to their different opinions regarding the Oppenheimer case and matters of the Atomic Energy Commission,* and their generally divergent political views.

Although the summer in Germany was to have been a vacation for Chandra, it became mostly a change of place and a change of study topics. He began to study group theory and quantum mechanics (beginning with Van der Waerden's two volumes of *Modern Algebra,* which he had bought on his arrival) in addition to completing the study of *Modern Analysis* (Whittaker and Watson) which he had begun in Cambridge. When the university was open during July, he attended Born's lectures on the applications of quantum mechanics to the problems of chemical

---

*Chandra, along with Robert Wilson and William Fowler, was on a committee for the Atomic Energy Commission in the mid-fifties to advise on the fusion project. At a meeting chaired by J. H. Williams, a member of the AEC, the directors of the national laboratories involved in the project, of which Teller was one, were to make presentations for their projected budgets. After the presentations, the committee was supposed to have an executive meeting with the chairman to discuss their recommendations. But Teller would not leave the session. When Wilson brought up the question of the executive meeting, Teller unceremoniously said, "Why do you want to have an executive meeting? What do you people know on the subject?" Thereupon Wilson naturally got quite upset and resigned from the committee; so did Chandra.

affinity. Chandra did not see Born often because Born was extremely busy, but, nevertheless, it was Born who had taken care of all the formalities regarding Chandra's stay and had provided excellent working conditions at the institute for Chandra. Besides mathematical studies, Chandra continued his researches, completing the unfinished computations he had accumulated. He produced two papers during the summer.[14]

He found some diversion, if it can be considered such, with his landlady's son, Eric Grunau, who was a friendly sort. He would come to Chandra's room and teach him conversational German and read with him the poems of Schiller and Goethe. By the end of two months' stay, Chandra was able to speak and understand German quite well. In return for Grunau's time and effort, Chandra arranged through his father to send him a "nice, brass, cigarette case (the freight and everything paid)," from India.

His summer in Germany came to an end with a ten-day trip to Berlin, which Chandra described as "a big city, clean, beautiful and magnificent." Along with tours of the zoological gardens, the National Gallery, and other sites, he spent a day at the Kaiser-Wilhelm-Institut für Physikalische Chemie, talking to Karl Bohnhöffer, A. and L. Farkas, and Paul Hartek, all physical chemists who were known for their fundamental contributions. He also visited the Potsdam Observatory situated in a suburb of Berlin. He was thrilled to meet Erwin Finlay Freundlich, an astronomer and astrophysicist of considerable renown. While Chandra was familiar with Freundlich's work, he did not expect the reverse to be true. On the contrary, as Chandra noted in a letter to his father on 22 September 1931, ". . . it was such a pleasant surprise for me when I found that Prof. Freundlich not only recognized me, but was familiar with my work! He wanted me to make a small 'Vortrag' (unofficial lecture) on my work 'Stellar Absorption Coefficients.' I spoke for an hour or more on the work I had done at Göttingen. . . . Prof. Freundlich congratulated me very much on my latter work and I also personally think it is the best I have done so far, and also important in that it gives (according to Freundlich) the 'Tod-Stop zu den Eddingtonische Theorie' (Death Blow to Eddington's Theory). Prof. Freundlich and his wife were 'at home' to me that evening."

With that happy ending to his summer sojourn in Germany, Chandra returned to Cambridge in early September to begin his second academic year. He immediately sent the paper on stellar coefficients of absorption, which he had completed in Germany, to Milne to be communicated to the *Monthly Notices of the Royal Astronomical Society* (*MNRAS*). While waiting for the Michaelmas term to begin in October, he continued his research on stellar coefficients of absorption in the theory of stellar atmospheres—work that involved "laborious numerical computations."

He adjusted back to the drudgery of Cambridge food. "I am at times vexed (i.e., feel helplessly cursed) at the bread, butter, jam and eggs, all insipid stuff," he wrote to his father. "I have, however, developed to a certain extent an almost 'Newtonian' mentality of not caring or allowing oneself to think of the tastiness or otherwise of the food one may be taking."

Lonely evening walks along Trumpington Road became a routine. Occasionally he would make Sunday a holiday. After breakfast, he would go for a day-long walk. He would first go along the winding Madingley Road leading to Oxford, with its many ups and downs and with poultry farms on either side, and then cut across open fields, enjoying the "solitary, lonely, quietness of the open air." He marveled at the sharp seasonal changes in the trees' foliage. He would return to his "digs" exhausted, but refreshed and ready to start working again.

In spite of his hard work and his almost continuous output of papers, Chandra was not happy with himself. As he approached his twenty-first birthday, he wrote to his father on 14 October, "I am almost ashamed to confess it. Years run apace, but nothing done! I wish I had been more concentrated, directed and disciplined in my work. He is wise, who strikes the exact balance between learning for oneself and attempting some really substantial things *consistent* with success. . . . Everybody has to strike a compromise, but the compromise will have to be fair. In my case it has been grossly unfair."

A conflict was brewing in his mind. Overly concerned with the limited duration of his scholarship, he was concentrating on what he could do. "In some ways, it is easier to write a paper than understand a subject," says Chandra. "There is always this conflict . . . Should you decide to do the things which you have known how to do or should you go into other things?" He was in astrophysics by sheer chance, because, on his own, he had found a problem to work on. He was convinced that he had made a fundamental discovery in the limiting mass of a white dwarf, but stalwarts like Fowler and Milne did not seem to appreciate its importance. What about his true love—pure mathematics? He had come into physics due to the insistence of his father, but he had hoped that he would not have to abandon mathematics. That hope had begun to recede.

Failing to pursue pure mathematics, Chandra would have been happier, he felt, if he could give himself to pure physics. He saw that physics was the frontier field in which fundamental discoveries were taking place. The star-studded Cavendish Laboratory was the center of activity. Dirac was his mentor. "He is just wonderful!" Chandra wrote to his father after he had a discussion with Dirac on 22 January 1932. "His philosophical insight into the general formalism of theoretical physics, his mathematical profundity which allows him to penetrate with ease any region of unexplored physical or mathematical thought, and with all this what humility! He almost represents to me the PERFECT MAN . . . 'almost' because of his utter unconsciousness of his own Depth."

But he could not bring himself to tell Dirac that he wanted to switch fields, that he would prefer to do work in theoretical physics. A feeling of loyalty to Fowler, who, Chandra felt, was mainly responsible for his admission to Cambridge, and the fear that he might not complete his degree if he switched fields, prevented him from making the switch from research in astrophysics to research in pure physics. This meant that if he was to follow his desires in the matter, pure mathematics and pure physics would have to be pursued on the side.

In the meantime, a further strong push towards astrophysics came unexpectedly. On 12 November 1931, at 10:30 AM, Chandra heard a knock on the door of his room. When he said, "Yes," the door opened, and he was amazed to find Milne standing in front of him. An Oxford professor at his door! Milne had come to tell Chandra that he had communicated his paper ("Stellar Coefficients of Absorption, Part II") to the Royal Society. He also had an interest in the work Chandra was engaged in at the time: ionization in stellar atmospheres. They had a long discussion, and Milne proposed that they write a joint paper, "Ionization in Stellar Atmospheres, Part III." It would be "Part III" because Milne had already published two papers on the subject. The new work Chandra was involved in was, in a sense, a continuation of that effort.

Chandra was thrilled by Milne's initiative and encouragement. The next two months he worked harder than ever; he was absorbed in numerical calculations and results which ran "through more than one hundred pages of closely written sheets." After completing the work and mailing the manuscript to Milne just before Christmas, he decided to take a week off in London. He had no choice. Cambridge was deserted. His landlady had an invitation from her friends to spend Christmas with them. "She requested me—if not forced me!—to go away for a week," he wrote. "During X-mas even all the restaurants are closed and I am practically at my wits ends." And in a rare instance of admission, he wrote on 23 December 1931, "I need to slow down a bit, since I have been working rather hard since I came back from Göttingen."

Another new year began. He might have smiled at his former optimism had he been able to reread what he had written on the previous New Year's Day before his mother's death: "One admits that this artificial demarcation of the ever ebbing tide of time is unnatural, strange . . . But it helps to review the past and look with bright hopes forward. It is to be hoped particularly that 1931 may see mother's recovering back to her normal health."

That hope shattered, on 1 January 1932 he could only philosophize, "As to what is in 'store,' experience alone in its rolling contact with time can reveal. From every point of view the conception of time has proved ununderstandable. From the point of view of science, even now we are not fully aware of the relations of time with space. From the point of view of philosophy, there seem to be very diverse opinions. From the point of view of Experience it seems to be the Death in Action or Action in Torture, (Definition due to Amiel[15])."

Workwise, however, his situation could not have been better. His joint paper with Milne and his own paper, "Model Stellar Photospheres," were being published in the January issue of *MNRAS*, occupying a major portion of the journal. He had every reason to be happy since he had completed the work he had planned six months before in Göttingen. "The Stellar Coefficients of Absorption," his joint paper with Milne, and the paper on model stellar photospheres all formed an interlocking chain of work. He was invited to present his results at the January meeting of the Royal Astronomical Society. After his presentation (of model photospheres only because of lack of time), he received high compliments from Eddington and Milne.

Nevertheless, he returned from the meeting rather depressed, sober, and "composed." "I do not feel happy," he wrote to his father on 12 January. "These 18 months of rather lonely life has slowly but steadily changed my whole attitude towards things in general. Chandrasekhar even of early 1931 would have jumped with joy. But he of 1932 feels only composed in an easy chair."

His mother's death had certainly cast a gloom over all his strivings and successes. Besides, the nagging doubts about the value of the work he was doing contributed to his depressed spirits and unhappiness. "My astrophysical work," he wrote, "has become largely a matter of numerical work and by such work one never learns anything new." Questions concerning the limiting mass were essentially ignored by others. Fowler was noncommittal, and Milne, who had initiated a bond of friendship and close association, was positively against his proceeding with this inquiry because it contradicted his own idea that *every star,* no matter what its mass, had a degenerate core.

Furthermore, although Chandra had greatly overcome the initial shattering experience of entering an environment which had been fraught with strangeness and he had adjusted to the idea that, unlike in India, he was not a celebrity, he still expected to be recognized along with his contemporaries, like Neville Mott, Norman Feather, and others who were "locally famous." That was not to be the case. They were in physics; physics was at the center, not astrophysics. He was alone. "My [first two years'] experience in Cambridge," recalls Chandra, "made me feel that I was essentially on my own and that whatever career I had in science must be due to my own efforts."

Chandra finally made up his mind to attempt to get into pure physics. He read some papers by Heisenberg and Pauli, which Dirac had suggested, as an introduction to quantum electrodynamics. He did not get very far in his efforts and, after considerable agonizing, he decided to discuss the matter with Dirac.

As Chandra recalls, it was a Saturday morning in May when he went to see Dirac in his rooms. He remembers telling him that he was extremely unhappy in Cambridge with what he had been doing. It didn't seem to him that he was accomplishing anything useful. He told Dirac that Cambridge was a place where if you did good work, you would get encouraged, and you would do better work, and in this way the spiral would build. But if you did nothing important, then you would get no encouragement, and you would do worse and worse.

Chandra considered himself to be in the latter category. Dirac, not being in astrophysics, was in no position to convince him otherwise. He was very nice and understanding, deploring the situation at Cambridge. Dirac then strongly advised Chandra to go to Copenhagen and study physics under the tutelage of the famous Niels Bohr. He told him that he would find a better climate there; the younger men were friendly, even though they were "big men" in physics. Dirac volunteered to write to Bohr on his behalf or, if Chandra preferred, he could ask Fowler to write.

The residency requirements of Cambridge University allowed a student to spend a year at another place, but Chandra needed permission from the office of the High Commissioner of India to continue his studies in Copenhagen under the support of the Government of India scholarship. Thanks to a strong recommendation from Fowler, the High Commissioner gave his permission. In due course, Bohr wrote and welcomed Chandra to his institute.

Chandra was delighted. He badly needed a change of surroundings. "Cambridge—in spite of Dirac!—gets on my nerves," he wrote. "The same little ten feet square—my garret!" Besides, it was always one of his "cherished" desires to study at Bohr's institute. He knew that it was a "very high-browed" place, to which "Fowler, Dirac, Heisenberg, Pauli, Jordan, Born, Klein, Darwin, etc., often went to consult and learn from Bohr." He wrote exuberantly on 15 June 1932 to his father: "It could be said *only* of Bohr that he is not only a great mind but one whose influence on the contemporary geniuses—particularly Heisenberg and Co!—has been colossal. In fact in the whole range of mathematical and physical history, it would be difficult to find Bohr's equal—I mean particularly in the greatness of his influence—at the moment I can think of only one name—Gauss."

As the opening date for the institute was 1 September, Chandra immediately began making plans for his trip, beginning with Danish lessons. In the meantime, he started working on a physics problem which Dirac had suggested.

Chandra left Cambridge on the evening of 15 August 1932 and reached Copenhagen on the morning of the 17th, crossing the North Sea between Norwich and Esbjerg on the steamer M.S. *England.* From Esbjerg to Copenhagen, he traveled by train; it was a unique journey for him because it was the first time he had been on a train which was carried on a ferry between the three islands that constitute Denmark. It was a delightful journey. "The sea was wonderfully calm, there was the full moon, and a lovely, gentle breeze," wrote Chandra on 23 August. "Though I had a 'sleeper' I was awake to see how the whole rake was transferred from land on to the ferry. The train is again transferred from the ferry on to the land by an engine which awaits it at the other end (the engines are too heavy to be transferred) and once again the engine's sharp whistle, puff-puff and the endless mimicry of the wheels and rails—rat tat."

While enjoying the unique nocturnal train journey, he was happy and at the same time quite apprehensive about what was in store for him. Three years before, while showing Heisenberg around Madras, he had heard about Copenhagen and Bohr's greatness as a teacher, philosopher, and "inspirer." Since then, rhapsodies about Bohr had reached him from many other sources. Now, the next morning, he was going to be in his presence!

Bengt Strömgren, whom he knew through correspondence regarding their scientific work, had come to the station to meet him. With his help, Chandra located Havas Pensionat, where arrangements for his stay had been made. Managed by Margere Have, this pension was situated within walking distance of the institute and was the more popular and, covertly, the more prestigious of the two pensions used by visitors to the institute.

As Dirac had indicated, the atmosphere in Bohr's institute was quite unlike Cambridge. It was extremely friendly and truly international in character compared with the dominance of the British in Cambridge. Chandra found himself in a group of enthusiastic young people, including among others, Max Delbrück, George Placzek, E. J. Williams, Victor Weisskopf, Hans Kopferman, and Leon Rosenfeld. There were also frequent visitors. Oskar Klein and Heisenberg came to the institute from time to time while Chandra was there. German was the "official" language of the institute. Chandra had no problem with that because of his previous summer's experience in Germany. He did continue, however, to take Danish lessons.

Weisskopf, Placzek, Williams, and Rosenfeld lived in the same pension. "We were rather a rambunctious crowd," recalls Weisskopf, and "he [Chandra] participated to some extent with us. We went to movies together, to Tivoli gardens often, enjoyed practical jokes . . . it was a strange group; and sometimes I remember this with a little bit of bad conscience. Whenever somebody came from some country, from America or from India, *we looked him over*—we decided whether he was an 'in man' or not. The poor fellow who was not accepted had a very bad life. I tell you this because he [Chandra] was 'in' in every respect . . . Shy and quiet compared to some of us. . . . But we all liked him very much. He was definitely an 'in man.'"[16]

Chandra also remembers those days of new friends and happy associations with some who were already famous and ahead of him on the ladder of their scientific careers. He was particularly drawn to Leon Rosenfeld of Belgium, a close associate of Bohr. Rosenfeld's fiancée was studying astrophysics, so he had more than a passing interest in the subject and discussed Chandra's work with him. Thus with new friendships, tea every Sunday at Bohr's house, walks and bicycle rides in the country, sometimes with Bohr and his family, Chandra's life took on a more communal dimension. He no longer felt so lonely and isolated. Bohr, of course, in and around the institute, was the center of attention and the main source of everyone's inspiration. His "acting, talking, living as an equal in a group of young optimistic, jocular, enthusiastic people, and approaching the deepest riddles of nature with a spirit of attack, a spirit of freedom from conventional bonds and a spirit of joy,"[17] made an extraordinary impression on Chandra.

He was also happy because he was working on a problem in physics suggested by Dirac. "I have been working at 'Dirac's' problem rather seriously," he wrote to his father, "and I have got out 'something' at least in which Prof. Bohr [is] very interested. Though I should not like to count the chickens before they are hatched, yet I rather think I have got hold of something not altogether trivial." His joyful moods are reflected in the rhapsodic way he wrote to his father about the autumn weather on 5 October:

Autumn! It is delightful here in Copenhagen. The autumn leaves, gay and vain in their changing hues of green, yellow, red and crimson, fluttering with excitement at every breath of the chilly Northern Winds—it is a drama of Nature revealing herself in flashes (and almost without modesty) the depth of her own Being.

Yet, as though to forewarn himself, he wrote further:

Poor autumn leaves! With all your vanity, with all your joy, you are little aware that the delightful Cold Wind which makes you faint with pleasure, is but the Ambassador of the ruthless Winter which will wither you, without feeling and without sympathy.

His sense of foreboding proved to be well founded. He thought he had solved Dirac's problem, so he wrote a paper entitled "On the Statistics of Similar Particles," and sent it to Dirac in Cambridge. Both Rosenfeld and Bohr read it and gave it their nod. Bohr communicated it to the *Proceedings of the Royal Society* for publication. But just as fall ended and winter set in, he received a short note from Dirac saying that there was an error in the paper. Chandra had not solved the problem.

At first Chandra did not believe Dirac. He wrote to his father on 21 October:

I have got into trouble with Dirac! He thinks my solution of his problem is wrong. In fact I think Dirac is wrong, and further he has not even read my paper. This means another long drawn-out controversy and I am rather certain that my case is perfectly clear cut and that Dirac will finally come around.

That was not to be the case. It did not take him long to realize that there was, in fact, a central mistake. "My paper sent to the *Proceedings of the Royal Society* is WRONG. That is all," he wrote to his father on 11 November. He withdrew the paper from publication.* Over four months of work was down the drain. "That was rather a severe blow to me," Chandra says. "I went to Copenhagen in 1932 with the hope that I would be able to integrate myself into some work in physics, in quantum theory. I tried this problem and failed. Then December came along. I had to return to Cambridge in June with a thesis ready . . . I went back to astrophysics."

As it turned out, Milne came to Copenhagen for a short visit in early December. He had been living in Berlin at the time. His visit gave Chandra an opportunity to discuss some work with him which he had set aside, his investigations of the equilibrium of rotating gas spheres. Chandra was soon able to complete a paper on the subject, and Milne agreed to report on it at the January 1933 meeting of the Royal Astronomical Society. At around the same time, he received an invitation from the University of Liège, Belgium, to deliver a series of lectures on miscellaneous topics in astrophysics.

Chandra's acceptance of the Liège invitation and his continued success in astrophysical research made his return to astrophysics complete. From then on he was to remain an outsider, just a friend to the "enthusiastic crowd" of physicists at the institute. He would only occasionally talk (rather mostly listen) to Bohr, who was not interested in astrophysics. "I cannot be really sympathetic to work in astrophysics," Chandra recalls Bohr saying to him one day, "because the first question I want to ask when I think of the sun is where does the energy come from. You cannot tell me where the energy comes from, so how can I believe all the other things?" Bohr was, at the time, occupied with questions of the measurability of electromagnetic fields and was working on this problem with Rosenfeld. "He [Bohr] is a very busy man," Chandra wrote to his father on 14 December 1932. "He scarcely discusses with anyone any problem unless he is sure he has time enough to go into the problem really deeply. He is not one who likes to talk glibly. He is Socratic in

---

* As Chandra recalls, Dirac had suggested to him the following problem: Ignore spin. Consider particle occupation numbers to be 0, 1, or 2. How would you then construct the wave function? What kind of statistics would you obtain?

his views with the difference that Socrates had time enough but Bohr has not."

All the same, Chandra recalls with a touch of poignancy the Sunday teas at Bohr's home, when Bohr would come down from his study. After tea, he would often take the physicists like Placzek, Weisskopf, and Rosenfeld upstairs to his study; Chandra would be left alone in the company of Bohr's young son, Aage Bohr,[18] with the feeling that what he was working on was irrelevant and not as important as what the physicists were doing.

Chandra occupied himself with the preparation for the Liège lectures, plans for writing his thesis, and of course continuing his research. He was supposed to go to Liège in the middle of February, return to Cambridge in the middle of May, and take his Ph.D. examination in June. The three-year scholarship would be over by then; he had to think of his future, about which his father was reminding him continually.

His trip to Belgium was delayed by a couple of weeks because Rosenfeld, who was responsible for getting him to Liège and was going to be his host, returned to Copenhagen to work with Bohr. On the last day of February 1933, they finally traveled together to Liège. Chandra stayed with Rosenfeld, Rosenfeld's mother, and his fiancée. Liège, a "pretty city with hills all around," was also an "industrial city with a little too much coal in the dust." But, as Chandra wrote on 2 March, "in one's home, in the presence of happy faces, and with a lively fire to keep one warm, who cares?"

During his first week's stay, he gave three of his six lectures. The first five were rather technical; he talked about stellar atmospheres, dealing with the theory of the chromosphere and the contours of absorption lines. The sixth and last lecture was a general, popular one on the internal constitution of stars. It attracted a large audience, including some bigwigs of the university. At the end, Chandra was honored with a bronze medal. Also, he was urged to publish his first five lectures in the Academy of Sciences proceedings.

These were happy days for Chandra. He was living with a happy family who included him and valued him. He enjoyed preparing lectures for an attentive and appreciative audience. He enjoyed the discussions with Rosenfeld and his fiancée. He had two "wonderful" Sundays in between his lectures; on one, he went for a long walk with Rosenfeld in the Valley of Ambleve—one of the most beautiful spots in Belgium; on the second, he took another walking trip through the "Fang"—one of the largest moors in Europe. He also squeezed in a trip to Antwerp, with its "neat (but narrow!) streets, beautiful harbour and its great church." While he was there he visited Antwerp's picture gallery, well known for its many originals by Rubens, Van Dyke, Jordain, and other Flemish painters and also paintings of the Belgian school. Altogether, he was pleased with his stay in Liège and left it with a feeling of "eminent success."

Chandra returned to Copenhagen on 16 March after spending a few days in Hamburg. Originally, the Hamburg trip was intended to be a pure holiday, a sight-seeing tour of the big city with its huge harbor filled with hundreds of ships steaming in and steaming out, with its zoological gardens where animals were not caged but left free in little islands, and with its art gallery filled with classical paintings and sculpture. Now that his lectures had to be prepared for publication, however, he had to spend some time working. He also spent a considerable amount of time with Professor Emil Artin, a very well-known algebraist, seeking his help and advice in studying modern algebra. Chandra had spent the summer in Germany studying the subject and wanted to see if he could make the transition into pure mathematics. "Artin was very kind to me," wrote Chandra to his father on 17 March, "and indeed [he] said that if I came to him in July, he would give me detailed instructions for a month."

During the next two months in Copenhagen, Chandra was preoccupied with work on distorted polytropes, which developed into a series of four papers and became more than adequate for his thesis material. He was also concerned about his future. While his father was urging him to return to India, Chandra was leaning towards extending his stay in Europe. But where was the support to come from? Whom should he approach? Should he write to the High Commissioner's office? Should he apply for a senior research fellowship in Cambridge? Or should he try for support at Copenhagen? With such questions on his mind, he returned to Cambridge in the middle of May 1933.

5

Fellow of Trinity College
Cambridge, 1933–1934

... to be a member of such a society, to be part of those ancient and restful buildings, and to have your name inscribed in the roll which counts in its past list so many names we all reverence.

E. A. Milne (in letter to Chandra)

If the conventional indicators were all that mattered, Chandra should have been extremely happy with his accomplishments during the year in Copenhagen. His Liège visit was a phenomenal success both socially and scientifically. He had enjoyed the warm hospitality of the Rosenfelds' home. His lectures were published in the form of a book, and within half a year (December 1932–May 1933) he had completed four papers on the equilibrium of distorted polytropes. He had more than adequate material for his thesis.

Yet, when he returned to Cambridge in May 1933, Chandra was unhappy with himself, suffused with a sense of failure. "I went there [to Copenhagen]," Chandra recalls, "to try to make a change. I wanted to work in physics, but I couldn't get anywhere. Any kind of work I thought I might be able to start in quantum theory or a related subject ended in failure. And when December came along, it became clear that I had to get my thesis ready in order to get my degree before the government scholarship terminated. So, I returned to astrophysics, and, after some thought, formulated for myself problems in distorted polytropes and began that series of papers."

The fact that he was compelled to do certain things in the interest of his career in the short range rather than pursue his innermost, strong desire to work in pure physics or pure mathematics left him dissatisfied. Physics was *the* fundamental science, so while he socialized with people like Victor Weisskopf, Max Delbrück, Hans Kopferman, and others who appeared to be at the hub of important discoveries, Chandra was not part of their science.

As Weisskopf recalls, "he [Chandra] was working in astronomy, astrophysics, spectroscopy, and related matters, which were not exactly the center of interest for the people who were there. Our center of interest at the time was Fermi's theory of beta decay, the beginnings of the field theory, the positron, the antiparticle, and the nuclear structure."[1] Chandra felt trapped in astrophysics because that was, at the time, the only area in which he could formulate problems on his own and solve them.

Furthermore, even accepting the short-range compromise, his future loomed dark and uncertain. The government scholarship would last only until the end of August, and according to the agreement Chandra had signed with the government of India, he had to return to India as soon as he completed his Ph.D. degree. Chandra's father was constantly urging him to begin applying for positions in India, sending him advertisements, asking him to secure letters of recommendation, and so forth. But Chandra was painfully aware of his fate if he returned to India without securing a situation in which he could continue his research and studies. He realized that he could become lost in petty quarrels and rivalries among people who could not "value scientific work and scientific pursuit (apart from scientific honors)." "A position similar to Dr. Vaidyanathaswamy's,"[2] he wrote to his father, "which would give me peace and quiet is the one I would positively want even if it were less remunerative." Such a position, however, did not exist. It had to be created. Candeth, the Director of Public Instruction, who had promised to create such a position for Chandra, was no longer in charge. The most likely thing that would happen if he just returned was that he would get no better job than that of a physics laboratory instructor. "If they want me to teach 'Wheatstone's Bridge' and supervise B.Sc. practical," Chandra wrote on 2 February 1933, "I would tell them I can as well lecture on the History of Mexico!"

In Europe he could continue his work more easily; a research atmosphere existed; opportunities abounded for discussions with experts in the field; and it was easier to get his papers published in first-rate scientific journals. He could perhaps once again try to do research in physics with Dirac's help. Milne wanted him to go with him during July and August to St. Andrews University, Scotland. Emil Artin had promised to instruct him in modern algebra personally if he went to Hamburg. A conference in Copenhagen was forthcoming in the fall of 1933, where there would be a confluence of all the great physicists. He would certainly want to attend that. Besides, he had been so preoccupied with his work during the past three years that he had had no time to travel. Chandra wanted to spend at least one more year in Europe. He would write to the High Commissioner's office and seek an extension of the scholarship for an additional year.

He met with a prompt rebuff. It was the old story again—bureaucratic stumbling blocks. "The High Commissioner people are damned fellows," wrote Chandra to his father on 27 April 1933, "who can cleverly place obstacles in one's way but can hardly raise a pin's head for others as a matter of common obligation." A year earlier he had written to his former teacher Parameswaran, inquiring about the possibilities of a job after his return. Parameswaran had replied that Madras government affairs were strange and that he had no say in the matter; however, he said (17 April 1932), "I do not think you need have any great anxiety about your future! With all the high class work you have done, with your reputation as a physicist well-established, with your uncle Sir Raman so well established in all higher councils where physics is concerned, you ought to be able to get a good footing easily."

These hopeful assurances were not of great help to Chandra, especially since he was absolutely determined not to base his future on his uncle's influence. Neither did he appreciate the further remarks by Parameswaran in a postscript of the same letter: "Perhaps you will be interested to hear that people are already busying themselves fixing up girls for you to marry on your return."

Chandra was determined to extend his stay in Europe. He would seek support in Cambridge and Copenhagen; if nothing materialized, he had enough savings to get by for at least six months anywhere in Europe. If he had to return to India, he could get by without a job for a year. "I do not see for the moment any dreadful hurry to get fixed up in a 'place,'" he wrote on 18 May 1933 to his father, who was urging him to take a temporary position at Dacca University.

With that determination, he returned to Cambridge on 17 May 1933 ready with his thesis. He expected Fowler to at least glance at it before he handed it over to the Registrar. But when he met with him, Fowler said, "No, no. Definitely not. There is no need. I have full confidence. Take it to the Registrar." Which Chandra did. He was informed within a few days that the oral examination would be on Tuesday, 20 June 1933.

Eddington and Fowler were to be the examiners and Chandra was to appear in the observatory rooms at noon wearing the traditional academic gown.

That it was going to be merely a formality was apparent to Chandra, as he had already, on invitation, presented a summary of his work on distorted polytropes (essentially his thesis) at the 12 June meeting of the Royal Astronomical Society. In the discussion that followed, he had received enthusiastic responses and praise from both Milne and Eddington. Henry Norris Russell, who attended the meeting, had said, "First, I must congratulate Mr. Chandrasekhar on the lucid presentation he has given of this intricate problem. His work is going to be of great practical value." After the meeting Russell had sought out Chandra to congratulate him again. In spite of all this, an examination was an examination, and Chandra was not totally without apprehension as he walked into the observatory room exactly at the appointed hour. As he recalls:

> There was nobody present. After about fifteen minutes, Fowler showed up in his old Ford and wearing his gown. He apologized to me for being late and dashed into Eddington's rooms. After a few minutes he called me in. Eddington was standing there in his slippers. Seeing us in gown, he said, "I am afraid I cannot rise to the occasion of a gown." We went into Eddington's study. Fowler and Eddington urged each other to begin. Fowler began. My answers to his questions were satisfactory to him, but Eddington was not quite happy with the way I answered them. Then Eddington asked questions. My answers were satisfactory to him, but not to Fowler. This went on for about forty minutes, when Fowler suddenly looked at his watch from his waistcoat and said, "Good heavens, I am late," and dashed out. Soon after that Eddington said, "That is all." And I walked out without knowing the outcome. It was not until I went to Fowler a few days later for a recommendation letter to extend my scholarship that I learned that I had passed. They [Fowler and Eddington] had sent a favorable report. The official declaration, signing the register, had to wait till October.

With the formality of his degree out of the way, Chandra was left only with the question of his future. He talked to Fowler and asked him whether there was any chance for him to stay an additional year in Cambridge. Fowler was not hopeful at all. "No, I don't think there is any chance," he told him. "You can try for a fellowship at Trinity, but the competition is quite severe. I doubt if you will get it." After some thinking, in spite of Fowler's discouraging remarks, Chandra decided to apply for a fellowship, the outcome of which would be decided in early October.

A Fellow of Trinity College! It was a wild dream. The only other

Indian who had been elected to a Trinity Fellowship was Srinivasa Ramanujan some sixteen years before. The fellowship would bring free rooms in the college, dining privileges at the high table, and an allowance of 300 pounds per year for four years! It would certainly solve his problem for the immediate future and put him in a strong position to secure the kind of job he aspired to on his return to India.

The competition, however, as Fowler had suggested, was formidable. It was open to candidates from all fields. Each candidate submitted a fellowship thesis, containing an account of his work during the previous years, and took two written examinations, one in general aspects of science and philosophy and the other in literature and the arts. Chandra submitted his thesis to Sir J. J. Thomson, the Master of the College, on 24 August and prepared himself for the written examinations on 29 September.

Although he went through all the required formalities, Chandra actually believed his chances of getting a fellowship were very slim indeed. As he recalls,

I was so sure I would not get the fellowship that I had made arrangements with Milne to spend the fall months in Oxford with him. I had saved enough money from my scholarship to spend an extra three months in England. In September I had gone to Oxford, rented a set of rooms for the fall, and paid off my landlady in Cambridge completely. I had even bought a bicycle because Oxford was a bigger place than Cambridge, but not big enough to use buses all the time.

On the day [9 October 1933] the fellowship was to be announced, I took a taxi with all my things to go to the station to go to Oxford, but on the way I stopped at the college to see who had been elected as fellows. *I was shocked to find my name on the list!*

When I saw my name, I remember telling myself, quite loudly, "This is it; this changes my life." I went to the taxi and asked the driver to take me back to my rooms.

Chandra promptly dispatched a telegram home:

ELECTED TO TRINITY FELLOWSHIP STOP WISH MOTHER WERE ALIVE STOP PLEASE NO PAPER PUBLICITY

The very next day he received a letter from Milne, dated 9 October:

It was with intense pleasure that I saw in the *Times* this morning of your election to a Trinity Fellowship and I hasten to send you my heartiest congratulations. I am very proud to have been associated with you in some of your work, and the satisfaction at your success is a very personal one.

I believe that the election to a Trinity Fellowship is one of the most important as well as one of the most gratifying events that can happen to one. I hope it will be a source of inspiration to you, as it certainly was to me, to be a member of such a society, to be part of those ancient and restful buildings, and to have your name inscribed in the roll which counts in its past list so many names we all reverence.

Further along in the letter, Milne had disclosed what was normally supposed to have been kept confidential—that he had been called in as a referee. He had provided a "long, careful and critical account" of Chandra's papers, "not always agreeing with them, but concluding that they showed a tremendous increase of power and maturity as the investigations mounted up." He was glad that Chandra had sent him his latest (unfinished) work on the chromosphere as he could comment on the real brilliance of the idea behind it. Milne wanted Chandra in no way to think that he [Milne] was claiming any part in his election. It was the intrinsic value of his contributions that had brought him the award. In fact, knowing something of the difficulties, he admitted that he had not been very hopeful of Chandra's chances. "I think," wrote Milne, "it is all the more credit to you that your work must have so impressed the electors. . . . It must be an added joy to you to know that you have been successful in a single attempt, in competition with people who might be considered to have an advantage in many respects."

Sir J. J. Thomson, the Master of Trinity, performed the traditional, colorful admission ceremony in the college chapel. Before the assembled college council, Chandra took the oath of loyalty to the college, loyalty to the cause of the furtherance of knowledge in whatever circumstances he was placed. "Each of us [the newly elected Fellows] had to kneel down before the master," wrote Chandra to his father on 12 October 1933, "who then, taking our hands in his, declared in Latin to the effect of our having been elected as Fellows of the College. In the evening there was the admission dinner when Sir J. J. made a speech for the health of the new Fellows. Naturally there was mention of Ramanujan having been my predecessor sixteen years ago. . . ."

Life indeed changed for Chandra. He could extend his stay in Cambridge by at least three years; the fellowship was actually for four years, but one year was granted as an "unconditional period." The Fellow could spend that year at any place of his choice. Therefore, Chandra could return to India during the fourth year and, while retaining the fellowship, look for a permanent situation. In this way he would be in a stronger position with the government "authorities." "Persons in authority always take advantage, when dealing with a person who is unemployed, of his unfortunate position," he had written to his father (15 June 1933). "So if I get a Cambridge Fellowship, the authorities would be prevented from playing that mean trick, and would be forced to treat me fairly."

The fellowship also served to some extent to counter the pressure from his father, who was not particularly pleased with Chandra's decision to extend his stay. He would rather have him return, take the best available job, marry, settle down, and share some of the family responsibilities as the eldest son. Chandra was fully aware of all this. He wrote on 12 October 1933, "I of course realize that this decision of mine would be interpreted as selfishness, lack of interest in home and so on. If one is convinced that way, then no amount of my assertion to the contrary would be of any avail, but I may as well mention (to be fair to myself) that this final decision has not been made without great pain and indeed during the whole of Sir J. J.'s speech last evening, and during all the time when so many were congratulating me there was only one picture before me—my mother in her silk sari . . . blessing me with all the force of her love to 'go forward.' Indeed I always have this vision which has been a great source of inspiration—intensely saddening, yet stimulating."

Further, he pointed out to his father that he had sought the advice of Eddington, Fowler, Dirac, and Milne. They were all unanimous; he should extend his stay to further his scientific research and career. Milne was candid. It would be stupid of Chandra if he did not utilize the excellent opportunity. In reality, Chandra did not need anybody's advice. He had made up his mind to extend his stay. Six years was a long time to be away from home, but as he wrote to Balakrishnan on 25 October, work was much more important and demanded attention to the exclusion of other, perhaps richer things in life. He adhered to his plans of spending the October term in Oxford working with Milne and returned to Cambridge in December 1933 to begin his term as a Fellow of Trinity.

Chandra's father had anticipated Chandra's decision; he wrote on 23 October, "What I feared, you have stated. Even when I got your letter that you sat for the written examination, or even previously, I told your aunt Sitalaxmi, that there was a possibility of your staying three years further, in the event of your election, and that she would have to recede from her wish that Sarada's marriage should take place only after your return and your marriage." Referring to Chandra's painful recollections of his mother, he continued, "You scarcely realise my own joys or pains in this matter. What you have achieved is only my dreams or hopes come true, of my son's obtaining intellectual 'laurels.' When your cable reached me I said to a few of my friends here that election to Trinity [Fellowship] and a further stay of three years was next door to F.R.S."

Having thus reconciled himself to Chandra's decision, Chandra's father proposed a short visit of his own to England and the continent. He had six months leave coming, and a fair portion of his expenses would come from tax savings if he went abroad. But Chandra was not very happy with the prospect. He was concerned about the difficulties his father, as an Indian, might experience in England. Outside the confines of the universities, it was not easy for Indians to find hotel accommodations, suitable meals, and so on. Besides, Chandra was well aware of his father's demanding nature and was concerned about himself—to what extent would he be able to cope if, for instance, he asked to meet Milne, Eddington, Dirac, and others? His father would not understand the deferential, distant relations between his son and the luminaries; he was likely to cause embarrassment, conflict, and unnecessary interference in Chandra's normal schedule.

Chandra did his best to discourage him from undertaking the trip. He wrote a long letter on 24 August 1933, in which he said, ". . . if you come here it would not be possible for you to stay with me or even with Vishwam [Chandra's younger brother, who was in London preparing for his ICS examination]; it depends on whether there are free lodgings at that time—which is unlikely—in the same house . . . also it would be difficult to get accustomed to English vegetarian food in a short time. Of course I am still vegetarian, but you would recollect how bad I felt in the beginning. . . . In London one could stay at 112 Gower Street but the half cooked rice and the frightfully hot curries completely upset my stomach during my four days stay there and I am positive (and Vishwam shares my view) that you can hardly stand even a single meal at 112 Gower Street. . . . Standard vegetarian restaurants (like "Churns" in London—where incidentally I have gone only once but as a guest of another!) cost easily 5 sh at a time." Chandra was not too encouraging about the continental tour either. "You are coming right in the winter," he continued. "It is not at all the cold of the winter which is trying in Europe. It is the *fog* and it is the *short span of four to five hours of visible atmosphere* (sunshine or even a passable clear atmosphere is a rare commodity during winter) and in addition the almost *continual dripping*.

Sight seeing under these conditions is pretty hopeless." He further warned his father about the language problem, the menu cards printed only in continental languages, and how difficult it would be to make sense out of them.

Chandra's father, however, was not someone who could be easily discouraged. "I shall certainly go to separate lodgings," he wrote in reply. "That I decided upon even without your letter and I have already obtained a list of all vegetarian restaurants on the continent. I know winter is not the best time to travel—but I do want a change badly and I have already booked passage. I do not mind the expenses nor the troubles I may have." He had his own plans for the tour. Since he had studied violin and the theory of Karnatic music for over ten years, he came to England armed with reference letters and introductions to several musicians interested in Indian music in London, Paris, and Florence. He arranged for himself private recitals in those places and also one over the British Broadcasting Service. He visited Chandra in Oxford and Cambridge, and Chandra escorted him on a week's tour in Germany. Altogether the six-month tour went well for him, and Chandra was happy at its conclusion in June, when his father sailed from Genoa, Italy. With a sense of relief, as it were, he wrote on 24 June 1934, on the eve of his father's sailing to India, "I have learned to know you better during your stay in England and I am sincerely glad that you finally came to Europe." He further added, "I must take this opportunity and be explicit regarding my real inner admiration of the high ethical standards which has shaped your and our—we your children's—lives. We are what we are made by our parents, and if you are convinced that I have kept steady, it is because of the sense of moral values which you and mother have by example taught me."

Chandra enjoyed some of the happiest times of his life from the fall of 1933 to the end of 1934. He moved from his "digs" into a nice set of rooms in Neville's Court (Trinity College) close to the famous Wren Library. He furnished the rooms with two oriental carpets bought at an auction[3] following the advice of Freye Gray. Dining privileges at the high table brought him into close contact with the scientific luminaries of the time. "I got to know the Cambridge scientific society," recalls Chandra. "Eddington used to dine at the college four or five times a week. Some other members of the high table were Rutherford, Edgar Douglas Adrian, Frederick Hopkins, J. J. Thomson, C. G. Darwin, and a great many others, including the mathematicians, G. H. Hardy, J. E.

Littlewood, and Harold Davenport. During the period I was there, there were occasions when I would sit next to one of them. Yes, Kapitsa was one of them; I got to know him well. I have wonderful stories about some of them."

Cambridge left a permanent imprint on Chandra. "He still talks like a Cambridge don," says Freeman Dyson.[4] He became permanently "anglicized" in his outward appearance and manners.

He was no longer as lonely as before. He felt assured of his work being appreciated. Astrophysics was going to be his predominant area of research. He would, at least for the time being, quit toying with the idea of switching to pure physics or pure mathematics. "I told myself," Chandra recalls, "I have four more years here, and I can do the work which I am able to do. I remember Harold Davenport saying to me, 'You can't teach an old dog new tricks. Why don't you continue in astrophysics? You got the fellowship on that account. There is enough time to do mathematics later.' So I did."

He felt less pressured and began to enjoy a more normal life. One of the things he did regularly during his fellowship years was to go for all-day walks on Sundays, covering 18 to 20 miles, to cities like Royston, Ely, and Saffron Walden. He would walk to a town, have tea in one of its small restaurants, and take a bus back to Cambridge just in time for dinner.

He also made friends. Besides the Grays, he became a close friend of Harold Davenport, a pure number theorist in mathematics, and David Shoenberg, an experimental physicist working in low-temperature physics. Occasionally they used to accompany Chandra on his Sunday walks and, as Chandra recalls, "Often, after 11 o'clock, after one had been working in one's rooms for two or three hours, you sort of walked around in the courts. If you saw a light in your friend's room, you went upstairs. David Shoenberg used to come up to my rooms, and we used to have tea. David was very politically inclined. So our conversations were not only about science. I recall, for instance, listening to Hitler on the radio in his rooms. He also used to be very interested in Indian political affairs, and we used to talk about Indian politics. Nehru's book *Towards Freedom* had just come out. I had a copy of that book. He borrowed it and read it, and we used to talk about it. Most of the people in academic circles were very conservative in Cambridge. Neither Dirac, Fowler, nor Milne were particularly sympathetic to the cause of India's freedom. Hence we avoided the subject. David was very sympathetic and was the only one with whom I could exchange views about India."

The RAS meetings added another dimension to Chandra's scientific life and allowed him to make a mark on the tradition-ridden, hierarchical scientific surroundings characteristic of the English. As a student, he had attended those meetings a few times as the guest of Fowler or Milne. Now as a Trinity Fellow, he could become a Fellow of the RAS in his own right. Milne nominated him and he became a Fellow without much ado. A trip to London to attend these meetings every second Friday of the month became a routine of Chandra's Cambridge life.

The meetings were held in Burlington House[5] on Piccadilly, not too far from Piccadilly Circus. Chandra would arrive in London before noon and go to visit William McCrea, whom he had come to know through these meetings. The two would have lunch at the Imperial College and walk over to Burlington House in time for that all-important British institution—Tea at 4:00 PM. People mingled fairly freely at these teas, though the young always displayed deference to their older and established colleagues. The meeting would begin at 4:30 PM and last until about 6:00 PM, with three or four fifteen-minute papers presented, each followed by discussion. The discussions were often quite lively, especially when controversies between famous figures, such as Jeans and Eddington, were involved.[6] Debating skills came into full play and no words were minced as Jeans, for instance, would say, "Well, I do not understand it, and I'm sure Sir Arthur also does not understand it." Sir Arthur Eddington would then meet the challenge. The attendance at these meetings exceeded sixty people, and as Chandra recalls, "[The sitting] was arranged in a very hierarchical order. In my time, the people who used to sit in the first row were Eddington, Jeans, and past presidents of the RAS, like T. E. R. Phillips, J. L. E. Dreyer. People like Milne and H. H. Plaskett sat in the second row. Bill McCrea and I used to sit in the last row. I remember once saying to Bill (when I got to know him well) that we could destroy the whole atmosphere of the RAS if we sat in the First Row. McCrea, the conventional man that he was, thought it was so audacious that I should even entertain such an idea."[7]

Papers had to be submitted a week before, by the first Friday of the month. Papers submitted by people like Eddington, Jeans, and Milne were always read, and they always came first. But young people had a fair chance to present their papers, as Chandra did. Beginning with his first paper at the June 1933 meeting, he read papers to the RAS six or seven times. People were extremely hospitable to his papers and all his papers were published promptly. He attracted attention and was introduced to people like Henry Norris Russell, Harlow Shapley, Sir Frank Dyson, the former Astronomer Royal, and many others. After one RAS meeting in 1934, Sir James Jeans came over and invited Chandra to his country home in Dorking. It was quite an experience for Chandra in his early twenties to be accorded such treatment by a world-renowned figure.

The Trinity Fellowship also brought an opportunity for Chandra to visit Russia during the summer of 1934 (7 July to 10 August). He owed this trip to B. P. Gerasimovič, the director of the Pulkovo Observatory near Leningrad. Chandra was still an undergraduate student in India when he had first received a letter from Gerasimovič, who was then in Kharkov, requesting a reprint of Chandra's first published paper in the *Proceedings of the Royal Society*. Young Chandra was thrilled at this recognition of his work by a foreign scientist. Since then he had kept in touch with Gerasimovič, sending him copies of his publications, and letting him know that he was in Cambridge. While in Copenhagen, Chandra heard a great deal about Russia from his friend Leon Rosenfeld, who was very enthusiastic about the revolutionary socialist experiment that was going on in Russia, and he wanted Chandra to visit Russia before returning to India. Accordingly, after he was elected to a Trinity Fellowship, Chandra wrote to Gerasimovič expressing his desire to see Russia. "I am very much pleased," Gerasimovič wrote, "to hear from you about your desire to visit us and see Russia. Our astrophysicists as well as myself will be very glad to make personal acquaintance with you and talk over many scientific questions." After some delay in getting the necessary visa, Chandra left Cambridge on 7 July 1934 for what turned out to be a memorable four-week trip to Russia.

He traveled by boat directly to Leningrad. As Chandra recalls, it was a pleasant five-day journey via the North Sea, through the Kiel Canal, skirting the northern coast of Germany. The warships of Germany were markedly visible. Most of the people on the boat were looking forward to their trip to Russia. Chandra, as almost all other young people of the time, shared their socialist views, as he was to write later, on 31 August 1934, to his brother Balakrishnan, "Russia appeals to me most tremendously. Russia gives the impression of a young man, full of ideals and who has such indomitable moral courage and indefatigable physical strength to go forward in spite of setbacks and who with his hopes derives consolation from his ideals during times of adversity."

This was still the period when there was no hint of the impending horror of Stalin's repression. All that began the subsequent autumn.

Chandra saw a striving, free Russia with friendly scientists who were kind and hospitable to him and who communicated freely with him. In Leningrad, for instance, he stayed with Gerasimovič in his official residence.[8] He met Lev D. Landau and Viktor A. Ambartsumian, who showed him around Leningrad, including the famous Hermitage.

During his week's stay in Leningrad, Chandra gave two lectures at the Pulkovo Observatory to large audiences. One of the two lectures was about his work on white dwarfs and the limiting mass, which had attracted little or no attention in Cambridge. Ambartsumian suggested investigating the problem in greater detail by avoiding some of the approximations Chandra had resorted to and working out the exact theory. As Chandra recalls, it was this remark of Ambartsumian, his interest and encouragement, that made him take up the subject again after his return to Cambridge and follow it to its conclusion.

Ambartsumian himself had done quite significant work on planetary nebulae. This and other topics of common interest drew Chandra to Ambartsumian and Nikolai A. Kozyrev, and he soon found himself spending more time with them than with Gerasimovič, the director who had invited him. Gerasimovič was not happy with this development; Chandra noticed the pronounced "generation gap" between the older Gerasimovič and the younger men, like Ambartsumian, Kozyrev, Landau, and others. Authority and seniority led to sharper lines of distinction and hence sharper conflicts here than elsewhere in the West. The upcoming younger scientists were not particularly charitable to older, senior scientists if their positions of authority lacked scientific merit. Gerasimovič, as a scientist, did not belong to the class of Ambartsumian and Landau, and the latter missed no opportunity to make this openly evident. Chandra recalls vividly a scene in which Ambartsumian said to him, "Look here, here is a set of papers by Gerasimovič. I turn to an arbitrary paper and to an arbitrary line. I am sure you will find a mistake."

Two years later, as Chandra would learn in Cambridge, almost all of the younger astronomers he met in Pulkovo were arrested, eliminated, or sent to Siberia. Gerasimovič was also arrested; he died in prison during the Second World War. Ambartsumian escaped a similar fate, since he was in the Crimea at the time. In the forties, Ambartsumian and Chandra did their monumental work on radiative transfer independently. In 1981, the Armenian Academy of Sciences arranged a special symposium at Yerevan to celebrate the fortieth anniversary of this work. It was there that Chandra met Ambartsumian again after forty-seven years.

After Leningrad, Chandra proceeded to Moscow to visit and lecture at the Shternberg Institute. A picture taken on the occasion of this visit has survived (see fig. 11c). It shows the director A. A. Kancheev, who invited Chandra to the institute, and the distinguished astronomer B. A. Vorontsov-Velyaminov. When Chandra visited this institute for the second time in 1981, the scientists at the institute retrieved the original picture and presented it to him with their autographs.

After Moscow, the next place on his itinerary was a visit to the Semeis Observatory in the Crimea, followed by Odessa on the Black Sea. Chandra vividly recalls his trip into Crimea:

> You get off the train at Sevastopol and take a bus to Crimea over the mountains. It's all perfectly green. At some point, people tell you to close your eyes— and then open them one second later when you cross over the ridge. The Black Sea is in front of you—just absolutely beautiful. Then you go down the mountains and see Yalta in the distance. You always think of Chekhov's story "Lady with a Pet Dog."

During the boat trip on the return leg of the journey, from Odessa to Istanbul across the Black Sea, Chandra found himself assigned to a cabin with three ladies, two Russian and one German.

> I was horrified that they [the boat officials] had asked me to share the cabin with three ladies. The two Russian ladies were totally upset about it. First of all [I was] a man and someone from outside Russia. As I could speak German and the German lady could speak Russian, we two made plans—the ladies would retire first while I waited outside the cabin. They would turn off the lights, and then I was to go in and get into my bed without turning on the lights. The first night when I went in I found the porthole open. I told the lady across from my bunk that the captain had told me that it should be closed; this was the time when long periodic waves—the solitary waves—were common. They could be so high that the water would come in. The lady said no, the porthole had to remain open. Hardly ten minutes had passed when a huge wave dashed against the porthole and poured water into the cabin. We all scrambled; the lights went on. No one worried about the plans we had made. We all had a complete shower—the only time I have ever taken a shower with three ladies. . . .

At the Turkish border, Chandra was mistaken for an Armenian with a British passport since Subrahmanyan had an Armenian sounding "yan" at the end. He was sent to the police, and it took three days to clear up the matter. But it gave him an opportunity to see Istanbul. "I had a very interesting time," recalls Chandra. "Then, there was a Turkish lady who got on the train at Istanbul (going to Paris). She only knew Turkish. Her husband who had come to see her off asked me to help her from station to station and across the borders. It was an ordinary train, stopping at all stations. I couldn't afford to travel by the famed Orient Express. Anyway, I missed many connections in trying to help the lady with her luggage. It was quite a problem. I was very relieved when we came to Paris and some friends of hers took charge of her."

On the whole Chandra returned to Cambridge in 1934 happy with himself and happy with what he had seen in Russia.[9] "Russia is quite different from the other European countries," he wrote to his father on 20 July 1934. "Women are employed as train-drivers, policemen— policewomen rather, in road repairs work, etc. I visited a glass factory. Attached to this factory is a theater, restaurant, technical school, playground—all meant only for the workers of the factory. In short, the provisions made for the workers are quite exceptional—though this Exceptional method is followed in every factory!"

He also wrote, "I have returned from Russia with the full determination to push steadily." Ambartsumian's suggestion that he should work out the exact theory of white dwarfs had made a deep impression on him. For the next four months he involved himself in detailed, tedious, numerical calculations in order to obtain as exact a theory of the white dwarfs as one could construct within the framework of relativistic quantum statistics and the known features of stellar interiors. He proved beyond any doubt that a limit on the mass of a star that could become a white dwarf was unavoidable. Through his work, he also thought that, once and for all, he would resolve a controversy between Milne and Eddington which had been brewing for a long time.

In order to understand the Milne-Eddington controversy and the link to Chandra's work, we must have a brief technical discussion of the ideas about stellar structure in the early 1930s.

The origin of the controversy was the nature of the boundary conditions one should use in determining the equilibrium configurations of stars. Eddington, the chief architect of the so-called standard model, believed that the conditions prevailing in the outer layers of the stellar mass were relatively unimportant in determining the central densities and temperatures of a star. Compared to the temperature of the order of a million degrees at the center, it did not matter whether one assumed the temperature of the outer layers to be zero or a few thousand degrees. Therefore, Eddington's perfect gas model of a star assumed, for mathematical simplicity, the boundary temperature to be zero and, based on certain assumptions regarding the central, internal conditions, the model predicted the physical properties (opacity, temperature gradients) of the outer layers.

Milne's contention was that such an assumption is unnecessary and should not be made for mathematical convenience. Since physical conditions in the outer layers of a star are observable, why not take them as they are, as boundary conditions, and ask whether the internal conditions that follow are compatible with the kind of internal structure one assumes.

"Framed this way, the question was a reasonable one," says Chandra. "Milne originally formulated the question exactly in this way: How can one determine the structure of the outer layers correctly from observations and use it as a proper boundary condition to determine the interior structure? Milne thought it conceivable that the outer boundary conditions suggested an internal structure which is different from what follows by assuming that the star is, for example, a complete perfect gas throughout."

Eddington would have nothing to do with these modifications of his standard model. The conditions at the center of the star were of paramount importance. That's all. He attributed to Milne the suggestion that the outer structure of the star determined its interior. "Does a dog wag the tail? Or the tail wag the dog?" he asked at one meeting, ridiculing Milne's idea. At another meeting he went on to say, "I have not read Professor Milne's paper, but I hardly think it is necessary, for it would be absurd for me to pretend that Professor Milne has the remotest chance of being right."

Subjected to such harsh public criticism, Milne reacted, rather unwisely, by departing from his original, reasonable hypothesis to take on an essentially extreme position attributed to him by Eddington, which held that one can never determine the interior without knowing the outer layers. He went on to make stellar models which *required* every star to have a degenerate core obeying one equation of state [pressure proportional to (density)$^{5/3}$] surrounded by outer layers of stellar material obeying the perfect gas equation of state [pressure proportional to (density)$^{4/3}$]. He had taken the idea of a degenerate core from Fowler's work on white dwarfs.

Now when Chandra discovered the limiting mass on his voyage from India and sent the short paper he wrote about his unexpected result to Milne, Milne saw that Chandra's work contradicted his idea that every star had a degenerate core. While acknowledging that Chandra had worked out the relativistic degenerate star "most beautifully," he also remarked:

Where, however, I must criticize your paper is in its conclusions. You conclude that a dense star cannot have a mass exceeding some value M_0. But the question then immediately arises:—What is the state of a mass which is very large ($>M_0$) for arbitrarily small L [luminosity]? *Certainly such a mass must have a configuration of equilibrium and the question is to find it.* The flaw in your reasoning is that you cannot prove that the solution appropriate to the outer parts of the relativistic degenerate core is Emden's solution $n = 3$ [leading to the perfect gas equation of state]; it may be one of the others. Your analysis simply proves that the relativistic equation of state cannot subsist right through the center when $M > M_0$. *Either a new centrally condensed "dense" configuration arises, or the new configuration requires a fresh intervening supporting surface (inside the relativistic degenerate core) and a new collapse is foreshadowed. You must investigate this to the bitter end and see what the final state really is. You may be able to prove that such a star must have an incompressible core at the maximum density of matter.* [Emphasis added.]

After some further discussion and suggestions regarding how to proceed, Milne went on to say:

I do hope you will carry the investigation to conclusion and meet the constructive criticism of mine. In its present form the conclusion about the mass of the white dwarf does not hold good, as my ideas go. . . . Your conclusion in its present form arises from the curious properties of "$n = 3$" but I think you have fallen into the Eddington error of inferring consequences from what can be only an incomplete mathematical treatment.

Milne was not willing to judge Chandra's results on their merit. It is apparent that he wanted Chandra to prove the validity of his own [Milne's] ideas. He put off further discussion with Chandra saying, in his letter of 12 February 1931, "Unfortunately I am rather pressed for time at present and have not the leisure I should like to think about your problem. But somehow there must be an answer to the problem: What is the physical state for arbitrary M and L?"

It was a good problem for which there was no easy answer. Chandra felt, rightly so, that his startling results merited publication, but there was no way to publish it in the *Monthly Notices of the Royal Astronomical Society*, since both Fowler and Milne effectively ignored it. Hence, he sent it to the *Astrophysical Journal* in America, and it was published there.* Subsequently, Chandra was able to show that not only was there

* Chandra sent his paper for publication on 12 November 1930. The referee of the paper, Carl Eckart, had some doubts about the derivation of the formula for the maximum mass. After they were clarified by Chandra, the paper was published in *Astrophysical Journal* 74 (March 1931): 81. It should be noted in this context that one year later Lev D. Landau published a paper, "On the Theory of Stars," *Physikalische Zeitschrift der Sowjetunion* 1 (1932): 285–88, in which he derived essentially the same formula for the critical mass as Chandra had

a limiting mass, but also if $M > M_0$ [the critical mass], degeneracy could not set in; the stellar material would continue to obey the perfect gas equation. Chandra says, "All of Milne's work was made to look ridiculous. I never said so explicitly, but I wrote to him what I thought the situation would be [if $M > M_0$], but he never believed that."

Chandra left it at that until he went to Copenhagen in 1932 and discussed the work with Leon Rosenfeld. Rosenfeld saw its importance and urged Chandra to publish it. Knowing Milne's objections, Chandra thought it wise to send it to *Zeitschrift für Astrophysik*. Erwin F. Freundlich, whom he had met the summer before at Potsdam Observatory, was the editor. But, as it turned out, Milne happened to be at Potsdam at the time, and he was asked to referee. The following excerpt from Milne's letter to Chandra of 1 October 1932 is self-explanatory:

The Editors of the *Zeits. für Astrophys.* have asked me to report on your paper. Unfortunately I have been unable to recommend acceptance, as the paper contains a mistake in principle, and in any case it would only do your reputation harm if it were printed.

After a long, "full" letter of explanation from Chandra, Milne somewhat reluctantly recommended that the paper be published. It is this classic paper in which Chandra wrote the following most often quoted statements:

For all stars of mass greater than M [1.2 times the critical mass], the perfect gas equation of state does not break down, however high the density may become, and the matter does not become degenerate. An appeal to Fermi-Dirac statistics to avoid the central singularity cannot be made.

Great progress in the analysis of stellar structure is not possible before we can answer the following fundamental question: Given an enclosure containing electrons and atomic nuclei (total charge zero), what happens if we go on compressing the material indefinitely?[10]

---

done earlier. But he went on to say: "For $M > M_0$ there exists in the whole quantum theory no cause preventing the system from collapsing to a point (the electrostatic forces are by great densities relatively very small). As in reality such masses exist quietly as stars and do not show any such ridiculous tendencies, we must conclude that all stars heavier than 1.5 $M_\odot$ certainly possess regions in which *the laws of quantum mechanics are violated*." Landau concluded that, "since we have no reason to believe that stars can be divided into two physically distinct classes according to the condition $M >$ or $< M_0$, all stars in great probability possess such *pathological regions* [emphasis added].

Remarking on this strange conclusion, Alan Lightman says, "It seems he found his theoretical result so preposterous, so disturbing to common sense, that he was willing to abandon the celebrated theory that produced the result" (Lightman, *Time Travel and Papa Joe's Pipe* [New York: Charles Scribner's Sons, 1984], pp. 34–35).

Not until much later did the far-reaching implications of this question become apparent. At the time, the white dwarf stage of stars was thought of as the end stage of all stars in the course of their evolution. Chandra had concluded that not all stars can become white dwarfs. Were there terminal stages other than white dwarfs? If there were, what were they? If not, and the collapse continued indefinitely, what happened to the stellar matter? Such questions should have been asked immediately and their answers should have been sought. If this had been done, neutron stars and black holes would have at least made their conceptual entrance into the astronomical world much earlier. That this was not the case, however, leads us to the ultimate frustration that came upon Chandra due to his totally unexpected encounter with Eddington.

During the intervening years, 1932–34, Chandra had occupied himself with other problems. He felt he had done as much as he could given the total lack of interest and encouragement from others. He had to worry about finishing his degree. In 1934, the situation was different. Chandra was relaxed, full of self-confidence, and ready to tackle the problem again. Ambartsumian's suggestion was a good one. If he succeeded in working out an exact theory of white dwarfs, Milne's objections regarding the approximations he was making would be completely answered and the matter would be settled once and for all.

The controversy between Milne and Eddington would also be settled. If Chandra could prove, on the basis of an exact theory, that not all stars can have degenerate cores surrounded by ordinary matter, Eddington's ideas would prevail. "I was very pleased," Chandra recalls, "because Eddington seemed to understand that. He took a great deal of interest in the day-to-day progress of my work. He even got me the only hand calculator, a Brunsviga, that was around and was being used by Gunnar Steenholt, a Norwegian visitor. Steenholt was not happy of course. During the three months from October through December, Eddington came to my rooms quite often, at least once, sometimes twice or three times, a week. As my numerical work progressed, I would show him the points on the emerging graph."

From what transpired later, what Chandra did not realize at the time was that Eddington apparently thought that the exact theory would also demolish Chandra's discovery of the limiting mass, and that Chandra would discover for himself that there was no such limit. Eddington was convinced that every star, no matter what its mass, could become a white dwarf. The white dwarf stage, the only known terminal equilibrium stage at the time, was simply a kind of peaceful, retirement stage for the star. Chandra's conclusion, namely, that massive stars (stars with a mass greater than the limiting mass) cannot reach such an equilibrium state, distressed Eddington, although he gave no indication of this to Chandra.

Chandra thus proceeded to work out a complete theory of white dwarfs. Accomplishing his task by the end of 1934, he submitted two papers to the Royal Astronomical Society. At their invitation, he presented a brief account of his results at the January 1935 meeting. Remember, he had already brought to the attention of Eddington, Milne, and others his belief that the fate of massive stars could not be identical to that of less massive ones. That conclusion, however, was based on approximate models. Now he had the exact solution to the problem, reinforced by extensive numerical analysis. Chandra did not doubt the validity of his results and the profound challenge they presented to those interested in stellar evolution.

In those days the society decided its program just a few days before the meeting, the list of papers to be presented being distributed at the gathering. Chandra remembers well the events leading to that historic meeting some fifty years ago:

I knew the assistant secretary, a Miss Kay Williams, rather well, and she used to send me the program ahead of the meeting. On Thursday evening I got the program and found that immediately after my paper Eddington was giving a paper on "Relativistic Degeneracy." I was really very annoyed because, here Eddington was coming to see me practically every day and he never told me he was giving a paper.

Then I went to dine in the college and Eddington was there. Somehow I thought Eddington would come to talk with me, so I did not go over to talk with him. After dinner I was standing by myself in the combination room where we used to have coffee, and Eddington came up to me and asked me, "I suppose you are going to London tomorrow?" I said, "Yes." He said, "You know your paper is very long. So I have asked Smart (the secretary of the RAS) to give you a half hour for your presentation instead of the customary fifteen minutes." I said, "That's very nice of you." And he still did not tell me that he too was presenting a paper. So I was a little nervous as to what the story was.

The next day at Burlington House, at the usual tea before the meeting, McCrea and I were standing together and Eddington came by. McCrea asked Eddington, "Well, Professor Eddington, what are we to understand by 'Relativistic Degeneracy'?" Eddington turned to me and said, "That's a surprise for you," and walked away.

The proceedings of the meeting, published in the *Observatory,* contain the following account of Chandrasekhar's paper:

Dr. Chandrasekhar read a paper describing the research which he has recently carried out, an account of which has already appeared in the *Observatory* 57.373, 1934, investigating the equilibrium of stellar configurations with degenerate cores. He takes the equation of state for degenerate matter in its exact form, that is to say, taking account of relativistic degeneracy. An important result of the work is that the life history of a star of small mass must be essentially different from that of a star of large mass. There exists a certain critical mass $\mathcal{M}$. If the star's mass is greater than $\mathcal{M}$ the star cannot have a degenerate core, but if the star's mass is less than $\mathcal{M}$ it will tend, at the end of its life history, towards a completely collapsed state.

After Chandra's talk, Milne made a brief comment, pointing out that he and another collaborator had similar results; their methods were more crude, but provided a better physical picture. The President then invited Sir Arthur Eddington to speak on "Relativistic Degeneracy." Eddington began by saying,

Dr. Chandrasekhar has been referring to degeneracy. There are two expressions commonly used in this connection, "ordinary" degeneracy and "relativistic" degeneracy, and perhaps I had better begin by explaining the difference. They refer to formulae expressing the electron pressure P in terms of the electron density σ. For ordinary degeneracy $P_e = K\sigma^{5/3}$. But it is generally supposed that this is only the limiting form at low densities of a more complicated relativistic formula, which shows P varying as something between $\sigma^{5/3}$ and $\sigma^{4/3}$, approximating to $\sigma^{4/3}$ at the highest densities. *I do not know whether I shall escape from this meeting alive, but the point of my paper is that there is no such thing as relativistic degeneracy!*

After making a few remarks about the history of the problem, the difficulty he had pointed out in 1924, and the way Fowler's appeal to Fermi-Dirac statistics had solved the problem, he went on to say:

. . . But Dr. Chandrasekhar has revived it again. Fowler used the ordinary formulae; Chandrasekhar, using the relativistic formula which has been accepted for the last five years, shows that a star of mass greater than a certain limit M remains a perfect gas and can never cool down. *The star has to go on radiating and radiating and contracting and contracting until, I suppose, it gets to a few km. radius, when gravity becomes strong enough to hold in the radiation, and the star can at last find peace.*

Dr. Chandrasekhar had got this result before, but he has rubbed it in, in his last paper; and, when discussing it with him, I felt driven to the conclusion that this was almost a reductio ad absurdum of the relativistic degeneracy formula. *Various accidents may intervene to save a star, but I want more protection than that. I think there should be a law of Nature to prevent a star from behaving in this absurd way!*

If one takes the mathematical derivation of the relativistic degeneracy formula as given in astronomical papers, no fault is to be found. One has to look deeper into its physical foundations, and these are not above suspicion. The formula is based on a combination of relativity mechanics and nonrelativity quantum theory, *and I do not regard the offspring of such a union as born in lawful wedlock*. I feel satisfied that the current formula is based on a partial relativity theory, and that if the theory is made complete the relativity corrections are compensated, so that we come back to the "ordinary" formula.

Eddington continued by making some vague remarks about two kinds of waves—progressive and standing—using a lighthearted analogy which made several people laugh:

I might compare the progressive wave with Professor Stratton and the standing wave with the President of the Royal Astronomical Society; only to make the analogy a good one, the Society would have to change its President gradually and continuously, instead of suddenly every two years. The formulae which apply to such a President would be different from the formulae which apply to an ordinary individual.

Eddington's comparison implied that Chandra had made a fundamental error of principle. The humiliating effect of having his thorough exploration of the problem tossed aside with such a loose analogy must have been chilling to Chandra. Before he could say anything, the President said:

The arguments of this paper will need to be very carefully weighed before we can discuss it. I ask you to return thanks to Sir Arthur Eddington.

He then called the next speaker to give his paper.

It would be an understatement to say that Chandra was left dumbfounded, shocked, and depressed. Instead of gaining recognition for having raised a challenging question, he found his years of hard work summarily, almost cavalierly, dismissed. Moreover, he, barely twenty-four years old and a relative newcomer to the research arena of astrophysics, confronted in Eddington a figure whose international prestige and authority could destroy him. He must have been overwhelmed by feelings of humiliation and helplessness when people came by after the meeting saying, "It was too bad, too bad," expressing their sorrow over what they believed to be Chandra's conceptual error.

In science, one has at times the inalienable right to be wrong. Certain kinds of mistakes could even be a source of pride. But this occasion was not one of those. Eddington had made Chandra's work appear to be based on a conceptual error, so that the profound conclusion Chandrasekhar had stated with so much conviction seemed wrong. If Eddington had thought this all along, why hadn't he said so? Why, in the privacy of

Chandra's rooms, did he encourage the young man to go on with the work that involved so much tedious numerical labor? Was his motivation only to discredit Chandra publicly, *or could Eddington be right?* Chandra must have entertained some of these thoughts at the conclusion of the meeting.

Chandra recalls having dinner later that evening with Harry H. Plaskett, a professor from Oxford. The dinner was quiet; neither of them said a word. No word of assurance came from Plaskett; his silence gave the impression that he too thought that Eddington was right. It was one of those things; mistakes happen. After the dinner, Chandra went to see him off at Paddington Station. He recalls, "Milne was there, and Milne was absolutely in a state of euphoria. Because Eddington's work had shown that my limiting mass was incorrect, his own idea that every star had a degenerate core must be valid. He told me that he felt it in his bones that Eddington was right. I was really angry at that . . . Well, I wished he had felt it elsewhere."

Chandra returned to Trinity at one o'clock the next morning, dejected and depressed. Later that morning, he saw Fowler and told him what had transpired at the meeting. Fowler offered some reassurance. In private, so did some others. But there was no doubt that Eddington's attack cast a dark cloud of suspicion and doubt over his work. The puzzling, yet simple and straightforward, discovery he had made on the long voyage from home, the subsequent effort he had put in to verify it, the disagreement with Milne, and the final vindication of his ideas after an enormous amount of work—all that was wiped away by a few flippant remarks. Eddington's style and prestige were sufficient to convince most that he (Eddington) might well be right. It seemed to Chandra that he had been bested through rhetoric; Eddington's remarks had little to do with the boundaries he had begun to probe and to believe he understood. He was left in a mood of utter frustration.

6

The Absurd Behavior of Stars:
Eddington and the White Dwarfs
Cambridge, 1934–1935

The subject is a fair field for the struggle to gain knowledge by scientific reasoning; and win, or lose, we find the joy of contest.

Sir Arthur Stanley Eddington, *The Internal Constitution of the Stars* (1926)

The combination room was deserted when Chandra returned to Trinity from London after the fateful meeting, but the coals in the fire were still glowing. Standing before the fireplace, he brooded over the unexpected course of events. He was totally puzzled by the behavior of Eddington. When had Eddington first begun to think that Chandra's work involved a fundamental, conceptual error? After all, hadn't Eddington been following his work almost on a daily basis during the past four months? If Eddington had thought he was in error, why hadn't he said so? Why did Eddington wait for such a public occasion to discredit him? What could be Eddington's motivation? What chance did he have to make other astronomers believe in the validity of his results when Eddington, who had such enormous prestige and established authority, said it was all wrong? And he recalls repeating to himself: "This is the way the world ends, this is the way the world ends, this is the way the world ends. Not with a bang but with a whimper."

Could Eddington be right? That was the question foremost in Chandra's mind. Since the issue was mainly a question of physics, the views of physicists were exceedingly important. Chandra thought that a definitive verdict from Niels Bohr, Wolfgang Pauli, or Paul Dirac would quickly settle the controversy and clear away the cloud of doubt that Eddington had cast over his work. The very next day, on 12 January 1935, he wrote to Leon Rosenfeld, his longtime friend, who was in

Copenhagen at the time working with Bohr. After some preliminaries, Chandra said,

> I feel very guilty to pass on immediately to a matter which is of exceeding importance to me and on which I would like you to consult Bohr as well.

After providing Rosenfeld with a short account of his work and the equations he had used, Chandra continued,

> Yesterday I gave an account of my work at the Royal Astronomical Society and after my paper Eddington sprang a surprise on everyone by saying the method of derivation of [equation] (1) was all wrong,[1] that "Pauli's principle" refers to electrons as being stationary waves and that the use of the relativistic expression for energy is a misunderstanding. . . . If Eddington is right, my last four months' work all goes in the fire. Could Eddington be right? I should very much like to know Bohr's opinion. Please consult him on the matter as soon as you possibly can and reply to me by air mail. You can understand my anxiety if you know that I have been working at the consequences of [equation] (1) for the last four months at an average rate of 12 hours a day.

Chandra wrote again within a few days after he had had a long conversation with Eddington. Eddington's interpretation of Pauli's principle was at variance with everybody else's. Chandra passed on to Rosenfeld Eddington's version of Pauli's principle. Rosenfeld replied on 14 January 1935:

> I may say that your letter was *some* surprise for me: for nobody had ever dreamt of questioning the equations [that Chandra used] and Eddington's remark as reported in your letter is utterly obscure. So I think you had better cheer up and not let you scare [sic] so much by high priests: for I suppose you know enough Marxist history to be aware of the fundamental identity of high priests and mountebanks.
>
> I submitted your letter to Bohr immediately, and I can state as follows the outcome of our examination of the question: In order to apply Pauli's principle to an assembly of electrons without interactions, one has simply to express that every non-degenerate stationary state of an electron in the external field considered may be occupied by zero or one electron. Now in your case (no external field), the stationary states are defined by the components of momentum and spin, and their energies by the relativistic formula (p. 2 of your letter). The eigenfunctions are plane waves, which may be defined either as stationary waves by a condition of reflection at the boundary of the volume V, or as progressing waves by a periodicity condition at the boundary; these two cases become equivalent in the limit, considered by you, of an (asymptotically) infinite volume, and both yield for the asymptotic density distribution in the phase space, precisely the expression you have used in your equations. Further this expression is relativistically invariant. These quite obvious arguments would seem to settle the question without any doubt.

In response to Chandra's second letter, Rosenfeld wrote again on 14 January:

Bohr and I are absolutely unable to see any meaning in Eddington's statements as reported in your second letter. The question, however, seems quite simple and has certainly a unique solution. So, if "Eddington's principle" had any sense at all, it would be different from Pauli's. Could you perhaps induce Eddington to state his views in terms intelligible to humble mortals? What are the mysterious reasons of relativistic invariance which compel him to formulate a natural law in what seems to an ordinary human being a non-relativistic manner.* That would be curious to know.

* It seems to us as if Eddington's statement that several *high* speed electrons might be in one cell of the phase space would imply that to another observer several *slow* speed electrons, in contrast to Pauli's Principle, would be in the same cell.

These private remarks reassured Chandra about the correctness of his ideas. But something more than private communication was needed to clear the cloud of doubt. In addition, further discussions with Eddington did not prove fruitful. Eddington had his own notions about the basic principles of quantum mechanics and about progressive and standing waves:

We cannot combine the wave functions (progressive waves) to produce standing waves. They are incoherent. If two functions are written with a + symbol, we only mean that both are present. We cannot combine progressive plane waves to produce standing waves in the quantum theory.[2]

Chandra struggled with alternative arguments and derivations that would make Eddington understand. He wrote to Rosenfeld on 19 January,

I am really sorry to trouble you, but Eddington is reading a paper at the Colloquium next Friday and I want to have real missiles to throw at him! Some really simple way of demonstrating that *any* theory which shows that $p \propto \rho^{5/3}$ is an identical proportionality for all densities [as Eddington believed] must be necessarily self-contradictory.

And Rosenfeld replied on 23 January:

I was very glad to read your preliminary note in the *Observatory*. It seems to me that your new work is very important indeed, and I think everybody except Eddington will admit it rests on a perfectly sound basis. As to the artillery fighting you are planning against Eddington, I could not imagine any missile more devastating than the one contained in my last letter. I feel a little dubious about the results of such a fight, since I do not expect Eddington, whatever the missiles, to collapse like a star with $\beta_M = 1 - \varepsilon$; it wouldn't be dignified enough for him to recant after he has gone so far as denying the existence of wave packets in quantum theory! Wouldn't it be a good policy to leave him alone, instead of losing one's time and temper in fruitless arguments? Nevertheless I wish you a great fun next Friday, and I even regret not to be there to enjoy the show.

More prolonged discussions with Eddington followed. Chandra wrote Rosenfeld again on 26 January:

I am afraid that I cannot leave Eddington alone! You can understand my disappointment. I have been spending months on my stellar structure work with the hope that for once there will be no controversy. Now that my work is completed, Eddington has started this "howler" and of course Milne is happy. My work has shown that his (Milne's) ideas in many places are wrong, but my work depends on the relativistic degenerate formula and Milne can now go ahead. . . . Already Fowler thinks that the relativistic formula (the one I have in my letter to you and also quoted in my note in the *Observatory*) cannot be regarded as proved.

So I have managed to get hold of Eddington's manuscript. He gave it to me and I am forwarding it to you for *you and Bohr alone* to read. I should be awfully glad if Bohr could be persuaded to interest himself in the matter. If somebody like Bohr can authoritatively make a pronouncement in the matter, it will be of the greatest value for further progress in the matter. It is terribly important to settle the matter as quickly as possible, otherwise intense confusion would result in astrophysics. . . . I am so sorry to trouble you over this but I hope you can understand.

Rosenfeld's response of 29 January said:

I vividly realize your troubles and feel very sorry for you. Bohr would be quite willing to help you, but he is very tired now and has to write two articles due for February 15th; after this is completed, he intends to leave to some rest resort to recover from a very strained semester. He therefore feels it difficult to concentrate himself on a new subject just now; but he has a proposal to you, which I think would meet your wish in the best possible way. Would you agree for us to forward confidentially Eddington's manuscript to Pauli, together with a statement of the circumstances and asking for an "authoritative reply"?

Chandra readily agreed, and the derivation of his relativistic formula along with Eddington's manuscript and other relevant material were sent to Pauli. About Eddington's manuscript, Rosenfeld remarked:

. . . After having courageously read Eddington's paper twice, I have nothing to change in my previous statements; it is the wildest nonsense.

Pauli's response was equally conclusive: Eddington did not understand physics. There *was* no ambiguity in applying the exclusion principle to relativistic systems. Eddington's principal error, according to Pauli, was in his wishful thinking, in his attempt to fit the exclusion principle to what *he* wanted in astrophysics. Chandra also wrote to Dirac, who was in the United States at Princeton at the time. He too thought that Chandra's treatment of the problem was flawless. But unfortunately neither Pauli nor Dirac was willing to enter publicly into the controversy and make an "authoritative statement" as Chandra had hoped. Chandra wanted this not so much to convince physicists of the correctness of his ideas, but to quell the doubts among astronomers.

Astrophysics was simply not part of the mainstream, frontier physics. Fundamental experimental discoveries and theoretical advances were taking place rapidly in the early and mid-1930s in physics. Dirac's theory of the electron had its triumphant vindication in the discovery of the positron, the first antiparticle to be discovered. The discoveries of the neutron, artificial radioactivity, Fermi's theory of β-decay, field theory, and so on dominated the interests of physicists. It appeared as though all the fundamental building blocks of matter had been found. It also seemed likely that the immensely successful ideas of relativistic quantum mechanics in unraveling atomic structure would soon be extended to bring nuclear physics into its fold. By contrast, most physicists thought that astrophysics was complicated, clumsy, and ridden with uncertainties and ambiguities. Physicists like Bohr wondered how one could make any progress in astrophysics when one has not even the vaguest idea about the energy production mechanism in stars. The consequence of all this was that Chandra's discovery remained unheralded among physicists and, as Chandra anticipated in his letter to Rosenfeld, astronomers continued to believe in Eddington.

Milne, for instance, forgot his long-standing controversy with Eddington over the conditions in the interior of a star. Without probing into Eddington's arguments, he accepted their conclusions because they fitted his own ideas. The very next month, in February, he published a letter in the *Observatory* declaring that his own work which led to conclusions similar to those of Chandra was *not* based on the relativistic considerations. He wrote to Chandra on 24 February:

Your marshalling of authorities such as Bohr, Pauli, Fowler, Wilson, etc., impressive as it is, leaves me cold. If the consequences of quantum mechanics contradict very obvious, much more immediate, considerations, then something must be wrong either with the principles underlying the equations of state derivation or with the aforementioned general principles. . . . Your family of curves, beautiful as it is, does not help me to *understand* stellar structure whereas my more limited program does so. To me it is clear that matter cannot behave as you predict in the domain $M_3 < M < \mathcal{M}$.

It was annoying and frustrating. Chandra had no recourse except to bottle up his feelings. "My last papers on stellar structure have involved me desperately in the rival jealousies of Eddington, Milne, and Jeans," he wrote to his father on 7 February. "I am taking care to be scrupulously polite to all of them. . . . It is the continuation of the history of my differences in attitude with Milne which has been brewing for the last three years. The explosion hasn't yet occurred. The whole thing would have been smoothed over had it not been for an awful howler Eddington has started. He thinks that Pauli's principle is wrong! I do not know what he is up to."

Eddington meanwhile kept up his attacks. No arguments, no alternate derivations were convincing.* "Eddington is behaving in the most obscurantist fashion," Chandra wrote to his father on 31 May. "Though Fowler, Dirac, Bohr, and many others agree with me, it is still very annoying to have such strained scientific relations with E. Indeed my differences with him (and also with Milne) have to a considerable extent made me unhappy during the past months. As Fowler said, they are being superstitious—and what havoc it makes in science!"

Chandra felt that the differences were not based on honest, scientific arguments. As he wrote to his father in a follow-up letter on 21 June, "the differences are of a 'political' nature. Prejudices! Prejudices! Eddington is simply stuck up! Take this piece of insolence. 'If worst comes to the worst we can believe your theory. You see I am looking at it from the point of view *not* of the stars but of Nature.' As if the two are different. 'Nature' simply means Eddington personified as an Angel!! What arguments could anyone muster against such brazen presumptuousness?"

The International Astronomical Union meeting of 1935 was held in Paris in July. There Eddington gave an hour-long talk, a considerable portion of which he spent proclaiming that Chandrasekhar's work was heresy; there was no such thing as relativistic degeneracy. Every star, no matter what its mass, had a finite state, and the idea of a limiting mass was an absurdity. Chandra, who was present at the meeting, tried to respond. "I sent a note to Henry Norris Russell who was presiding, telling him that I would like to reply to Eddington," Chandra recalls. Russell

---

* Chandra persuaded Christian Møller, a physicist known for his contributions to quantum mechanics and relativity, to write a joint paper with him in which they derived the formula for normal modes, under contention by Eddington, using Dirac's relativistic wave equation, showing that there was no mistake. Eddington followed it up with a rebuttal which made little sense.

sent a note back saying, "I prefer you don't." "And so I wasn't even allowed to reply to Eddington . . . Well, that was the protocol. That gives you a picture of that time. Eddington had talked, and who was this young man to come and say he was incorrect? Russell refused that I give any reply to Eddington. Privately though, when I met him earlier, he was frightfully enthusiastic about my work and had whispered to me, 'Out there we don't believe in Eddington.'"

There was no simple way to prove that Eddington was wrong, no direct observational tests. After Chandra's somewhat unsuccessful attempts to enlist the public support of the physicists, he had very few resources left to counter Eddington's influence. He wrote to the astrophysicist Svein Rosseland, whose reply of 18 May 1935 declared, "Personally I do not like to take part in what, according to the Irishman's definition, must be regarded as a private English fight. But it is interesting to watch the game from a safe distance." The theoretical reasoning of Eddington was too complicated for most astronomers to follow, but his style and prestige were sufficient to convince most of them that he might be right. The following excerpt from a letter by William McCrea, a relativist and a friend of Chandra, illustrates the situation clearly:

I remember one RAS meeting when Eddington produced what struck me then as an unanswerable argument. As I recall, it went like this: Only *special* relativity is involved. Basically in order to obtain the so-called "relativistic" form of any property, we require the fundamental equations of the subject to be *Lorentz-invariant,* i.e., the same for all inertial frames. If for any reason some particular frame of reference is uniquely defined by the physics of a problem, then for that problem it is absurd—said Eddington—to require a lot of other frames to be of the same status. In the case of a spherically symmetric star, the rest frame of the star is a uniquely preferred frame. In discussing the physics of the star we are not justified in claiming that any other frame is equivalent to that unique frame. So Eddington said that there can be no requirement of Lorentz-invariance in such a case. It was therefore not surprising that insistence upon that requirement led to a manifestly absurd result. . . .

When I listened to Eddington on this occasion I could not immediately weigh up all the implications of what he said, but my instinct seemed to tell me that he might be right. . . .

What I am ashamed of is not having tried to get to the bottom of the sort of argument Eddington produced. Had anyone other than Eddington produced such arguments, I suppose I should have done so. But they were superficially satisfying to me, and since they satisfied Eddington, I confess that I was content to let it go at that. In any case I was not working at stellar structure. However, I did profess to know something about special relativity and I ought to have gone into the subject from that side.[3]

Rudolf Peierls, a noted physicist, did try to get to the bottom of things. He recalls,

Eddington's arguments tended to shift with time. At one point he claimed that relativity and quantum mechanics could not be combined. When it was pointed out that the combination of relativity and quantum mechanics had proved itself in Dirac's theory of the hydrogen atom, he argued that this was a question of symmetry. Relativity was all right for a problem with spherical symmetry, such as the hydrogen atom, but in statistical mechanics you were considering a rectangular volume element $dxdydz$, and that had a different symmetry (I would have thought a star was more like a spherical box than a rectangular one!). He also brought in the difference between standing and progressive waves.

I did not know any physicist to whom it was not obvious that Chandrasekhar was right in using relativistic Fermi-Dirac statistics, and who was not shocked by Eddington's denial of the obvious, particularly coming from the author of a well-known text on relativity. It was therefore not a question of studying the problem, but of countering Eddington.

It was for this purpose that I wrote my paper in the *Monthly Notices*. The simplest way to derive the equation of state is to use cyclic boundary conditions, which allow the use of progressive waves. This was one of the points criticized by Eddington, and therefore I looked for a simple proof that, for a large enough system, the use of the cyclic boundary condition was justified. When I found it, I sent it to the *Monthly Notices* because the point would be obvious to the physicists, but not necessarily to astronomers. . . . I do not believe Eddington ever took any notice of my paper.[4]

Faced with such a situation, Chandra had to make a decision. Should he continue to argue and confront Eddington, or should he leave Eddington alone? "I was extremely enthusiastic about my work before the ill-fated meeting," recalls Chandra, "and had planned ahead quite a bit. I had planned to work out the theory of rotating and pulsating white dwarfs. Obvious problems, because the only thing I had to do was to use my exact equations of equilibrium in place of the Emden equation that people had used in other connections. Since I was familiar with the theory of rotating polytropes, it was quite easy to see how to extend that work for rotating white dwarfs. Likewise, I was quite familiar with the theory of pulsating stars and had a clear idea regarding what to do in the case of pulsating white dwarfs. But after the January meeting, I lost the necessary enthusiasm. . . . I felt a psychological impediment to go on working and publishing things which, I knew, people were going to ignore or consider wrong. Still I worked out a number of things; for example, the theory of rotating white dwarfs, which I didn't publish at that time, but later gave an account of at the Paris meeting." That meeting was four years later, in 1939. It was an international meeting in Paris devoted to the special topics of novae and white dwarfs. Both Chandra and Eddington were invited to give talks. Having had the previous experience of not being allowed to reply to Eddington, Chandra had made sure that he would be giving his paper first on Friday. Eddington was scheduled to give his paper the following, final day of the meeting. But by that time Chandra had given a complete account of his theory with full numerical details; he again enunciated the significance of his results on the prevailing notions of stellar evolution:

It should be clear now that a discussion of the role which the White Dwarfs are likely to play in any theory of stellar evolution must be necessarily linked with the evolutionary significance which we attach to the two critical masses, M_3 and $\mathcal{M}$.

For stars of mass less than M_3, we can tentatively assume that the completely degenerate state represents the last stage in the evolution of the stars—the stage of complete darkness and extinction. These completely degenerate configurations with $M < M_3$ are of course characterized by the finite radii.

For $M > M_3$ no such simple interpretation is possible. The problem that we are faced with can be stated as follows:

Consider a star of mass greater than $\mathcal{M}$ and suppose that it has exhausted all its sources of subatomic energy—hydrogen in this connection. The star must then contract according to the Helmholtz-Kelvin time scale. Since degeneracy cannot set in, in the interior of such stars, continued and unrestricted contraction is possible, in theory.

However, we may expect instability of one kind or another (e.g., rotational) to set in long before, resulting in the "explosion" of the star into smaller fragments. It is also conceivable that the star may decrease its mass below M_3 by a process of continual ejection of matter. The Wolf-Rayet phenomenon is suggestive in this connection.

For stars with masses in the range $M_3 < M < \mathcal{M}$ there exist other possibilities. During the contractive stage, such stars are likely to develop degenerate cores. If the degenerate cores attain sufficiently high densities (as is possible for these stars) the protons and electrons will combine to form neutrons. This would cause a sudden diminution of pressure resulting in the collapse of the star onto a neutron core giving rise to an enormous liberation of gravitational energy. This may be the origin of the Supernova phenomenon.

The above remarks on the evolutionary significance of M_3 and $\mathcal{M}$ are made with due reserve and no definiteness is claimed for them.

There was some discussion of his paper, but nothing of significance happened. Eddington, although present, did not raise any objections. But as Chandra recalls, "I came on Friday, and I talked about my work, and then on Saturday, at the morning session, Eddington came. He started by saying, 'Our beliefs on Saturday must be different from our beliefs on Friday.' And then he went on to give his paper."

Eddington, in his paper, elaborated on his earlier ideas, why standing waves should be used instead of progressive waves and so on, and said, "Chandrasekhar's investigation must be rejected because his mathematical formulation does not in any way correspond to the physical problem." But this time Chandra was prepared, and during the discussion he made it clear that Eddington and he differed in their fundamental premises. Since Eddington had argued that the relativistic degeneracy formula reinstated the paradox which Fowler had removed, Chandra explained that no such reinstatement took place in his formulation of the problem. Eddington, however, persisted in his views and went even to the extent of saying that Fowler himself did not understand his own formulation of the paradox. "Fowler is a mathematician—he does not understand physics," Eddington said. The following excerpt from the summary of the discussion typifies Eddington's remarks:

. . . in stars of mass greater than the critical masses mentioned by Dr. Chandrasekhar there is no limit to the contraction, so that if the star is symmetrical and not in rotation, it would contract to a diameter of a few kilometers, until, according to the theory of relativity, gravitation becomes too great for the radiation to escape. This is not a fatal difficulty, but it is nevertheless surprising; and, being somewhat shocked by the conclusion, I was led to reexamine the physical theory and so finally to reject it. Whatever Fowler's view of the paradoxes may have been, he eliminated the difficulty by showing how the contraction could be stopped and how the star could become cool again.

Finally, in response to Gerard Kuiper (an observational astronomer, not a theorist), who asked whether there were any observational tests that would permit a choice between the two rival theories, Eddington insisted that there were not two theories; "Observation can decide between rival hypotheses, but not between rival conclusions which profess to represent the same hypothesis." "At this point, I got very angry," Chandra says, "and I asked, 'Well, Professor Eddington, how can you say that? Last week we were at a discussion together with Dirac and Pryce, and Dirac and Pryce told you they did not accept your theory of degenerate matter. And if Dirac does not accept your theory, I do not see how you can claim, with respect to an observational astronomer, that there is only one theory.'"

This meeting, as the one in 1935, was being presided over by Henry Norris Russell. After this bit of heated exchange, he closed the discussion and left the matter unsettled. Chandra had already decided to change his area of research. He had decided not to publish the work he had done in connection with white dwarfs and the effect of rotation and pulsation on the equilibrium state, all of which would be reapproached in the sixties, over twenty years later, in the context of the theory of pulsars.

The July 1939 Paris meeting was the last time Chandra and Eddington met face to face. Soon after that the war broke out, and Eddington died in 1944. This last meeting was a poignant one.

After the meeting ended, there was a gala lunch at the City Hall. The mayor of Paris was there, and all the Paris greats were at the high table; Eddington, Russell, [Louis] de Broglie, [Irène] Joliot-Curie. I was sitting way in the corner somewhere, really extremely upset and annoyed, because of the way in which the whole discussion had gone. After lunch, I was standing entirely by myself waiting to leave in the next hour to take a train to Cherbourg where I was to catch the steamer *Normandy*. (In fact, it turned out to be its last voyage at the time because of the war.) Eddington suddenly appeared next to me. He said, "I am sorry if I hurt you this morning. I hope you are not angry with what I said." I said, "You haven't changed your mind, have you?" "No," he said. "What are you sorry about then?" I said and turned away. Eddington sort of stood there for a few moments and walked away.

From aboard the *Normandy,* Chandra wrote his father, "I had a nerve-racking controversy with Eddington who showed himself thoroughly opaque to reason. I had a trying time avoiding him, but he was equally after me—we were playing blind man's bluff!" But "I regretted later, I still do," says Chandra. "I was rude, was unforgiving when he came . . . essentially to apologize."

After the traumatic 1935 meeting and the subsequent years during which Eddington's attacks on relativistic degeneracy continued unabated, one would certainly expect that the relationship between Chandra and Eddington would have become strained forever. On the contrary, strange as it may seem, their feelings towards each other retained a resilient continuity. "It never destroyed my respect for him," Chandra says. "It never gave me a feeling that I was not on speaking terms with him. You may be surprised . . . during the spring of that year [1935], we went on a bicycle trip together and Eddington took me to the Wimbledon tennis championship match. I recall the match on the center court between the Englishman Fred Perry and a Frenchman, whose name I do not recall. Eddington was for Perry while I liked the Frenchman." And the following year [1936] when Chandra returned from India after getting married, Eddington invited the couple for tea. When he learned that they were leaving for America soon, he asked Chandra to his rooms one morning. "Let us not talk science," Chandra recalls him saying. "That's what we have done all along." Eddington then talked about his early years, the poor circumstances he grew up under, his living alone, and the loneliness of an intellectual life. He then brought out a map of England on which he had pinned all the places to which he had bicycled and marked the routes he had taken. "You are the first person to see this map," he said to Chandra. Chandra was obviously moved. "I sort of felt," says Chandra, "that Eddington was trying to add to our professional relationship a personal dimension. The enormous respect I had for him made me feel grateful, grateful that I had such an opportunity to know him."

During the subsequent years, they corresponded with each other mostly, but not entirely, about scientific matters. Eddington's letters[5] are full of warmth, humor, and affection. He wrote about his travels, his reactions to various conferences, his triumphs and failures with his cycling. "My cycling 'n' is still 75. I was rather unlucky this Easter as I did two rides of 74-3/4 miles which do not count;[6] I still need four more rides for the next quantum jump. However, I had marvelously fine weather, and splendid country—chiefly south Wales." Further on in the same letter he notes: "Tomorrow I have to put on a weird costume— knee breeches and silk hoses—and get my Order from the King. Then holiday—then the I.A.U.—then the British Association at Cambridge which will make my fourth conference this year." In another letter, after discussing some scientific questions, he writes, "As a lighter matter the following incident may amuse you. The Master recently had to give lunch to the Duchess of Kent. He applied to the local food office for extra rations in order to entertain H.R.H. suitably. They rose to the occasion. Yes, he might have one extra pennyworth of meat, and some rice." Still in yet another letter during the height of the Second World War, he wrote, ". . . I often get inquiries after you from your former colleagues in Trinity. The Observatory Club has had to lapse. The almost complete absence of research students makes a great difference to the atmosphere of the University. On the other hand the undergraduates have suddenly developed great enthusiasm for astronomy, and last year founded an Astronomical Society with more than 200 members . . . 'n' is now 77. I think it was 75 when you were here. It made the last jump a few days ago when I took an 80-mile ride in the fen country. I have not been able to go on a cycling tour since 1940, because it is impossible to rely on obtaining accommodations for the night; so my records advance slowly." And finally, in the last letter he wrote in 1944, he said, "I have finished my new book on Fundamental Theory. I have only two more sections to write. I have not been feeling well. I am going to Evelyn Nursing Home this evening and when I come back I will finish the book." But in the nursing home they found that he was in an advanced stage of cancer; he died the following week.

Chandra, though he remains puzzled over Eddington's behavior in his encounter with him, retains the highest admiration for Eddington's extraordinary scientific achievements, his charm and wit, and his great influence in many fields of human endeavor. He speaks movingly of Eddington's pacifism and internationalism, often quoting from Eddington's obituary for Karl Schwarzschild:[7]

The war exacts its heavy toll of human life, and science is not spared. On our side we have not forgotten the loss of physicist Moseley, at the threshold of a great career; now, from the enemy, comes news of the death of Schwarzschild in the prime of his powers. His end is a sad story of long suffering from a terrible illness contracted in the field, borne with great courage and patience. The world loses an astronomer of exceptional genius, who was one of the leaders in recent advances both in observational methods and theoretical researches. . . . It would be paying an ill-service to the memory of one who gave the final proof of devotion to his country, to seek by these traits to dissociate him from the rest of his nation. We would rather say that through him a new spirit was arising in German astronomy from within, raising, broadening and humanising its outlook; and we are confident that he has left behind an influence which will enlarge and revivify the old traditions.

In turn, in the obituary speech Chandra gave on Eddington at the University of Chicago, he said that posterity may rank Eddington next to Karl Schwarzschild as the greatest astronomer of his time.[8] Perhaps by reading a part of what Chandra said in that tribute we may understand why they were able to keep such good relations.

I believe that anyone who has known Eddington will agree that he was a man of the highest integrity and character. I do not believe, for example, that he ever thought harshly of anyone. That was why it was so easy to disagree with him on scientific matters. You can always be certain he would never misjudge you or think ill of you on that account. This cannot be said of many others.

In scientific matters, Eddington of course could be harsh and brutal without any regard to how it affected others' lives. Milne, as we have seen, suffered from the brunt of his attacks. Jeans and Eddington swapped blows during almost all of their scientific lives.*

Whatever Eddington's intentions, there is no doubt that those whom he attacked were not immune to feeling stung and put off balance by his professional candor, even though they might recover with a feeling of respect for his forthrightness later. Thus, in spite of all his personal warmth and affections, Eddington did not spare Fowler or Chandra when it came to science. In his address on the "Constitution of Stars," given at the Harvard Tercentenary Conference of Arts and Sciences,[9] Eddington discussed white dwarfs, and with the preface, "I will say a word or two about Professor Fowler's explanation," he then proceeded to disarm Fowler's proof with what one might almost call relish:

My colleague Fowler was in his youth a pure mathematician, and I am afraid he has never really recovered from this upbringing. Consequently, although his paper contained reassuring equations, it did not clearly reveal the simple physical modification of ideas which wave mechanics brought about. He proved that the star would manage all right. But, as you may have inferred from Professor Hardy's revelations,[10] I am not an extreme worshipper of proof. I want to know why; a proof does not always tell you that. As Clerk Maxwell used to ask, "What's the go of it?" Well, in this case the "go of it" was that whereas the older theory said that atoms could only be ionized by high temperature the new wave mechanics said that high temperature was not essential because they could also be ionized by crushing them under high pressure . . . that was what Fowler's rather mysterious result really meant; but I think that it is still not at all known.[11]

Then Eddington introduced further developments, about the need to take relativistic effects into account, and went on to rehash his disagreement with Chandra in a tone of amused tolerance for the misguided:

All seemed well until certain researches by Chandrasekhar brought out the fact that the relativistic formula put the stars back in precisely the same difficulty from which Fowler had rescued them. The small stars could cool down alright, and end their days as dark stars in a reasonable way. But above a certain critical mass (two or three times that of the sun) the star could never cool down, but must go on radiating and contracting until heaven knows what becomes of it. That did not worry Chandrasekhar; he seemed to like the stars to behave that way, and believes that is what really happens. But I felt the same objections as

* Chandra cites the following example: After seeing the announcement that Jeans was to present a paper on radiation viscosity, Eddington sent Jeans a postcard with the formula for radiation viscosity. "Sort of childish," Chandra says. "What was important was the concept of radiation viscosity; once the concept was formulated, the derivation of the formula was a simple matter. But Eddington wanted to deride Jeans."

twelve years earlier *to this stellar buffoonery;* at least it was sufficient to rouse my suspicion that there must be something wrong with the physical formula used.

He then repeats what he had said at the January 1935 meeting, that the relativistic degeneracy formula was the result of a combination of relativity theory with nonrelativistic quantum mechanics. "I do not regard the offspring of such a union as born in lawful wedlock. The relativistic degeneracy formula—the formula currently used—is *in fact baseless* [emphasis added]; and, perhaps rather surprisingly, the formula derived by a correct application of relativity theory is the ordinary formula, Fowler's original formula which everyone had abandoned. I was not surprised to find that in announcing these conclusions I had put my foot in a hornet's nest; and I have had the physicists buzzing about my ears, but I don't think I have been stung yet."[12]

Eddington, who had said, "There is a kind of sureness which is very different from cocksureness,"[13] had become a victim of "cocksureness" himself. "I don't think Eddington's tirade against me was derived from any personal motives," insists Chandra. "You may attribute it to an elitist, aristocratic view of science and the whole world. Eddington was so utterly confident of his views that as far as he was concerned he was a Gulliver in a land of Lilliput. He was not affected in the least by what other people said or did not say."

More importantly, the fact of it was that his views on relativistic degeneracy were a part of his entire philosophy of the universe. Around 1937, Chandra asked Eddington to what extent his ideas on relativistic degeneracy were relevant to his theory of electrons and protons. "Absolutely fundamental," was Eddington's reply. "If my ideas on relativistic degeneracy are wrong, my entire theory of electrons and protons is wrong." When Chandra kept quiet, not saying anything in response, Eddington wanted to know why he had asked the question in the first place. Chandra said, "I am sorry to hear what you just said." Up to a point, Chandra meant what he said. He did feel sorry that Eddington's whole effort in that area should depend upon ideas which were to Chandra so obviously wrong.

It is indeed one of the strangest ironies in science. Eddington, whose bold embrace of Einstein's general theory contributed so much to its early acceptance in the English-speaking world,* failed to see the far-reaching consequences of a very simple and straightforward application

---

*According to Eddington, no one else understood the significance of the Lorentz transformation. In a 4 July 1938 letter to Chandra, Eddington said, "One almost wishes that the

of the special theory. "Kamesh, suppose, just for a moment, Eddington had accepted my result," Chandra says. "Suppose he had said, 'Yes, clearly the limiting mass does occur in the Newtonian theory in which it is a point mass. However, general relativity does not permit a point mass. How does general relativity take care of that?' If he had asked this question and worked on it, he would have realized that the first problem to solve in that connection is to study radial oscillations of the star in the framework of general relativity. It's a problem I did in 1964, but Eddington could have done it then in the mid-1930s! Not only because he was capable of doing it—he certainly had mastered general relativity— but also because his whole interest in astrophysics originated from studying pulsations of stars. And if he had done it, he would have found that the white dwarf configuration constructed on the Newtonian model became unstable before the limiting mass was reached. He would have found that there was no *reductio ad absurdum,* no stellar buffoonery! *He would simply have found that stars became unstable before they reached the limit and that a black hole would ensue. Eddington could have done it.* When I say he could have done it, I am not just speculating. It was entirely within his ability, entirely within the philosophy which underlies his work on internal constitution of stars. And if Eddington had done that, he would stand today as *the greatest theoretical astronomer of this century,* because he would have predicted and talked about collapsed stars in a completely and totally relativistic fashion. It had to wait thirty years!"

Such an exploration was not outside Chandra's ability either. He reported some of his work on rotating white dwarfs at the 1939 Paris meeting, and the paper in the Synge volume published in 1972 contains an almost verbatim account of the work he had done in 1935.[14] Also, John von Neumann was visiting Cambridge in the spring of 1935. "He was rather lonely," Chandra recalls. "He used to come to my rooms often. Naturally we discussed Eddington's objections. John said, 'If Eddington does not like stars to recede inside the Schwarzschild radius,[15] one probably should try to see what happens if one uses the absolute, relativistic equations of state.[16] We started working on that together,[17] but to go on we had to study equilibrium conditions within the frame-

---

Lorentz transformation had never been invented. Nearly every paper I attempt to read on the structure of the nucleus tries to invent Lorentz-invariant formulae—which is nonsense. Clark and I have just published a paper on a double-star problem in relativity; neither we nor Einstein, Levicivita [Tullio Levi-Civita] and others who have treated the problem dreamt of using Lorentz-invariant equations in it—it would have been altogether wrong to do so."

work of general relativity. Soon John left Cambridge and forgot the problem, and I got sufficiently discouraged with the situation to leave the problem alone."

A few years later J. Robert Oppenheimer and George M. Volkoff did study the problem and wrote their classic paper on neutron stars.[18] It is impossible to know, of course, what some individual would or would not have done, or what would have been the consequences if a particular development had or had not taken place, but there is no doubt that there is a kind of poetic justice in the fact that it was Chandra who worked out the theory of pulsation of spherical stars in the framework of general relativity in 1962, proving their relativistic instability. The work had enormously significant consequences for the theory of pulsars and made Chandra's entry into his "seventh period" all the more dramatic. "The moral," Chandra says, "is that a certain modesty of approach toward science always pays in the end. These people [Eddington, Jeans, Milne], terribly clever, of great intellectual ability, terribly perceptive in many ways, lost out because they did not have the modesty to say, 'I am going to learn from what physics teaches me.' They wanted to dictate how physics should be."

Nevertheless, it is conceivable that Eddington, in his tenacious opposition to Chandra's discovery, was motivated by what he conceived to be honest scientific disagreement. But what about the others, the physicists, the astronomers who could have remedied the situation? Physicists did not intervene because not only was astrophysics simply not at the center of their interest, but they had stopped taking Eddington seriously. He had more or less abandoned the well-trodden empirical path to science; he was advocating a purely mathematical, epistemological approach and was engulfed in deriving the total number of protons and electrons in the universe and the value of the fine structure constant. However, in the astronomical community, to which Chandra's discovery should have mattered most, Eddington still enjoyed an almost mystical admiration, and his authority reigned supreme. None other than the great Henry Norris Russell had prevented Chandra from responding to Eddington's "wildest nonsense" in the open 1935 Paris meeting. Further, in his article in the *Scientific American*,[19] where Russell gives an account of Chandra's discovery and the theory behind it, he finds it necessary to warn the reader that *Eddington has criticized the theory*. Although he goes on to say that Chandra's theory is generally accepted and Eddington's paper criticizing it was comprehensible only to a small group, the harm was done. Likewise R. H. Fowler, who *knew* Eddington's arguments were baseless, points out in a footnote in his book on statistical mechanics that *Eddington says* that the relativistic degeneracy formula is wrong, but he does not come out and say that he (Fowler) does not think so.[20]

It took nearly three decades before the full significance of the discovery was recognized and the Chandrasekhar limit entered the standard lexicon of physics and astrophysics. Five decades would pass before he was awarded the Nobel Prize. "It is quite an astonishing fact," Chandra says, "that someone like Eddington could have such an incredible authority which everyone believed in, and it is an incredible fact that in the framework of astronomy there were not people who had boldness enough and understanding enough to come out and say Eddington was wrong. I don't think in the entire astronomical literature you will find a single sentence to say Eddington was wrong. Not only that, I don't think it is an accident that no astronomical medal I have received mentions my work on white dwarfs. I was given the Gold Medal of the RAS in 1952; the citations and the speech made by the president refers to my work on stellar dynamics, refers to my work on radiative transfer, but not a word about my white dwarf work. I was given the Bruce Medal a year later in 1953; Otto Struve made the point that I was the youngest to receive the Bruce Medal, but did not refer to my white dwarf work. In fact, none of the recognition that I was given, up to the Dannie Heineman Prize in 1974, mentions my white dwarf work. The Heineman Prize in physics was the first one to mention it. It is ridiculous to talk about these things, but I personally believe that the whole development of astronomy, of theoretical astronomy, particularly with regard to the evolution of stars and the understanding of the observations relating to white dwarfs, were all delayed by at least two generations because of Eddington's authority."

Of course one could turn around and ask Chandra why he did not pursue his ideas and, if he was so convinced, press the issue to its bitter end. It was not because he was unaware of this possible alternate course. He wrote to his father on 14 February 1940, "If I had followed the traditional methods (set by Eddington among others), I should have made a real capital out of his errors." Chandra says, "It was a personal decision I made at the time. I felt that astronomers without exception thought that I was wrong. They considered me as a sort of Don Quixote trying to kill Eddington. As you can imagine, it was a very discouraging experience for me—to find myself in a controversy with the leading figure of astronomy and to have my work completely and totally discredited by the astronomical community. I had to make up my mind as to what to do. Should I go on the rest of my life fighting? After all I was in my middle twenties at that time. I foresaw for myself some thirty to forty years of scientific work, and I simply did not think it was productive to constantly harp on something which was done. It was much better for me to change the field of interest and go into something else. If I was right, then it would be known as right. For myself, I was positive that a fact of such clear significance for evolution of the stars would in time be established or disproved. I didn't see that I had a need to stay there, so I just left it."

People did come to know Chandra was right in time. In retrospect, this traumatic event became a turning point in Chandra's life. It had a "sobering effect"; he became inward bound. He developed a distinctive style dominated by aesthetic considerations. The relentless mastery of a certain area and, once mastered, the ability to leave it entirely for another became Chandra's hallmark in his scientific pursuits.

Years later, in 1983, while much was being said publicly and privately among the circle of his students, colleagues, and admirers about the fifty-year slumber of the Nobel Committee in belatedly awarding him the prize for his white dwarf work, Chandra remained lighthearted and could say that he was happy that this most prestigious prize came after all the other prizes and awards. Otherwise, he says, his story would have been like that of a certain general in the army who attended a dinner with rows and rows of pins and medals on his well-starched uniform. When a lady at the table asked him, in awe and admiration, what all those honors stood for, the general pointed to the top medal and said, "Dear lady, this one, this top one was awarded to me by mistake. The others followed as a domino effect."

7

"I Must Push On in My Directions"
Cambridge and Harvard, 1935–1936

It is a great pity that human beings cannot find all their satisfaction in scientific contemplativeness.

Niels Bohr

Niels Bohr made this statement when his friend and junior colleague Leon Rosenfeld announced his engagement. In a letter to his father on 5 October 1934, Chandra said he agreed with Bohr, yet he felt he must do the contrary and did not know how to apologize for it. "Lalitha and I," he wrote, "have come to a definite mutual understanding that we should get married on my return."

Chandra was in an ebullient mood during the fall of 1934. In spite of immediate uncertainties, he was looking forward to what he felt would be a bright future. The first year of the Trinity Fellowship proved to be a triumphant year. He was at peace with himself for having remained in astrophysics and no longer suffered from doubt or conflicts about whether to go into physics or pure mathematics. His trip to Russia and the welcome he received there both scientifically and otherwise had only boosted his confidence further. His research on stellar structure was on a well-charted course with Eddington's encouragement and interest in the outcome. The Cambridge University Press (on the advice of G. H. Hardy) wanted him to write a small book on stellar atmospheres in the "Cambridge Mathematical Tracts" series. "No commercial value for this enterprise," he wrote to his father on 16 November 1934. "I would be paid only 10 pounds, but one has the satisfaction of 'registering' oneself as a serious Cambridge mathematician; almost all the eminent Cambridge mathematicians have written in this series." Furthermore, he was presented with a possible opportunity to spend a year in America at Yale or Harvard. Four "Henry Fellowships" were going to be awarded for the year 1935–36. Harlow Shapley, director of the Harvard College Observatory, had written to Chandra, wanting him to apply. Milne and Eddington were also pressing him to do so. There was little or no doubt Chandra would get one of the fellowships if he just took the trouble to apply, since he was ideally qualified for it. He was less than twenty-five years old, distinguished, and a British subject.

But should Chandra accept the fellowship, which was to begin towards the end of July 1935, he would then face a difficult situation. "I should [then] not be able to come to India earlier than in the spring [of 1936] to get married," he wrote to his father. "For one thing, one can be a candidate for the Henry Fellowship if one is unmarried—the fellowship will have to be vacated on marriage. Secondly, I am to attend the International Astronomical Union towards the end of June and this combined with the fact that I am lecturing in Cambridge during January–April [of 1936] makes it hardly worthwhile to pay what should in effect be a 'Flying Visit' to India." After considerable deliberation, and in spite of pressure from Milne, Eddington, and his father,[1] Chandra decided not to apply for a Henry Fellowship. "Now that Lalitha and I have an understanding of our mutual attitudes," he wrote to his father, "it does make it difficult for both of us. I feel I am pretty deep in a mess. I should feel much worse if it were not that I am now passing through an essentially successful period in my scientific life."

Chandra's father, who was becoming increasingly apprehensive that Chandra was not paying attention to the "other sides of his nature" and devoting himself so exclusively to astrophysics that he was becoming an "astrophysical animal," was extremely pleased with the news of his son's engagement and plans to return home and get married. "I am happy and sad at the same time," he wrote to Chandra on 19 October 1934. "Sad because your mother is not alive to welcome home your bride-elect. I pray to God that the union you propose be blessed and happy." At first he had some unfavorable thoughts regarding the alliance,[2] but he had met Lalitha a few times before and had been favorably impressed.

Lalitha had completed her M.A. in physics and was the headmistress of a school in Karaikkudi at the time. Chandra's father invited her, along with her physics professors, Appa Rao and Parameswaran, for dinner at Chandra Vilas. He noticed how Lalitha mixed only with Chandra's sisters; she stayed inside and had her dinner with the women in the family rather than with the men. Lalitha, as he put it, was a "modest, quite reserved young lady." "Hope that your bride-elect would be good to you," he wrote to Chandra, ". . . just as your mother was good—as I now realize—how your mother was different in the mental make up from that of her other sisters and brothers in all matters, combining in herself only the good qualities of those amidst whom she was born, the intellectual capacity of her father, the good meekness of her mother, the charitable nature of her eldest brother and so on. On this day [19 October 1934], you begin your 25th year of your life . . . I do hope and pray that 'this maiden passion for a maid'[3] will bring all happiness that a father can wish for his son."

Four unmarried daughters were also on Chandra's father's mind. Sarada, the oldest, was eighteen years old. Chandra's father thought that, rather than attend school, she would do better to get married "and be a delightful hostess running a home for a young man." He was, therefore, in search of a suitable bridegroom for her and had written to a former classmate of Chandra, who had been to England, asking him whether "he would consider Sarada as one of the possible brides." But all educated, eligible young Indian men were not like Chandra. Most followed the normal, overwhelmingly prevailing custom; their marriages were in the hands of their parents, who were particular about caste, subcaste, subsect of subcaste, and the amount of dowry in arranging the marriages of their sons. Sarada's marriage, in any case, had to await Chandra's, since Chandra's father had "vowed to himself" that he would not get any of his younger daughters married "until he had a daughter-in-law at home." Chandra, being the eldest son, should marry before his brothers. "I do earnestly wish," he wrote to Chandra on 11 March 1935, "that you had better leave in Xmas week of this year 1935, stay with us until the spring term—i.e. very nearly up to Easter: i.e.—I expect you to get married early in January 1936. (This would enable me then to seriously consider Sarada's marriage also)."

The commitment of this much time appeared impossible for Chandra. The worsening controversy with Eddington aside, "The amount of work that has to be done would take me well into the next year," he wrote on 1 March 1935. "I have hardly begun my tract, which is due before July, and I must continue my stellar structure work energetically—already other people in Cambridge (Fowler, Wilson, and a young man Pryce) have gone ahead some distance with my work but in directions in which they will do more completely than I. But I must push on in *my* directions."

He had different ideas regarding Sarada's marriage as well. "I most certainly think that Sarada herself should be consulted (*not* advised) on the matter," he wrote to his father in his letter of 1 March. "She is surely old enough to understand what marriage means and she should be asked to express her feeling frankly on her marriage. . . . She should be asked to think for herself and be given time to do so. She is barely seventeen and starting married life in her present health—unless the husband is particularly considerate—might prove disastrous." His father's response was the expected one. "India is not England. The girls must get attached to somebody and 18 years is a good age for a girl in India to be so. What I have said regarding Sarada will apply with equal force to your intended bride . . . Do not be perverse in saying to yourself that I shall postpone my marriage so that my sisters may not get married early."

Thus Chandra, by announcing his "bride-elect" and his "mutual understanding" with Lalitha, put himself on the hook of his father's desire to reel him home to India. His confrontation with Eddington had left him depressed, discouraged; he had stopped dining regularly at high table, which prompted Milne to write him in June, "I was most distressed to hear from you that you have discontinued dining regularly in the Hall at Trinity owing to the damping influence of Eddington. I think we must all struggle against this feeling of discouragement. We have a responsibility to our scientific posterity not to give up our efforts merely because of the dominating influence of one man. Posterity will put us all in our proper, humble places, and will judge us chiefly for our courage in stating the truth as we see it. Did you ever read how Newton was discouraged by the adverse opinions of his all-knowing confreres? It has been the same all along, in all departments of activity, esp. creative activity, that are worthwhile; to be in fashion is really to be behind the times. I for one hope that you will continue to write papers on stellar structure . . . I beg you not to discontinue dining in the Hall."

Meanwhile, Chandra also had begun to feel unsure of the situation between Lalitha and himself. How well did they know each other? True, they had corresponded during Chandra's first year abroad, but then had stopped, and had only started writing to each other again after Chandra became a Trinity Fellow. But it was all in "a sort of vacuum." "I began to feel," Chandra recalls, "[that] we had not at all communicated what was central to our lives. Lalitha did not know much about my work, the people I was dealing with. And I did not know much about her life, her aspirations."

With these doubts and uncertainties, Chandra went to Copenhagen for a six-week visit to the Niels Bohr Institute. There, in the company of his friend Leon Rosenfeld and away from Cambridge and Eddington, he felt relaxed enough to sort things out for himself. He came to the conclusion that any thought of marriage was premature. Lalitha and he should wait until they had an opportunity to see each other, talk to each other, and find out whether they were prepared to make a "go at it"; marriage was a serious step to undertake, especially since they would have to depend entirely on themselves.[4] His job outlook was uncertain. He might have to live abroad. Was she prepared for that? He wrote accordingly to Lalitha, and they came to a new "mutual understanding" to completely free each other from any kind of commitment and let the future take its course.

Having arrived at this new definite understanding with Lalitha, Chandra wrote his father promptly after his return from Copenhagen (22 April 1935): "Since you left England, I had been frightfully busy—first my visit to Russia, then my long work on stellar structure, then my discussions with Eddington, Milne, my lectures, etc. I had my first breathing time in Copenhagen. I realized there that my relations with Lalitha were purely illusory. I really had not known her at all. I had seen her in college, but that was five years ago. I was simply deluding myself. So I have written to her breaking off our understanding . . . I only wish to say that my rather lonely life in England has made me center my interest almost solely in my studies and as such I have no desire or interest to get married. But that is only for the present. In any case, I think that if any suitable offers come for Sarada, then she should be married, I mean that their lives should not be made to depend on the fact that I continue single."

If Chandra, by summarily dismissing marriage and his engagement with Lalitha in this letter to his father, intended to ward off further pressure from him, he failed utterly. Chandra's father became more concerned, more worried; he hazarded guesses as to the cause of the disengagement; he thought that Lalitha was responsible for it. He wrote, "Miss L has not sufficiently *known* her father[5] to be able to transfer that love to a husband." He imagined Chandra moody, disappointed by the incident. He advised him "not to worry at all—there are ever so many girls who are reading in the B.A. class, much younger, who are better for family life." He asked Chandra to seek consolation in reading great literature, Tennyson, Browning, Shakespeare. He wanted him to come home even earlier than his previous proposal. "Come in October, stay at least three months with us to see if you can not come to a decent understanding ending in marriage—or choose from some other girls reading in the B.A. class, respectable, healthy."

Brushing aside feelings of annoyance, Chandra continued to frame respectful replies to his father; he responded calmly that Lalitha was not to be blamed for the changed situation. He had thought over the matter carefully and decided to remain single, and his decision regarding his way of life *must not* affect the marriage of his sisters. There was no way he could return to India before April or May of 1936. He was to lecture again in early January, and the work on the tract had to be finished. He had started taking French lessons in preparation for his trip to Paris to attend the International Astronomical Union meeting in early July. He had made plans to spend a few weeks in the fall at Oxford working with Harry H. Plaskett with the idea of working out a "full fledged physical theory of turbulence to apply it to stellar atmospheres"; he was busy studying "some kinetic theory and hydrodynamics."

Still, work notwithstanding, he felt an intense longing to return to India. As he wrote to his brother Balakrishnan on 26 June 1935, he felt homesick. "I am returning surely before June next year, i.e. in less than a year! How I look forward to it! I will be different and so will you and others. Six years! How long! . . . I can hardly imagine how things are at Chandra Vilas. Memory fails not because I cannot recollect but because memory recalls what is no more. Whenever I think of Chandra Vilas, the scene I almost always recollect is mother in a red saree resting in the easy chair in front of our house. All gone! I recollect too our reading together—how I used to bore you reading all my texts—Ruskin especially . . . I want to be so back with you all, *but, but* . . . As you wrote once, for work we sacrifice the richer things in life. But is it worth it after all? The world judges not by what we cherish, but what we accomplish and even that only to forget."

A week in Paris in early July was a respite from such conflicting emotions as well as from work. He flew from London to Paris, his first experience in an airplane. "I had an exciting flight from London," he wrote to his father from Paris on 15 July, "especially exciting as we crossed a thunderstorm. Higher and higher to avoid the clouds—the pilot said we went as high as 7,000 feet." The plane flew even higher, 8,000 feet, on the return journey, and the view of the coasts of England and France from the air was something at which to marvel; so was the shadow of the plane on the clouds. "It is almost incredible," he wrote after his return to Cambridge on 19 July, "[that] I was walking in the streets of Paris at 11 o'clock and I was in Cambridge well before 5 o'clock!"

In spite of Eddington's tirade against his work at the conference, Chandra enjoyed the meetings, the garden parties, and the discussions with "some great astronomers, H. N. Russell, H. Shapley, Adriaan van Maanen, Bertil Lindblad. . . ." With his friend McCrea, he enjoyed the sights of Paris, carnival rides, and even went to a demonstration on Bastille Day, which McCrea thought was a bit dangerous. And it was during this meeting in Paris that Shapley raised the possibility of Chandra lecturing at Harvard College Observatory during the following year. After his return to the States, he would send Chandra a formal invitation if Chandra would think such a visit possible. This presented further complications in his plans to return to India! But how could he forgo the opportunity to visit the United States? He agreed to consider the invitation, thinking that, if things were arranged, he could still return to India, at least for a short visit, during the spring of 1936.

In the meantime, Chandra had to continue to contend with the concerns of his father, who was now urging him to plan ahead and look for a job in India. The director of the Kodaikanal Observatory, one Mr. Royds, an Englishman, was to retire within the next year and a half. The Government of India would advertise the position. Chandra, according to his father, had "really no good competitors" among Indians; the proposed salary scale (400–900 rupees per month) was not good enough for a European to seek the directorship. He could make the job sure for himself with some practical training at the Oxford observatory and a recommendation letter from Plaskett. Therefore, he should "register" himself quickly as a prospective candidate. But Chandra had neither the inclination nor the desire for an administrative job in an observatory.

After his return from Paris, Chandra had gone to Oxford for a few weeks with the intention of doing some observational astronomy. The previous year a new telescope had been installed and Plaskett was using it to make some notable solar observations. Chandra soon found, however, that he would not have the opportunity to work with the telescope; Plaskett wanted him to analyze the solar spectra he had obtained. Measuring wavelengths was not the type of observational work Chandra had in mind, and he returned to Cambridge rather disappointed. "My inclination is for a chair in mathematics with a congenial atmosphere," he wrote to his father on 16 August 1935. "Also apart from the Government finding me good enough for Kodaikanal, I should first have confidence that I am—which I do not feel now. My efforts to do practical work so far have turned into fiascos! . . . If no Englishman would go over to India, the Government may quietly offer the place to [K. R.] Ramanathan who I think would only be too glad to get it, in spite of his supposed statement to Krishnan that he would rather have me at the observatory. The only occasion I met him he slighted me!! But I *do* think Ramanathan is very good."

A more surprising matter for Chandra was a letter from his famous uncle Raman offering him an assistant professorship at the Indian Institute of Science, Bangalore. "The salary of the post is Rs 500–600 together with a free house, a 10% retiring allowance and other privileges," Raman wrote on 8 August 1935. "The appointments of this Institute are usually made for a term of five years but may of course be renewed thereafter by the Council. . . . I do not wish to make any rash promises, but I have a feeling that at the end of your 5-year tenure you can look forward to a full Professorship being created for you. . . . The climate of Bangalore is the best to be found anywhere in India, and a very excellent house and garden goes with the appointment. . . . I shall be very glad if you will kindly send me a brief cablegram in reply—'Reply affirmative' or 'Reply negative' so that I may know what to do. It is necessary that the incumbent should join his post not later than April next [1936]."

Raman had sent a copy of his letter to Chandra's father, who reacted by sending this cable to Chandra:

MY ADVICE KEEP OFF HIS ORBIT.

In his next letter he explained why he was strongly against Chandra's accepting Raman's offer: "I ardently desire that one *Madrasi* at least would be able to do something in the world of physics by his own efforts and not be indebted to Raman who thinks he is such a 'king maker' of Professors and Doctors of Science. . . . I do not wish it to be said of *you* particularly or that the world should think of you that you stood to gain because of his *relationship*. If you will *honor* my *sense* of *self respect* at least if not your own, you may not accept the offer." Chandra did not need this strong admonition from his father, but he was happy that, at least in this instance, he and his father were of like mind. While Chandra respected Raman's brilliance in physics and his sensational discovery, Raman was not a role model for Chandra. Because Raman was given to sensationalism and reveling in controversies, and prone to speak in contradictory terms, he annoyed Chandra thoroughly. Just a few months before this job offer, Raman was reported to have said that Chandra was wasting his time in astrophysics, in the backwaters of science, and he would not have such a person miles near Bangalore. He had also said that if Chandra changed his field and worked in nuclear physics instead, he might be able to get somewhere. Such reports had thrown Chandra into a rage. "I wonder if anything could be done to stop him talking about me," he had written to his father (14 March 1935). "He can be assured anyway that I will not get so much as scores of miles near Bangalore." He knew that in Bangalore he would not find the quiet, congenial atmosphere he needed to go about his own work. "With C. V. R., I would be expected to turn miracles, make splashes continuously," he had written to his father on 24 July 1935, before Raman's offer. "Apart from my general inability to do those somersaults I would prefer a position where nothing would be demanded or expected of me. In this respect my present position at Trinity is an ideal one. No master, no responsibilities personal or otherwise. I am my own master."

The rejection of Raman's offer was, therefore, a foregone conclusion in Chandra's mind. Yet, the response had to be clothed in polite terms; Raman was after all an elderly uncle and a preeminent Nobel laureate physicist in India. Chandra did a masterful job in rejecting the offer. His letter of 18 August 1935 began with his expressing how surprised and flattered he was by the offer and his thanking Raman for considering him a "suitable candidate for the post of the Assistant Professorship." He went on to say that his first impulse was to send a cable accepting the offer [obviously far from the truth]; but on second thought, he became diffident as to whether he could do justice to "the dignity of the position" Raman had so generously offered him, realizing particularly that his studies and research centered around "the remote astrophysical and stellar problems" and were "outside the central arena of modern physics." Chandra went on to say that he had studied theoretical physics (one could not help it at Cambridge!) and perhaps could lecture on general principles of wave mechanics, quantum mechanics, and other branches of pure and applied mathematics, but he could not be expected to come to grips with problems of molecular structure, chemical physics, and the like, in which he presumed the experimental physicists in the institute would be most interested. He was afraid he would prove a disappointment for them and that Raman himself would begin to wish that he [Chandra] was more of a "genuine physicist" than he was.

In addition to all these considerations, there was the problem of taking up the position by April of the following year. He was invited (and he had essentially accepted the invitation) by Professor Shapley to spend some months at Harvard. At the Harvard Observatory there were quite a number of people who were actively interested in his work. So for the widening of his own scientific outlook on such problems as he was interested in, it was an opportunity he could not afford to miss. In conclusion he wrote, ". . . one rather delicate matter. I have had reports that a certain part of the press (influenced by persons jealous of your achievements) is violently antagonistic to you. I realize, of course, that such attacks are launched under cover by people who have discovered their inability to make any further genuine contributions to science and who, as I said, are jealous of the powerful school you are building around yourself. I am, therefore, a little afraid that that particular part of the press to which I have referred to might start a fresh scheme of scandals, if I were to be appointed to a position at your Institute. Your scientific reputation is so unique that such attacks from the press can hardly affect you personally, but you would agree that they might have quite unpleasant reactions on a beginner in research as I am."

The matter of Raman's job offer thus ended, Chandra set forth for a brief vacation in Scotland. A Scottish friend, Gordon Sutherland, had arranged for him to stay at a farmhouse at Kirkmichael in Perthshire. Chandra needed rest. He needed to get away after eight months of harrowing stress. The Eddington episode, pressure from home to return, uncertainties in his relations with Lalitha—all of these had led to many sleepless nights. The farmhouse at Kirkmichael was situated in a "wonderfully beautiful hilly country." With nothing but miles and miles of moorland around, it was ideally suited for long walks in invigorating fresh air. Within a few days in the quiet surroundings he felt refreshed and rejuvenated, and when he was not on the moors taking in "violently the fresh air and the cool breeze," he found himself immersed for a change in literature instead of equations. He read Dostoevski's *The House of the Dead* and *Crime and Punishment.* He read Honoré de Balzac's *The Wild Ass's Skin* and started reading *Kristin Lavransdatter* by Sigrid Undset before the three-week holiday ended. Bohr once had spoken of Undset as the greatest Norwegian writer. "Dostoevski's *Crime and Punishment* is almost Greek in its conception," Chandra wrote to his father on 4 September 1935. "Sometimes Thomas Hardy has been compared to Dostoevski, but Dostoevski is an incomparably finer artist. If Dostoevski's works could be compared to some of say, Rubens' paintings, then Hardy's are neatly finished diagrams drawn on graph paper with ruler and compass! Compare for instance the tragedy of Tess with that of Sonia, or the culmination of the forces in Raskolnikoff's confession with the melodramatic ending of *Return of the Native. Crime and Punishment* is more superb in its conception than even Hugo's *Les Misérables.* Great as is the tragedy of Raskolnikoff, greater still is the tragedy of Sonia. She reminds us of Fantine when poverty and starvation forces her to a prostitute's life, of Ophelia in her tragic devotion to Hamlet, of Cosette in her simplicity, but to whom could we compare her when, for instance, she reads out the resurrection of Lazarus to her lover. Dostoevski himself characterizes her most delicately in the words of Raskolnikoff as he threw himself at her feet, 'I do not bow to you personally, but to the suffering humanity in your person.'"

In this foray into literary studies, stars were not totally forgotten. "I have been able to *think* more about stars than I would have," he wrote to his father on 11 September, "because of the lack of 'paper and pencil.' Paper and pencil necessarily land one in the intricacies of the calculation! I have a number of ideas and I am getting almost impatient to get back to Cambridge to work them out."

Refreshed and invigorated, he returned to Cambridge during the last week of September. Soon thereafter he received a formal invitation from Shapley. He was offered a lectureship at Harvard in "Cosmic Physics" with a remuneration of $550 a month. He would be expected to stay for a minimum of three months. The appointment could begin in December or January.

Chandra decided to leave Cambridge towards the end of November, reaching the States in early December. He planned to return to Cambridge in March of 1936 to lecture during the Easter term. This meant he had to postpone his visit to India until the summer or the beginning of autumn of that year. In the meantime, he would also keep looking for a suitable position in India. He did not have to worry too much since he still had one year of the Fellowship to fall back upon. Also he had saved enough to live without any job in India for a year if he was forced by circumstances to do so.

As it happened, a position in India that was considerably to his liking appeared. In October, after his return from Scotland, he saw an advertisement in the London *Times* for a mathematics professorship in the Government College of Lahore. He promptly wrote to the Public Service Commission of the Government of India declaring his intention to apply for the position. But Chandra's father was ahead of Chandra in this matter. He not only secured the necessary formal application form from the commission to send to Chandra, but he also forwarded to the commission a copy of the form which he had duly completed along with a full list of Chandra's publications. (A list of thirty-seven papers which he had prepared himself from the copies of papers Chandra had sent to him; he provided a copy of this list to Chandra to make it easy for him to apply for the position.) "I am forwarding this application," he wrote to the commission, "so that it may reach the Public Service Commission within the due date. I have forwarded one form to my son for him to fill in and post directly to you. It is presumed necessary action will be taken on receipt of the application form signed by me, pending receipt of my son's, so that the application *may not be thrown out on merely technical grounds,* since there has not been sufficient time between the date of receipt of the forms from you, and the due date, for a signed application from him."

In the meantime, however, Chandra had decided to withdraw his application after learning from Littlewood that Chowla was a candidate. "Chowla is one of my best friends," he wrote to his father on 7 November 1935. "I admire his work and person. I feel that it is unfair for both of us, if I should press forward. . . . Prof. Littlewood informs me that Chowla is extremely anxious to get the Lahore job and in consultation with Littlewood I have now decided to write a letter to the Public Service Commission withdrawing my application in favour of Chowla."

Chandra's father was terribly disappointed and upset by Chandra's decision. "Please yourself," he wrote to Chandra on 15 November. "You have made a laughingstock of me before the Commission. You have thrown away all your chances of coming to India. You could have withdrawn later. If you were selected and if you were not able to accept the offer . . . even then, it would have gone a long way to secure other jobs in India. That was one reason why I wanted you to apply." But Chandra felt differently. When he had thought of applying, he had no idea that Chowla would be a candidate as he had a good position at Andhra University, Waltair. Further, since he had already committed himself to the trip to America and it was extremely unlikely that he could return to India in time to accept the job, he felt it was unfair to compete with Chowla. "For the sake of personal vanity," he wrote on 7 November, "I do not want the ill feeling of one who is not only my personal friend but one whose work and abilities I greatly admire." His father needed to be appeased, consoled, so he told him that Chowla was more likely to get the job in any case. Chowla would receive strong recommendations from Professors Hardy and Littlewood. Fowler and Milne would certainly support him, but perhaps not as strongly as he would wish. "I can not really pretend to Fowler and Milne," he said, "that I am really anxious to get the post for I am definitely *not* anxious." Chandra asked his father not to worry about being made a laughingstock; the commission surely realized the difficulty in communicating overseas and was aware that the application was made in good faith to meet the deadline before Chandra's intentions were fully known. Chandra also pleaded with his father not to worry about him throwing away all his chances for a return to India. Chandra reminded him that he still had the Fellowship for two additional years. Further, he had decided to take his Sc.D.[6] in 1938—the earliest time when he would be qualified to take it—and that certainly would restore, even enhance, his chances of getting a job.

This matter settled, he left Cambridge on 29 November to sail on 30 November from Liverpool to the "New World." He spent the night at Liverpool with his friends the Feathers. The voyage on the 27,000 ton ship, the Cunard *White Star Britannica,* began in a stormy way. The seasickness he experienced during the first forty hours was worse than that which he had experienced when crossing the Indian Ocean on his maiden voyage five years before. "It was a most distressing experience," he wrote after the Atlantic quieted down, still feeling as though he had somehow managed "to come out alive from a grave." But as soon as the journey became tolerably comfortable, he took out a copy of the collected plays of Henrik Ibsen which he had brought aboard. Ever since his mother had translated *A Doll's House* into Tamil, Ibsen had remained prominent on his reading list. In quick succession he read the well-known plays: *Ghosts, An Enemy of the People, A Doll's House, The Master Builder, The Pillars of Society,* and *Hedda Gabler.* Rather disappointed with the melodramatic ending in *The Pillars of Society,* Chandra wrote to his father, "Ibsen is striving for justice, so he describes a world where justice eventually wins. But that is not the real world—the poor are always punished, the idealists are always stoned, and students always caricatured, and Justice never wins. And mountebanks always masquerade as High Priests or should I say the Pillars of Society!" On the other hand, while Ibsen had unmercifully brought about the logical conclusion of Nora leaving her husband in *A Doll's House,* in *Pillars of Society,* Chandra felt Ibsen had allowed his optimism to run away with him.

With such reading the week-long voyage across the Atlantic was a real escape from his usual world of stars and the complexities governing their evolution and destiny. Instead, Chandra plunged into Ibsen's world of ordinary mortals with their sufferings and delights, their ambitions and frustrations.

The New World made its appearance early on Sunday morning, 8 December. Along with other passengers, Chandra rushed to the deck at the first sight of land. By 1 PM that day, he had disembarked at Boston and was pleasantly surprised to find that Shapley himself had come to the pier to receive him. Shapley took him to his digs near the observatory grounds, where he would sleep the next three months. Later that evening, Shapley invited him for tea at his home. Chandra met several astronomers there, some of whom he already knew by correspondence. Among them was Gerard Kuiper, who had just made some observations of white dwarfs which had bearing on Chandra's work on highly collapsed configurations.

The following day, Chandra was shown to his office in the observatory, an entirely new experience for him. In Cambridge, be it the digs of his student days or the rooms in Trinity, he had lived, studied, and worked there. In America it was different. The "digs" was where one slept; the "office" was where one studied. "I do my 'office work,'" Chandra wrote to his father. "Indeed, I feel quite flattered to have my first 'office'!" He also had to go out for all his meals—including breakfast. That was a bit troublesome, but he had to get used to it. Soon he found himself a member of a "bachelor-club," composed of three other astronomers, Drs. Fred Whipple, Jerry Mulders, and Kuiper. They went to eat together and that made it more pleasant, but it also meant "too much shop talk." The observatory atmosphere was extraordinarily friendly and informal, totally different from that of Cambridge. No high and low dining tables, no combination room, no gowns. None of that cultivated aloofness typical of the British. Yet, not unlike other students and scholars from India, Chandra felt a special affinity to England, especially to Cambridge, where he had by then spent five years. "I prefer Cambridge, England [to Cambridge, Massachusetts]," he wrote to his father on 18 December 1935. "There I felt practically at home. Here in spite of the extraordinary friendliness of the people, I feel a stranger."

He learned that he was expected to give eight to ten lectures, at the rate of two lectures a week, on stellar structure. In addition, he would be requested to give one or two public lectures. But all this would begin during the second week of January, after Christmas vacation. Therefore, he had some free time in which to travel. He visited New York, "the city of cities." It was not a particularly pleasant day since it was raining all the time. Still, it was worthwhile just to have even a "glimpse of the staggering sky-scrapers and of the activity and rush in the most modern of the modern cities of the world."

He spent his first Christmas in America with Paul S. Donchian[7] in Hartford, Connecticut. Immediately after Christmas Day, he proceeded to Princeton to attend his first American Astronomical Society meeting.

"We had a fairly interesting meeting," he wrote to his father. "I had some conversations with Russell and got to know a number of American astronomers. On one of the days, Shapley arranged a show—'The follies of 1935'—and my part in it was to give a 'scientific contribution' and I took my 'revenge' on Eddington by telling some funny stories about him which were quite well appreciated." He also had a second chance to see New York. It must have been a better day than his first visit, since he went up the Empire State Building, "the tallest building in the world, 1265 feet high with 102 stories," and enjoyed the "magnificent view of New York and its skyscrapers from the top of the world."

After his return to Harvard, he was busy with his lectures, which were well attended and well appreciated. The winter was more severe than in England. The perpetual snow and the resulting slippery roads left little to do outside since they made impossible the long walks which had been the only relaxation Chandra religiously practiced. His friends urged him to learn ice skating. "I am not very keen," Chandra wrote to his father during January 1936. "I once tried it in Copenhagen and I fell so frequently that I gave more amusement to others than to myself." But there was no problem with keeping himself busy. "There is too much rush," he wrote, "too many things going on. Last week Lawrence—a very eminent nuclear physicist—gave a course of five lectures. Now Dingle—from England—is giving a course of ten Lowell lectures on philosophy and so on." There was very little time to do his own work. His lectures would end by the middle of February, and he looked forward to some traveling in the States before returning to England by the end of March.

It was now certain that he would not be able to return to India before August. Having given up the chance for the mathematics professorship at Lahore, he vied temporarily for a readership at Andhra University in Waltair. He had expected Andhra University to look for a replacement for Chowla if he was selected for the Lahore position. However, Chandra learned from his father that Chowla faced an unexpected threat to his chances of getting the job he wanted so badly. Raman and Max Born, who was a visiting professor at Bangalore, had proposed Nagendra Nath for the professorship. Since Nagendra Nath was a theoretical physicist primarily, he was no match to Chowla in merit for a professorship in mathematics, but given the political situation and the high prestige Raman commanded in India, no one could foretell the outcome. The news about this move by Raman and Born so infuriated Chandra that he wrote to his father, "I am very disappointed to learn that Raman has gone so far as to use his position to jeopardize the chances of truly meritorious persons. . . . Because Nagendra Nath is at Bangalore, he has convinced C. V. R. that he is the modern Ramanujan of India. . . . Hardy and Littlewood who are more personally interested in Indian mathematics than any living Indian rank Chowla the second best (the first being Vijayaraghavan). Of course they don't count. . . . I am so disgusted with the whole situation that my desire to settle finally in India and be of some service to Indian science seems to dwindle day by day [22 January 1936]."

In the meantime, a future for Chandra in America seemed to chart itself without any effort on his part. His eminently successful lectures prompted Shapley to nominate Chandra as a candidate for election to the "Society of Fellows." If elected by the Harvard faculty, of which there was little doubt, he would be a Fellow for three years, starting January 1937. The fellowship would pay $2,700 per year, inclusive of rooms in the college and dinner, and it would allow him to pursue his research in complete freedom. Concurrently, he received an invitation from Otto Struve of the Yerkes Observatory to visit Yerkes and "give some lectures under the auspices of the University of Chicago."

Before Chandra could respond to this welcome invitation, which would make it easier for him to see a bit more of the United States, he received another offer from Struve of a research associateship at Yerkes with a starting salary of $3,000 per year. "It seems to me," Struve wrote, "that your brilliant theoretical work could be made even more valuable to astronomers if it could be combined with practical investigations by the observers. This would require a close cooperation between yourself and the astronomers at the great observatories of America and might lead to new types of investigations of a non-routine character that would be especially valuable in the present state of astrophysics."

For Chandra, this sudden interest in him and his work among American astronomers was surprising. But Struve's arguments in favor of close cooperation between a theorist like him and observational astronomers made a great deal of sense. Chandra had experienced firsthand the value of such interaction at Harvard. Due to his discussions with Kuiper, whose observational work closely followed and preceded his theoretical studies, he had gained considerable new insights into astrophysical problems, which he had not had before. Further, Kuiper was going to join the University of Chicago, and so was Strömgren, his good friend from

Copenhagen days. Struve was doing something new in America. He was bringing young theorists and observational astronomers together in a place given predominantly to observational work. Thus, apart from choosing between the Harvard Society of Fellows and a research associateship at Yerkes, Chandra felt his immediate future had been virtually settled for him. He decided that he would work in America for at least three years.

This unexpected turn of events, he knew, would certainly not be welcomed by his family. In anticipation of at least some resistance from his father, Chandra wrote to him on 27 February 1936, "I have learned to realize more and more that scholarship demands its own sacrifice and the intellectual interests outgrow—rather, necessitate the sacrifice of— other considerations." However, that did not mean, he explained, that he was selfish and only concerned with the furtherance of his own ambitions. A life of scholarship demanded personal sacrifices as well. "Without attempting to justify myself," he said in the same letter, "I may state simply that the reason why D. L. [Lalitha] and I agreed to break our 'understanding' was that the extra-intellectual considerations like love [and] marriage . . . put a restriction on the mutual freedom required for my studies—others may be different from myself in this respect, but I cannot help being myself." Further on, looking at his future from another angle, he wrote, "I feel under the circumstances of intrigue and underhand dealings going on in Indian scientific circles, it may after all be best that I spend some time in America. But I have decided to come home, at least for two months, before I come to America again."

Letters from America took three weeks to reach India. Therefore, he had a considerable interval of time before he would receive his father's response. In the meantime he planned to leave Boston on 7 March for Chicago and Yerkes Observatory at Williams Bay, Wisconsin, with a brief sight-seeing side trip to Niagara Falls on the way. He booked his return passage to England on the *Berengaria,* a ship twice as big as *Britannica,* sailing from New York on 18 March. This gave him a week or so at Yerkes and also a few days in Washington, D.C., for sight-seeing. The week prior to his leaving Harvard was busy with a "host of parties" to celebrate his stay there. He had made many new friends despite the short duration of his visit. It was not England; it was America, and American hospitality lent itself to quick friendships. Some friends even came to the station to bid him farewell, which would have been unusual in England.

Chandra was no different from any other tourist when it came to

Niagara Falls, which he saw once during the day and once at night when it was illuminated by floodlights. The first impression of slight disappointment turned into one of awe for one of "Nature's most stupendous extravagances," as he described it in his letter from Yerkes. "The last lingering, almost plaintive look with which I tore myself away from the Falls created in me a sense of void which only nature can produce in her most magnificently malignant moods—what else, other than Malice can produce that awful roar, that continuous streaming flood of staggering pillars, all in their strength only to break into a million droplets and in its foam, create the fathomless castle of rainbow colors as the sun sternly shines on?" Nothing he had seen in nature or of monumental human dimensions until then equaled the intensity of feeling evoked by the falls. The only other impression that came close was that of Red Square in Moscow at midnight with "the clock chiming the International, the whole black tomb of Lenin bathed in gorgeous red, and silky white of the Kremlin in the background contrasting the wild contortions of the Byzantine church built by Ivan the Terrible." But that intensity was of an altogether different kind, created at least in part by his political leanings at the time. "The difference is only one of kind," he wrote. "Nature frightens what Man emboldens, and while She destroys with malice, He creates with delight. . . ."

In Williams Bay, Chandra stayed with Otto Struve. The observatory and its grounds, with beautiful lawns in front and wild woods in the back touching the shores of Lake Geneva, which was frozen white at the time, made a good impression on Chandra. Yerkes was one of the great observatories of the world, enhanced by the international character of its staff. Struve and the others received him cordially. Chandra's lectures went quite well, and he had a very enjoyable time.

When Chandra left Williams Bay for Washington, D.C., on 13 March he was strongly inclined towards accepting Struve's offer. After a weekend of sight-seeing in the capital, two days in Princeton with Russell, and a day in New York, his first visit to America was at an end. He was back in England on 25 March.

As expected, Chandra's father had serious reservations about Chandra's accepting either of the two offers from America. Distance from India, higher cost of living, and racial discrimination aside, America was perceived as shallow and inferior when it came to academic standards as compared to Britain. "I know, as a matter of fact," wrote Chandra's father on 20 March 1936, "that American text books are much *shallower* than the British books on the same subject." He warned Chandra to beware of American salesmanship, advised him to consult with Eddington and Milne, and wondered whether going to America jeopardized Chandra's chances of getting a Cambridge Sc.D. degree and of becoming a Fellow of the Royal Society. "It is these British titles which may get you a decent job in India," he wrote to Chandra, "since I believe you will after *all,* return to *India.*"

But, Chandra, during his return voyage, had already decided to accept Struve's offer of the research associateship at Yerkes.[8] He did, however, consult with Eddington and Milne. Both advised him to choose Yerkes over Harvard. "Your astronomical reputation and work would gain very much," Eddington said, "by being attached to one of the great observatories of the world." Chandra too was convinced that at Yerkes he would be able to "contribute more solidly" than he had done previously. "I think that is sufficient ground," he wrote to his father in reply on 1 April. "F.R.S. etc., are completely trivial. I may get it sometime within the next five or ten years; the chances are not affected by my going to America. In any case I am quite indifferent to it." Besides, America had been first to recognize his scientific potential sufficiently to consider him "worthy of an annual salary with a definitely senior position in a university." He felt, paradoxically, that in India not a soul was aware of his existence. In England, although Fowler, Eddington, Milne, and Plaskett treated him well and had such high regard for his merits, there was still some prejudice against giving Indians a permanent position. In any case, none of them came forth with an alternative offer. Chandra sent his acceptance letter to Struve on 22 April. He would join Yerkes on 1 January 1937 as a research associate. The appointment for one year in the first instance would become a regular faculty position later on. Struve expected Chandra to stay at least five years at Yerkes.

Now that his immediate future regarding his scientific career was settled, Chandra began to worry about home, his short visit to India, and preparations to go to America. It was a time of great stress in the family. Chandra's father was trying to find suitable husbands for his daughters. Chandra's brother Vishwam was staying in London to complete his degree in metallurgy. That meant a strain on the family income. Chandra's other brother, Balakrishnan, was studying medicine, but that vocational decision was made under the pressure of their father. Balakrishnan still wanted to be a writer and philosopher. After four years of hard work and study, he was on the verge of quitting medicine altogether. He needed Chandra's attention, advice, and encouragement.

It was all extremely depressing to Chandra—"the so-called educated men demanding 25,000 rupees as bridegroom price, and otherwise sensible men like Ramakrishnan sticking to minor subcaste partitions." Chandra felt he was running away from it all and not facing his responsibilities. Perhaps he had been too selfish in accepting Struve's offer.

Yet, he was fully convinced that under the prevailing circumstances his scientific life in India would be one of continual torture. The rivalries between Saha and Raman had divided the country's scientists into two camps. Provincialism and favoritism were rampant. A committee,[9] appointed to look into the matters of the Indian Institute of Science at Bangalore of which Raman was the head, had criticized Raman totally unfairly, which eventually led him to resign and start his own institution. It was all exasperating for Chandra. Chandra had met Saha briefly when he was in England in June 1936. "I could not help feeling his scheming, sneering attitude—an attitude which desires to destroy all things productive—an attitude which blurts out when in inviting me to come back to India he asks me to 'join in' the politics [5 June 1936]." In spite of all this, he felt that if only he had returned he would have been happier. "I have been unfair to you and all," he wrote to his father. "Things could be simpler, but there needs to be change—social conditions in India *must* change. It cannot go on eternally as it is—bridegrooms demanding 24,000 Rs."

As for his own marriage, Chandra's father had not given up. "You think seriously of marrying when you come here," he wrote. "Get in touch with Miss L. I may tell you by your not marrying or delaying it till 1941, the younger sons will find it difficult to get *married*—since parents of girls will not consider marrying them to younger sons while their elder sons remained unmarried. Some decent girls were offering, when I suggested to the parents Balakrishnan as a possible bridegroom, they refused to think of the proposal." Chandra waivered a bit on feeling this renewed pressure. He was considered not only an impediment to his sisters' marriages, but also to his younger brothers'! He would think about it, although it seemed so impossible and impractical to get married in two months and transport the bride and himself to America in the dead of winter to a life that was so altogether new to both. Williams Bay would be as different again as Cambridge had been when he first went there from India. Besides, Lalitha "was a dream," he wrote, "and I dare say that I imagine her quite differently from what she probably is now, six years is a long time. Even those whom one had known well change, and those whom one had only begun to get to know . . ."

He would have to wait and see. He booked passage on the *Conte Verde,* sailing from Genoa on 31 July and return passage on the *Conte Rosso,* leaving Bombay on 13 October.

8

Lalitha
Madras, 1936

Children, it is not enough to love much; you must love well. Great love is
good, undoubtedly; Wise love is better. . . .

From Anatole France, *"Bee," the Princess of the Dwarfs*

The journey home on the *Conte Verde* was uneventful. The seas were
relatively calm, and Chandra experienced none of the violent seasickness he was subjected to on his journey to England six years before.
"I recall very little," Chandra says, "except one or two minor things.
I sailed from Genoa, having taken the train through France, through the
Alps, as I had done on my journey to England. I remember to have mis-
calculated the time interval and had to rush in a taxi from one station to
another in Paris to make the connection. Then, I remember not shaving,
letting the beard grow on the eighteen-day journey. And when I shaved
the day before we landed in Bombay, some of the returning Indian students whom I had befriended said that I looked more handsome with
beard on than without it."

Chandra's father, his sister Rajam, and her children were at the pier
in Bombay harbor to welcome him. After a few days stay in Bombay,
Chandra headed alone to Madras to see the rest of his family. But before
he left Bombay, he wrote a letter to Lalitha; he was coming to Madras,
and he would like to see her and talk things over. Lalitha was in Bangalore
working in Raman's laboratory at the time. A year had elapsed since
Chandra's previous letter to her when she was in Karaikkudi, but she
had known about Chandra's arrival in India from a brief newspaper account which included Chandra's picture. While she was wondering
whether she would hear from him, the letter arrived bringing new expectations and joy. "I could not believe this was happening," Lalitha recalls. "I wrote to him immediately to say that I would meet him in
Madras soon and took the next train to Madras."[1]

Chandra Vilas had changed during the six years of Chandra's absence. A new extension, Parvati Bhavan, had made its appearance. "The whole place is now grown up," wrote Chandra to his father on 27 August 1936. "The coconut trees, the mango trees, all green and cool give the place an almost idyllic appearance!" Along with Chandra Vilas and its surrounding environment, Chandra's younger sisters had changed. "All the girls, Sarada, Vidya, and Savitri," he continued, "have developed into quite individual characters; Vidya passionately fond of her music and looking very calm has really a great deal of fire inside her; Savitri—to spell her name without an "h" as she does!—all delicate like tinsel frame with an extremely pleasant laughter—like the tinkling of silver bells; and Sarada, a little too serious for her age. Balakrishnan, I find, is in a very critical period of his life. He will of course complete his medical course, but right now he is consumed by his 'artistic' pursuit. It is best that he is left *entirely* free."*

It was good to be home, in Madras. "It was still very much like when I had left it," recalls Chandra. "There were so many friends remaining. I went to Presidency College and saw my former teachers, Appa Rao and Parameswaran. I also visited, I remember, Queen Mary's College; Savitri was a student there at the time and my former classmates, Ammani Ammal and Rajesvari, were lecturers. I quickly got back into the spirit and even began to feel very disappointed that I couldn't get a position in India."

As "Foreign Returned," a phrase often used in those days, Chandra was a celebrity of sorts; he was accosted for talks and speeches by all sorts of organizations. Chandra managed to get out of most of such requests. He did give a talk at Queen Mary's College and before the Physics Association of Presidency College. He also made a trip to Calcutta to meet his friend K. S. Krishnan and managed to make him listen to a talk concerning his work rather than on the grand personalities in Cambridge, England, and Cambridge, America.

Chandra's earlier decision to postpone his marriage indefinitely wilted away rather suddenly when he saw Lalitha again after six years. She was more than a dream, she was quite real. There was not even the slightest uncertainty regarding their mutual feelings. If they were ever to marry, it would be to each other and to nobody else. Lalitha shared Chandra's dedication to science. He became convinced that she would be a help

---

*Balakrishnan went on to combine a distinguished medical career with that of a successful writer.

rather than a hindrance to his single-minded pursuit. His main concern thus quickly laid to rest by their discussions, Chandra sent a telegram to his father in Bombay announcing their decision to marry. Because he had only two months in India, he asked his father to write to Lalitha's grandfather, seeking his formal approval. All the customary formalities were to be side-stepped. There was no need, for instance, to consult an astrologer about the compatibility of their horoscopes. "Lalitha's marriage to me," he wrote to his father, "is independent of the relative 'arrangements' of our two horoscopes." What was more important was that he needed to borrow from his father the passage money for Lalitha. He had already changed his reservation on the *Conte Rosso* from a single berth to a double-berth cabin.

After deciding to get married, Lalitha and Chandra began to discuss their future. They walked along the beach as Chandra articulated his concerns about the difficulties they would have to face. He saw little hope in ever securing a position in Madras. The Madras Government, while it had obligated him to return and serve for at least five years, did little to create a suitable position. He had the equivalent of $1,000 saved; not enough to pay for an additional passage and the start of a new life. "Chandra was overly concerned with money matters," Lalitha recalls. "I was not. I was, of course, concerned with going so far away from home, leaving behind my mother, grandparents, sisters, aunts, and uncles. I did not know when we would return to see them all again. Although the responsibilities of setting up a home in a totally different culture did not alarm me, it meant I had to give up, to some extent, any aspirations I had about myself. However, I did look forward with excitement to the opportunity of going abroad. I knew a little bit of Chandra's life, his friends, and the eminent scientists he came in contact with. That was all very exciting."

The wedding took place on 11 September 1936. There was no time for the customary elaborate four-day marriage ceremony. Besides it was a long-standing decision of his father that his children would not have that kind of conventional wedding, which involved considerable expense on the part of the bride's family. The dowry system and the expenses of his daughters' marriages had turned him against such conventional extravaganzas. A one-day ceremony, fulfilling all the essential precepts of Hindu rites,[2] was performed at the Tirupathy Temple at Tiruchanur near Madras. Lalitha's family took care of all the necessary arrangements for the ceremony, attended by the couple's immediate family members.

A general reception for all the relatives and family friends was given the following day at Lalitha's family's house, Sri Vilas, in Madras.

A marriage like that of Lalitha and Chandra, based on mutual choice, colloquially termed a "love marriage," was, and still is, extremely rare in India. The custom of arranged marriages still predominates, along with the practice of asking astrologers to determine whether the horoscopes of the prospective bride and groom match properly. Of course if the right family connections or a good piece of land or a good dowry is in the picture, there are ways to modify what is ordained by the heavens. Seeking a second opinion with a good, hand-warming *daksina* (bribe) from a second astrologer always helps. If not, there are many well-prescribed ways (gifts to be given to the poor, priests to be fed, help sought from a particular God at a particular temple, etc.) to cure the bad influences and assure a long, happy married life. After "horoscopic" considerations come caste, subcaste, sub-subcaste restrictions, family situations, and the past and present histories of the families. Marriage is a serious business; it requires the blessings not only of all the living, but also of all the dead in the family.

That Lalitha and Chandra did not have to worry about such matters says a great deal for their exceptional family backgrounds. We have seen how Chandra's family, beginning with his grandfather, was transformed from a village-based, land-cultivating, and religion-centered family to one that is urban educated, westernized, and government service oriented. That alone, however, was not sufficient reason for such a radical departure from the established ways, since a majority of the families with a similar history followed the well-trodden path of tradition when it came to family and social arrangements. Chandra's family had gone a step beyond. Raman, Chandra's uncle, had set a precedent by choosing his own bride at the age of eighteen. Therefore, in Chandra's family there was no objection to an individual (a male individual at least) choosing a suitable partner, provided it did not interfere with the marriages of others. However, caste was still a factor. Fortunately for Lalitha and Chandra, they both belonged to the same brahman caste, posing no major threat of opposition within the families.

Likewise, Lalitha came from a family in which education took precedence over other matters. She was the third daughter in a family of four daughters. Lalitha's mother herself was the third daughter in a family of five daughters. Three of Lalitha's aunts were teachers. Her two older sisters studied medicine and became doctors; her younger sister took her master's degree in Sanskrit. Lalitha had acquired hers in physics. Remember that it was customary, even mandatory, for parents in those days to marry off their daughters long before they attained puberty. The education of women consequently was very limited and offered only to the fortunate few who could afford it. Even then they would likely attend just a few grades in elementary school. How is it then that in Lalitha's family all the women pursued higher education and careers in public service? What triggered this change? Who was responsible? It is a story worth telling.

Lalitha's grandparents were Visalaksi and Subramania Iyer. Mr. Iyer was trained in civil engineering and, like Chandra's grandfather, was among the first few in his family to depart from the traditional brahmanic village life of his parents and ancestors. They came from the same general area as Chandra's ancestors, the Tanjore district of Madras Presidency. After a few years of service as a civil engineer in the Madras Government, Mr. Iyer, finding the physical exertion more than he could bear, began to teach engineering in an agricultural college in Saidapet near Madras. Along with his wife and five daughters, Mr. Iyer's household included the children's grandmother and an aunt, whom everyone called Chithy.

Subbalakshmi was their oldest daughter, born on 18 August 1886. She had a normal, happy childhood in a loving family with a grandmother whom everybody obeyed and respected, and whose fascinating stories from the great epics were a delight. Her mother and Chithy were quite learned in Tamil and Sanskrit, and they made sure the children were familiar with lots of poetry, devotional songs, and music. Subbalakshmi herself was extremely good at school, and after she finished her fourth grade of elementary school at the age of nine, she competed in the public examination of the Madras Presidency; she ranked first in her district. If she had been a boy, she would have continued her education in an English school with a scholarship. But since she was a girl, according to the dictates of the time, her education had to be discontinued and a husband had to be found for her. She had to be prepared to be a daughter in another family. In due course, Subbalakshmi's parents found a suitable match and she was married at the age of eleven with all the pomp and rituals of a customary Hindu wedding. The astrologer who had studied the horoscopes had predicted a long and happy life for her. As she went around the fire seven times holding the hand of the boy to whom she was being married, she behaved as she was supposed to—shy, keeping her eyes down, and managing not to even glance at the boy. She was delighted with the fabulous new Banaras silk sari she had received for the occasion. She dreamt of wearing it, as her mother wore her own wedding sari, on many special occasions. The ceremony over, she returned home. Only after a later ceremony when she reached puberty would she go to the home of her husband.

Just a few months had passed when the family received word that Subbalakshmi's boy-husband had died. Shock was felt all around the house. Sympathizers, family friends, and relatives streamed in. With every new visitor a new wave of tears and sobbings erupted from her mother and Chithy. When Subbalakshmi was told the news, it did not make much of an impression on her. Why were they all crying and weeping so much? They did not know the boy very well. She had never even seen his face.

She had often wondered why Chithy dressed so differently than her mother. Chithy always wore a coarse, white cotton sari, like her grandmother. But she was not old like her grandmother; she was young and energetic like her mother, who garbed herself in gorgeous silk saris with her long thick hair combed and tied into a big bun behind her head. Chithy's sari went tightly over her head, and it was evident that she had no hair underneath. She wore no blouse, and her bare back showed at times when she was working in the kitchen. When everyone else feasted, she generally ate only simple food. All this did not make sense to Subbalakshmi, but when, after wondering about it for a long time, she had asked her mother why Chithy acted so strangely, she had encountered only a flood of tears. Another time, when she was visiting relatives in the village, she had witnessed a little girl, two years old perhaps, being teased and harassed by some older girls. They were calling after her, "*Widow! Widow!*" and trying to snatch the necklace with a *thali* around her neck saying she had no right to wear a *thali*. She had no idea why they were saying such things to a small child.

Subbalakshmi was too young, too innocent to realize the full consequences of the tragedy that had been wrought upon her. If tradition prevailed, she would be denied remarriage and motherhood and would remain, like her aunt Chithy, a virgin widow for the rest of her life. When she came of age, her head would be clean-shaven; she would wear no more silks, skirts, saris, and blouses; no perfumes, no adornments whatsoever were allowed. Wearing a coarse, white cotton sari tightly wound over her head, living on only the plainest food, she would have to make herself as invisible as possible, confined to a relative's kitchen or backyard lest her budding youth and beauty tempt some passing male. Her lovely wedding sari would be snipped into pieces and made into skirts for her sisters. She would be barred from all auspicious occasions and ceremonies. People on important business would avoid her, because seeing a widow would bring bad luck.[3] She would live the rest of her life as a house servant, at the mercy of others for two cotton saris a year and two meals a day.

But Subbalakshmi was fortunate. Her parents, deeply religious though they were, could not bring themselves to inflict such hellish punishment on their twelve-year-old, vivacious and charming, obedient and intelligent daughter. When she reached puberty, instead of bringing in the barber, Mr. Iyer appeared before Subbalakshmi and told her, "You will begin your education again. You will learn English." Those were sweet words to someone who had loved school and had regretted that her education had been stopped. Her father began the lessons the next morning from an old English primer and an English dictionary. Before long Subbalakshmi had made up the lost time in all the subjects and was ready to join the public secondary training school in Madras.

From then on, no force in the world was strong enough to stop Subbalakshmi from continuing her education, from secondary school to high school, from high school to a preparatory convent school for her college education. She graduated in 1911 with a B.A. degree, which brought her national fame. Newspapers all over India reported the success of the young brahman widow in Madras who had won first-class honors, outshining all the men of her year.* Every step of the way she had faced obstacles, of course, but her parents and Chithy were always there to help. It required immense courage and sacrifices on their part to stand up against orthodoxy and tradition and to bear the financial burdens of her education.

It was not long before Subbalakshmi became aware that she was not alone in her tragic situation. Child widows abounded throughout India. There were countless widows in their full youth; there were also pre-teenage and teenage wives with irreparable damage done to their minds and bodies by older, but immature husbands; girls mauled by men three or four times their age and then abandoned. Subbalakshmi, who became Sister Subbalakshmi in the convent school, decided to dedicate her life to the liberation of such unfortunate women through education. She received a teacher's training degree from Presidency College and then went about tirelessly working for her avowed cause.

* Most of the information on Subbalakshmi's life given in this chapter is from Monica Felton's *A Child Widow's Story* (New York: Harcourt, Brace, and World, Inc., 1967).

She was fortunate in having strong support from "Christina Lynch, an Irishwoman—tall, blonde, formidably handsome—an ardent feminist who was passionately in love with India."[4] Miss Lynch was an Inspectress of Women's Education in the district, and during the course of her government service had realized how inadequate the educational opportunities were for Indian women and girls. Miss Lynch and Sister Subbalakshmi worked together as pioneers in the field of women's education in South India. To begin with, Subbalakshmi's house became a Widows' Home. Chithy cooked, cleaned, and became the housemother. As more and more unfortunate ones came to the door, they moved to larger and larger places, till they settled in the so-called Ice House,[5] a palatial site across from the Marina Beach in Madras. They then pressured the government into building a new school (Lady Willingdon's School and Teacher's Training College) adjacent to the Ice House and a new college for women (Queen Mary's College). In a matter of a few years, their efforts produced a cadre of women teachers. Unfortunate widows, who otherwise would have slaved miserably without honor the rest of their lives, became independent through their education and led lives that were examples to other women. Some became doctors. Sister Subbalakshmi became a household name in South India. She represented South India in the first All-India Women's Conference in Poona in 1927. She testified before the so-called Age of Consent Committee, appointed by the government in 1928 to investigate possible legislative measures to abolish child marriages altogether and to set a minimum age for women before they can be married. Through the Sarada Ladies' Union and other women's associations, Subbalakshmi championed the cause of women's education long before Mahatma Gandhi made it a national goal of the highest priority.

Needless to say, all this effort had a profound effect on her own family. Her sisters, although they married young, continued their own education as they reared their families. For the next generation, which included Lalitha, early marriages at the age of ten or twelve were out of the question. Completion of high school and, preferably, college before marriage became the normal, expected behavior. Marriage was no longer the supreme goal for the women in Lalitha's family. Subbalakshmi's tragedy and her parents' courage had brought about a revolutionary social transformation in one generation.

Lalitha grew up in this transformed atmosphere in her grandparents' home, Ananda Vilas, in Triplicane, Madras. Lalitha's father, Doraiswamy, was a medical officer in the Indian Army. The First World War had taken him away from home when Lalitha was three years old, therefore she has only a hazy and distant memory of her father during the war years; her grandfather, however, provided all the love and affection she needed. Lalitha's mother took the opportunity of her husband's absence to continue her own education. After the war, he was promoted to the rank of captain and posted in Bangalore. The family was happily together again.

One day about a year later, however, Captain Doraiswamy came home with a very bad toothache. It soon spread to his other teeth and, upon examination, it was found that his entire lymphatic system had collapsed. This was attributed to the shell shock he suffered at the war front in Mesopotamia. The ailment developed rapidly and brought his life to an end rather suddenly when Lalitha was only ten years old. Lalitha's mother, with a new life inside her (Lalitha's younger sister Radha was born soon after her father's death), returned to her parents' home with her three daughters. Under the circumstances, it was the ideal place for her children to grow up. They would get a good education in Sister Subbalakshmi's school and they would be surrounded by the affection and influence of their grandparents.

Three generations lived together in their large household. Lalitha and her three sisters, their six cousins, and their youngest aunt belonged to the newest generation. Lalitha's mother and her sisters formed the tier above. Then, there were the grandparents. As Lalitha has written,[6]

Ammamma [grandmother] was the pivot and around her the family acquired a sense of unity. She had roots deep in Hindu culture and tradition and introduced us to the beauties of the *Ramayana* and the *Mahabharata* through the stories she told us and the songs she taught us. . . . She offered her daily prayers to Goddess Lalithambal after whom I was named. She also had a Shiva *lingam* that her father had given her. This she placed on a brass plate and when her prayers were over the *lingam* was surrounded by circles of flowers in white and red and yellow that came out of our yard. . . . In this home Thatha's [grandfather's] influence was like a gentle breeze which blew in to give us our good health through the daily walks he insisted we take to the beach. If Ammamma was the center of the family, then he was the circumference, which not only protected the family, but through which we emerged to see the world at large every summer during the holidays; for, it was he who planned these trips to different parts of South India, and always to a place where there was a river, fresh air, and plenty of opportunities for long walks and places to see. The whole family, all three generations of it, would set out on our annual expedition, with our suitcases, bedding and cooking equipment (I have counted as many as sixty items on the platform before we boarded the train for our place of destination). Thatha also gathered all the children around him every night after dinner in the front verandah of the house, and entertained us with jokes and puzzles. He would tell us about his childhood and growing up in the village of Rishiyur. He would teach us the names of the constellations. In this way, he was not only the grandfather, but a father, a brother, and a playful companion. He filled the void created by the loss we had experienced at Bangalore.

Along with the grandparents and Sister Subbalakshmi, Lalitha's mother played an important role in the family. Chinnakka (Little Sister), as she used to be called by all the younger ones in the family, including her own daughters,[7] was shy and reserved, a model of patience and fortitude. Putting aside her personal tragedy, she cared for all the children, attended to their education, and imbued in them what she considered as the best in Hindu tradition and culture. "With Ammamma, she observed all our religious rites," writes Lalitha. "[She] learned and sang the Tamil and Sanskrit songs and *slokas* (verses), read widely and deeply into the Tamil (and to some extent into the Sanskrit) literature of our poets and philosophers, and also observed the many charming festivals of our Hindu culture. She was a great cook and her specialities were the unusual dishes she prepared on such occasions." She also managed to learn the Persian script and the Urdu language on her own. When she discovered that she had talent in painting, she took lessons at the Fine Arts School and began to paint seriously in her spare time. She also collected examples of the immense variety of *Kolam* (designs) that women in South India draw every morning outside the front entrances of their homes with powders made from chalk, lime, or rice. Because her husband's death was caused by the war, she was left with a small pension and an allowance for her children from the Government of India. She used these funds well to educate her daughters, two of them in medical school and Lalitha in college. With frugal spending, she managed to save enough to buy a larger house in Mylapore for the entire family. This new house, with a beautiful garden of flowers and fruit trees, was adjacent to Chandra Vilas. Lalitha's mother named her new home Sri Vilas, and the entire family moved there when Lalitha was in college.

Thus Lalitha, like Chandra, inherited an exceptional family background. "We were both scions of families that had acquired national attention," says Lalitha. "Chandra's family due to his uncle C. V. Raman, and mine due to my aunt." While advanced degrees in western education were the dominant aspiration for everyone in the family, it did not mean a total departure from the traditions and ideals of the past. On the contrary, it meant a new awareness, a new enlightened commitment, to what was considered correct and right in one's values. Lalitha, like

Chandra, was an avowed vegetarian; other than that, she was prepared to be "modern," to live in the West. She was ready to be the ideal wife for Chandra.

Lalitha and Chandra met when they were students in the physics honors course. As a girl of seventeen, Lalitha was the only "lady student" designated to sit in the front row. She had noticed with interest the young man with a crew cut, always sitting behind her in the second row. "Chandra was brilliant in the way he chose the subject and presented it in a colloquium in 1928," Lalitha recalls. "He knew his subject, was completely involved in his thoughts, and also showed a measure of excitement which reached across to all who listened to him." Once Lalitha asked Chandra whether she could see his laboratory record book, and he readily agreed. The record book went back and forth between them through the intermediary of the laboratory attendant. At an annual party where a group photograph of the physics honors students was taken, Chandra gave Lalitha a large sweet-smelling rose to pin on her sari. And one day, seeing Lalitha take her bicycle out to ride back home from college, Chandra followed her all the way on his bicycle. He persuaded her to enter an essay competition that their teacher Appa Rao had set up. Finally, just a few days before he left for England, he came to her house with a list of books that she had asked for. Alone in Lalitha's study, awkward silence engulfed both of them, each hoping the other would break the ice and say something. Someone in the family saved the situation by bringing in tea and refreshments.

Later, from Cambridge, Chandra wrote to Lalitha regularly during his first year. While Chandra was struggling on in Cambridge and continuing his research career, Lalitha, who had similar aspirations, was not so fortunate. After her graduation in 1931, she too wanted to continue her research in physics, but she was given little or no guidance in Presidency College. She wanted to go to England and study in Cambridge, but her mother said no. Progressive as she was, she was not yet ready to send her unmarried daughter alone to Cambridge to live so close to her possible future husband in total freedom. "Get married to Chandra and then go," she told Lalitha. As that was not possible under the circumstances, Lalitha had to remain behind. She taught for a year in a high school in Madras and then for a year in Lady Harding's Medical College in Delhi. She next became the headmistress of a middle school at Karaikkudi for some time, and then returned first to Presidency College and later to the Indian Institute of Science in Bangalore, hoping to take up physics research. Neither Parameswaran in Madras nor C. V.

Raman in Bangalore helped her find a meaningful research problem. She had finally settled down to study on her own when Chandra's letter arrived. During the six years while Chandra was away, she had spared him of her frustrations and waited patiently for him to decide their future. Now they were happily united.

After the wedding, the couple spent a few pleasant days in Bangalore, enjoying the climate and sights of the beautiful city. En route to Bombay, they visited Chandra's sister in Nagpore and then sailed from Bombay Harbor on 13 October for England. Only two months at home after six years abroad was too short a stay. Chandra was sad to be leaving home again for an uncertain future.

"I am afraid I could not speak much to either of you during the few days of your stay in Bombay," Chandra's father wrote to them soon after their departure:

I was too suffused with emotion that my voice refused to speak and got choked. I wanted to give you some advice, but am unable to express it in better terms than in the words of Anatole France: "Then King Loc spoke again in these terms: Children, it is not enough to love much; you must love well. Great love is good, undoubtedly; Wise love is better. May yours be as mild as it is strong; may it want nothing, not even indulgence, and may some pity be mingled with it. You are young, beautiful and good; but you are human and, for that reason subject to many miseries. This is why if some pity does not form part of the feelings you have for each other, these feelings can not be adapted to the circumstances of your common life; they will be like holiday clothes, which are no protection against the wind and the rain. You only love those securely whom you love even in their weaknesses and meannesses. Mercy, forgiveness, consolation, that is love and all its science."[8]

Chandra also felt that his stay in Bombay was too rushed. He did not have time to say all that he wanted to say to his father. He had sensed his father's unhappiness and loneliness; he was staying alone in Bombay because of his job, while the rest of the family was in Madras. His responsibilities were heavy. "I feel almost that you are bearing the 'cross' now," Chandra wrote from aboard the *Conte Rosso* on 23 October. "I feel only too much, that I am neglecting my duty towards my home and am selfish with regard to my work. But, as I said, I shall not be longer in America than I really need be—not more than three or four years. By the time you retire, I shall have returned too, and we can all have a happy time together in Madras."

This time the seas were calm. Lalitha, who had traveled some years before to Rangoon by sea and had experienced rough weather, was happy that this journey was smooth. They arrived in Cambridge on

30 October. David Shoenberg provided accommodations for the couple in the garret above his rooms. Trinity rooms were not for married couples; however, Chandra was allowed to keep his rooms for study and work till the end of the term. The attic room had no cooking facilities except for a hot plate to make tea, soups, and one-dish meals. A small restaurant called Whim became their regular eating place. There Lalitha was introduced to Welsh rarebit for the first time (it has remained one of the few favorite vegetarian dishes she can count on when eating out in restaurants). "Cambridge was no surprise to me," Lalitha recalls. "I knew the kind of life Chandra was leading from his letters. I was look-ing forward to experiencing it: seeing Trinity, walking in the backs along the Cam, meeting his friends. I liked Shoenberg. I thought he was charming."

There was only a month to spare in Cambridge. Chandra had booked passage on the *Laconia,* sailing from Liverpool for Boston on 5 De-cember. The days became crowded with packing, travel formalities, and invitations for tea and lunches. "During the past week," Chandra wrote to his father on 20 November, "L and I had tea with the Eddingtons, Redman, Shoenberg, and lunches with the Smarts, Beers, and Suther-lands." But amidst all the pleasantries of farewells, Chandra had some anxious days fraught with the difficulties of getting visas to America for Lalitha and himself.

Those were the days when there were no student exchange visitors, and no F-, J-, or H-type visas for an extended stay. There was only an immigration visa. When Chandra had returned to England from Amer-ica and received the appointment letter from Yerkes that spring, he had contacted the American consul in London and appeared before a man "who was really very rude," Chandra recalls. "He was writing some-thing; I stood there quietly without his greeting me or looking up at me, or even asking me to sit down. When he finally looked up, he asked me,

'What do you want?'

'I have an appointment at the University of Chicago. I would like to get a visa,' I said.

'What kind of passport do you have?'

'A British passport given to an Indian,' I replied.

'You cannot get a visa.'

'You mean I cannot accept this position?' I asked.

'I do not see how you can accept it because you cannot get a visa to go to the U. S. Indians have no quota. They cannot immigrate to the U. S.'

That was that."

Chandra then wrote to Struve, who contacted the legal counsel at the University of Chicago. It turned out that there was a section of the Immigration Act, 4D, under which permanent resident visas could be granted to missionaries from countries which had no quota. For the purposes of the law, the legal counsel said, a person could be considered a missionary if he had taught in a school, college, or university. When Chandra went back to the American consul with this information and a letter from Sir Arthur Eddington attesting to the fact that Chandra had taught in Cambridge, the consul had agreed to grant him the visa, but Chandra had to wait to file the application, which could only be done two months before his departure date.

Now that Chandra had returned from India married, Lalitha needed a visa too. This created a new series of problems. The record of her birthdate in her passport was not acceptable; a birth certificate was needed. Independent attestation from the principal of Presidency College from the registration records in the college had to be obtained. A marriage certificate had to be presented, although no such certificate existed since Lalitha and Chandra were married according to Hindu rites in a nongovernmental ceremony. "The only way we can resolve this problem is for her to get a character certificate from someone responsible in this country who knows her personally," the consul declared. Fortunately, Sarvapalli Radhakrishnan, a professor at Oxford, was also a family friend of both Chandra and Lalitha. He wrote a good character-certificate letter for Lalitha, and his signature was accompanied by his full title. That solved the immigration problem.

The *Laconia,* scheduled to leave Liverpool on 5 December, stayed in the harbor for two days with all the passengers aboard. "The boat was overloaded with whiskey and the tide was low," recalls Lalitha. "They had to unload cartons of whiskey before the boat could move, then load them again. In the meantime, we were treated to all sorts of goodies." After a comfortable Atlantic crossing, they landed in Boston on 16 December. Whipple was at the harbor to meet them. They spent a few days in Boston and Cambridge before heading for Williams Bay. Lalitha met many of the friends Chandra had made during his previous year's stay. "It has been a race with time," Chandra wrote his father on 29 December from Williams Bay. "A few days at Boston and Harvard packed with meeting many of my last year's friends and discussions and parties. Then to Chicago and finally we came to Williams Bay on Monday 21 December."

The unfurnished house on the observatory grounds which Chandra and Lalitha were supposed to occupy was not ready. While the cleaning went on, they were the guests of the Kuipers for a few days. "It was only last Saturday that we have moved into 'our' house," wrote Chandra in his first letter from Williams Bay (29 December) to his father. "The house is far from 'ready' yet, but we have all the essential furniture—one table, two chairs and a studio-couch (which can be converted into a double bed), and the china and the cooking utensils."

The young newlyweds from Madras were told that the prevailing near-zero-degree temperatures were mild compared with what was in store for them in January: twenty below zero! "But we are not afraid," said Chandra in his letter. They quietly set themselves to the "hundred odd things" that had to be done in furnishing the house and settling down. "We have an oil furnace in the house which keeps us warm," wrote Chandra on 14 January 1937, "but with only two upright chairs and a table which has to serve all purposes it has been pretty difficult. Further I have started my lectures here—I began my 'course' yesterday—I am giving a course of lectures on Thermodynamics. L comes to lectures as well."

In due course the new furniture ordered from Chicago arrived, and Lalitha and Chandra settled down to make Williams Bay their home. Chandra's colleagues at the observatory were friendly and helpful to the newcomers. The director, Dr. Struve, and his wife were "almost paternal" in looking after the younger men of the staff. The assistant director, George Van Biesbroeck, was also kind and helpful. Kuiper, though primarily an observational astronomer, was keenly interested in theoretical developments; Chandra found it extremely stimulating to have discussions with him. And then there was Strömgren, who was to join the faculty soon, whom he had known since 1932 in Copenhagen. They shared some common ideas on properties of matter in stellar interiors and had plans to begin a collaboration. As for social life, there was not much. "Williams Bay is too far from civilization for anything to happen," wrote Chandra to his father on 4 February. "But the people around us are very obliging and now we feel that we can conquer at least half the world if not the whole of it!" Indeed people were nice. When Chandra and Lalitha were both bedridden with colds and fever in January, someone or other came to their house to help, brought meals, and took care of them. The setting was casual and informal among the community of scientists. They visited each other in their homes and had long pleasant evenings.

As for Lalitha, there was a lot to learn and cope with. "People were friendly, but the main thing for Chandra and me was to understand

16. Chandra and Lalitha after their marriage, Madras, 1936.

17. Farewell from family and friends, Madras railway station, October 1936.

8. Meeting of physicists in Washington, D.C., 1938, organized by George Gamow and
Edward Teller. *Left to right:* (*seated*) D. Inglis, . . . , L. H. Thomas, George Gamow, S. Rosseland, Harlow Shapley, J. A. Fleming, . . . , R. Gunn; (*standing*) R. Seeger, K. R. Herzfeld,
Willis E. Lamb, Jr., C. L. Critchfeld, Wolfgang Pauli, M. Schönberg, S. Chandrasekhar,
. d'E. Atkinson, . . . , Jean B. Perrin, Edward Teller (*last three unidentified*).

19. Chandra and Lalitha at the dedication of the McDonald Observatory, Mount Locke, Texas, July 1938. Photograph by Elwood M. Payne.

20. Henry Norris Russell and Chandra at symposium to commemorate the fiftieth anniver-
ary of the University of Chicago, September 1941.

21. Chandra with George Greenstein, son of his colleague, Jesse L. Greenstein, Williams Bay, c. 1945.

22. Chandra with Michael Lebovitz, son of Norman Lebovitz, Chicago, 1981.

23. Chandra with all his brothers and sisters on visit to Madras, 1961. *Left to right:* (*front*) Vishwam, Rajam, Chandra, Bala, Balakrishnan; (*back*) Savitri, Vidya, Ramanathan, Sarada, Sundari.

24. Lalitha, 1956.

25. Paul A. M. Dirac and Chandra, Williams Bay, 1958.

26. Chandra with Igor D. Novikov and Ya. B. Zel'dovich, General Relativity Conference, Warsaw, 1962.

27. Royal Medal recipients, 1962. *Left to right:* Sir Landsborough Thomson, Sir Cyril N. Hinshelwood, H. J. Emeléus, and S. Chandrasekhar.

28. Chandra receiving the National Medal of Science from President Lyndon B. Johnson, Washington, D.C., 1967.

29. Prime Minister Indira Gandhi and Chandra on the occasion of the Second Jawaharlal Nehru Memorial Lecture which Chandra delivered, New Delhi, November 1968.

30. Chandra receiving the Nobel Prize in physics from King Carl XVI Gustav of Sweden, Stockholm, 1983. Photograph by Jan Collsiöö.

31. Copley Medal of the Royal Society recipients, London, 1984. *Left to right:* (*front*)
S. Chandrasekhar, Sir Clive Sinclair, J. R. Mallard, P. Mansfield, Sir Sam Edwards,
H. H. Hopkins; (*back*) E. Mayr, A. L. Cullen, Mary F. Lyon, R. P. Kerr, A. R. Battersby,
R. Bond, J. F. Davidson, J. M. S. Hutchison.

each other," she says. "The first priority for him was his science. There was not much time for other things, which I well understood and appreciated." Besides attending astronomy and astrophysics lectures at the observatory, she joined local women's groups. "I belonged to a Tuesday evening study club of women from various walks of life and the local chapter of the American Association of University Women," Lalitha recalls. "There was a traveling library department; we got books from Madison library on various subjects, read and reported on them. I really enjoyed interacting with women from diverse backgrounds—some high school teachers, some farmers' wives, some grocers' wives, etc. We enjoyed a good rapport, and although I was new to this country, I could communicate with them. Of course there were problems. Some of them had such set ideas about India. They thought only of poverty, sickness, snake charmers, rope tricks, and so on. You try to correct them by telling them that is not everything, by telling them other aspects of India. I used to give talks about India's history, civilization, and culture. I also used to play on the *vina* in those days and had brought a *vina* with me from India. I played it for some music groups and gave talks on Indian music."[9]

As for a serious scientific career for herself, she had to give up the idea. Although she attended lectures, took notes, and studied them seriously, the responsibilities of taking care of the house and their life together came first. "I had to give up the idea of further studies," Lalitha says. "Chandra was not too happy about it; he felt that I was ready to undertake a research problem, but I made the decision not to continue, since I felt I would not be able to devote my full time. I understood that Chandra had to give most of his time to his science. That is the way a scientist is made."

Chandra, in the meantime, had to contend with increased pressure to return to India. Hardly a month had elapsed since they had arrived in Williams Bay, when he received a telegram from C. V. Raman asking him to reply by cable whether he would accept a readership in applied mathematics in Andhra University. Then in May 1937, the directorship of Kodaikanal Observatory became open. Chandra's father wanted Chandra to apply for the position. He even sent him the application form. Chandra had no desire to try for an administrative position, let alone the fact that Lalitha and he were just getting settled in the United States at considerable investment in travel expenses, new furniture, and household appliances. Besides, he had just been promoted to an assistant professorship with the prospect of tenure. "I think it would be a real shame," Chandra explained to his father, "if I should have to leave Yerkes so soon after being made a permanent member of its staff and without even staying the minimum time required to make even a partial use of the advantages of being attached to one of the greatest observatories of the world." Lalitha too was just as determined as Chandra about staying at Yerkes and taking full advantage of the opportunities there.

Thus Williams Bay, Wisconsin, became their home, and Chandra launched his scientific career at the Yerkes Observatory of the University of Chicago. Chandra hoped that, after a few years, with his stature as a research scientist firmly established, he would have an opportunity to return to India to a position of his liking, in which he would be "free, free from any administrative encumbrances, and free to learn and teach advanced mathematics and theoretical physics."

For the observatory itself, Chandra's presence along with the other newcomers, including Kuiper and Strömgren, heralded the beginning of a new phase in its forty-year history.

9

Scientist in the Midst of Political Turmoil
Williams Bay, Wisconsin, 1937–1952

I have always been proud that I had a part in bringing you to the University of Chicago.

Robert M. Hutchins, 22 September 1971

Actually, Williams Bay, Wisconsin, became the home of Chandra and Lalitha for twenty-seven years. It was also the home of Yerkes Observatory, which was equipped with the world's largest refracting telescope when it was founded by George Ellery Hale in 1897. Otto Struve, the director of the observatory in 1937 when Chandra arrived on the scene, has written an account of the observatory's beginnings.[1] Apparently, during the summer of 1892, Hale took a break from his summer vacation in the Adirondack Mountains to attend the meeting of the American Association for the Advancement of Science at Rochester, New York. There, "one hot evening," writes Struve, "a group of delegates were sitting in front of the hotel, trying to keep cool. The young man [Hale] was in a receptive mood and his ears readily caught a tale that the famous Cambridge telescope builder, Alvan G. Clark, was telling to a group about him." In 1887, Lick Observatory on Mount Hamilton near Santa Cruz in northern California neared completion and began to attract worldwide attention. A long-time civic booster and trustee of the University of Southern California, Edward F. Spence, appealing to regional pride, demanded an even larger telescope in the Los Angeles area. The people of southern California, in the midst of the uproar and sudden wealth of a land boom, responded to his appeal, especially since it was backed up by Spence's own personal contribution of $50,000. Enthusiastic support from newspaper editorials, railroad companies, and developers came forth. "This was ample warrant," says Struve in his account, "in the judgement of the hour, for ordering a pair of 40-inch glass

185

discs, which in the course of three years had been successfully made by M. Mantois in Paris. Unfortunately, however, the land-bubble had meanwhile burst, the gift was worthless,[2] and Mantois was vainly seeking payment of the sixteen thousand dollars at which the discs were valued. Here was a great opportunity, said Clark, for someone to get a large telescope without loss of time. He had tested the glass and found it perfect, and nothing would please him more than to figure a forty-inch objective."

Hale hurried back to Chicago, where, less than a month before, he had been appointed an associate professor of astrophysics in the newly founded University of Chicago under its first president, William Rainey Harper. Hale was a talented and dedicated astronomer. His wealthy and doting parents[3] had given him an early start by providing him with every educational advantage; he had a private tutor and attended a private school that specialized in science training. When he was fourteen, he persuaded his father to get him an excellent Clark-made four-inch refractor telescope so that he could watch the transit of Venus on 6 December 1882. A few years later, Hale persuaded his father to buy him a professional Browning spectroscope so that by his freshman year at MIT, he had already begun his work on solar phenomena, particularly the large irregular structures known as solar prominences seen on the limb of the sun at solar eclipses. Before he graduated, he would invent the spectroheliograph, an instrument designed to permit the study of the prominences without waiting for an eclipsed sun. When the facilities at Harvard College Observatory proved inadequate for his work with the specially built spectrograph, he acquired through the resources of his father a handsomely built private observatory of his own in the backyard of the Hales' home in Kenwood, a wealthy suburb of Chicago at that time. Completed in 1891, the Kenwood Observatory, with a twelve-inch refractor telescope and a laboratory for spectroscopic work, was better equipped than nearly any other American university observatory.

In appointing Hale to the faculty, Harper had struck a bargain. Hale's father agreed to give the entire observatory—telescope, instruments, and all—to the University of Chicago. In turn the university would pay the operating expenses and also raise $250,000 to build a larger observatory. Here then was a golden opportunity, not only to build a larger observatory, but also to build the world's largest telescope with the forty-inch lens so readily available. Young Hale's enthusiasm infected Harper. Together they approached Charles T. Yerkes, the street-railway magnate, from whom Harper was already trying to solicit large contributions to the university. Yerkes, described as a flamboyant character,[4] became interested when he read Hale's eight-page letter describing the exceptional opportunity to get the forty-inch lens for Chicago and to build a telescope far superior to the Lick Observatory telescope. The "biggest telescope in the world" would, he knew, both attract visitors to Chicago and bring fame to the university. One meeting of Yerkes, Hale, and Harper after this 2 October 1892 letter was enough to induce Yerkes to come forth with full support for the enterprise of building "the largest and best observatory in the world," which, of course, would be known as the Yerkes Observatory. Plans were completed in 1894 and the construction of the observatory building, as it stands now, was begun in 1895 at Williams Bay, Wisconsin, on the northern shore of Lake Geneva.[5] The first astronomical observations with the completed telescope were made by Hale and his associates in the summer of 1897. It became the most modern and complete observatory of its day, and Hale naturally became its first director.

In 1903, Hale left Yerkes for Pasadena, California, where he built the Mount Wilson Observatory with his own sixty-inch reflecting telescope[6] and took with him "some of the ablest members of the Yerkes staff, notably Messrs. Adams, Ellerman, Pease, and Ritchey."[7] Edwin Brant Frost became the director of Yerkes in 1905, and a period of retrenchment followed (1905–32) because of the gradual transfer of the Carnegie Institution's financial support from Yerkes to Mount Wilson and the rigid economy imposed by the presidents of the University of Chicago.

In 1932, Otto Struve became the director of the Yerkes Observatory. A period of revival began in concert with a dramatic change in the university under its most influential president, Robert M. Hutchins. Struve began to assemble a brilliant observatory staff. In 1932 he already had C. T. Elvey, William Morgan, and George Van Biesbroeck and was looking for others. In 1935, on the occasion of the annual dinner given by the trustees of the University of Chicago for members of the faculty, President Hutchins, greeting Struve on the receiving line, asked him to pay him a visit in his office. "When I did so, a few weeks later," says Struve, "he asked me whether I had given some thought to the question of increasing and improving the staff of the astronomy department and he outlined to me his own policy of building up weak departments through the appointment of only relatively young first-class research workers. Were there such workers available in astronomy? I replied that I could name three or four immediately: Kuiper, Chandrasekhar, Bengt Strömgren, and one or two others."[8]

A few more visits to the president's office and Struve was ready to offer positions to all three of the people he had mentioned on the spur of the moment. Kuiper and Strömgren, along with Chandra, joined the Yerkes staff in 1936–37. This marked the beginning of a new era at Yerkes and also in American astronomy, which then was still dominated purely by observational studies. As Martin Schwarzschild says, "It was quite novel to have a pure theorist like Chandra in an observatory devoted to pure observations."[9] European born and trained, Strömgren and Kuiper brought the prevalent strength of Europe, that of mixing theory with observations and placing theoretical considerations in the context of observations. Chandra's Cambridge training made him more of a mathematical astronomer, but unlike his mentors Milne and Eddington, Schwarzschild says, "Chandra had no snobbishness in regard to his mathematical work. He did not shy away from numerical, computational solutions. He mixed rigorous analysis with numerical calculations, as the problem required."

However, according to Schwarzschild, this kind of work was so contrary to the general attitudes in the country that Chandra's presence as a theoretician had little or no impact. On the contrary, especially on the West Coast, Chandra's style of astronomy was not even considered proper astronomy. A whole new generation of astronomers would have to come along before theoretical astronomy would become a part of astronomy proper.

The groundwork for such a change began at Yerkes itself. Chandra's principal assignment when he joined Yerkes was to develop a graduate program in astronomy and astrophysics. Along with Kuiper, Chandra devised a sequence of eighteen courses spread over two years; courses in stellar atmospheres and interiors, stellar dynamics, solar and stellar spectroscopy, solar systems, and atomic physics. During the years that followed, he taught twelve to thirteen of the eighteen courses, teaching at least one, sometimes two courses each quarter. He was put in charge of the library (ordering books and journals), of advising students, and of arranging the weekly colloquia. Soon "with Struve, Kuiper, Chandrasekhar, Morgan, and Hiltner, Yerkes became a leading institution in every respect," says Schwarzschild, "including the development of one of the most outstanding, if not the outstanding graduate school in astronomy and astrophysics in the country. . . . Chandra was by far the most active member of the group. He just loved to give lectures and was very demanding of his students, many of whom felt enormous loyalty to him."

Schwarzschild recalls his own experience when he first met Chandra in 1937. Schwarzschild was a guest of the Kuipers, who had invited him to spend the Christmas holidays, and he met Chandrasekhar at a small party at Kuiper's house. "I remember that evening vividly," says Schwarzschild. "It was the first time I was in personal contact with a dark-skinned person, and I am embarrassed to say, I recall feeling an instinctive aversion throughout the evening. Then and there I made up my mind that this was something I had to overcome extremely strongly. When we met again a few days later, whatever that aversion was, it had completely disappeared. Even though I was a guest of the Kuipers, I spent most of each day with Chandra. We had terribly important things to discuss about stellar structure. . . . We very fast found our way to each other; he soon, jokingly, gave me home assignments to read at night. One evening the Kuipers decided it would be fun to go to a movie, but I said, 'I cannot. I have Chandra's homework to do.' Mrs. Kuiper insisted, and we went to the movie when she offered to sign a letter of excuse of why I couldn't do my work. I presented that letter to Chandra the following morning."

Chandra's reputation as a teacher, and his youth and enthusiasm for research, soon began to attract students from all parts of the world: Paul Ledoux from Belgium; Mario Schönberg, Jorge Sahade, and Corlos Cesco from Argentina; Gordon W. Wares, Ralph E. Williamson, Wasley S. Krogdahl, Margaret Kiess Krogdahl, and Louis R. Henrich from the United States. These were some of Chandra's students and associates during his early years at Yerkes. Later on, after the Second World War, Guido Münch, Arthur Dodd Code, Donald E. Osterbrock, Esther Conwell, Jeremiah P. Ostriker, and many others were to follow.[10] Chandra's own research and writing continued, unabated by the teaching, the advising, and the weekly colloquia he instituted.[11] For instance, during his first year at Yerkes, in addition to writing half a dozen research papers, he completed the manuscript of his first monograph, *An Introduction to the Study of Stellar Structure.*

This pattern of teaching, research, and writing was to persist from 1938 to 1944, a period in which his researches encompassed work on stellar dynamics, dynamical friction, stochastic problems in physics and astronomy,[12] and the negative hydrogen ion. From 1944 to 1949, Chandra's preoccupation was with radiative transfer and, as he often says, it was the happiest period of his scientific life. "My research on radiative transfer gave me the most satisfaction," says Chandra. "I worked on it for five years, and the subject, I felt, developed on its own initiative and momentum. Problems arose one by one, each more complex and difficult than the previous one, and they were solved. The whole subject attained an elegance and a beauty which I do not find to the same degree in any of my other work. And when I finally wrote the book *Radiative Transfer*,[13] I left the area entirely. Although I could think of several problems, I did not want to spoil the coherence and beauty of the subject [by further additions]. Furthermore, as the subject had developed, I also had developed. It gave me for the first time a degree of self-assurance and confidence in my scientific work because here was a situation where I was not looking for problems. The subject, not easy by any standards, seemed to evolve on its own."

The onset of the Second World War, and his anguish and concern for India and the world at large, dampened Chandra's exuberance for science. "With the War dispersing all values," he wrote to his father on 26 December 1940, "the pursuit of science and of study and scholarship seems almost futile and unimportant. If I have continued, it has been a matter of habit and if I persist to continue, it is without zest or enthusiasm." He felt that "the combined population of the 'greater' Germany and the Soviet Republics (2×10^8 human beings!) were the pawns of two (three?) irresponsible men," and Europe and its civilization faced a disastrous transformation. And in the enshrouding darkness, whatever the past follies and crimes of Britain with respect to India had been, Chandra felt India should fight with the British. If there was any future for India, it was *with Britain*. When Holland, Belgium, Norway, and France went down in quick succession, and Greece and Turkey were trampled, he was alarmed. "The security of India is threatened," he said in his letter to his father on 28 October 1940. "Under the circumstances, India should wholeheartedly support the British instead of playing politics. India under Germany and Italy would be far worse than anything we have known." He was impatient with America, which, prior to the Pearl Harbor disaster, watched the war from the sidelines and showed no willingness to back up strong sympathy for Britain with action to defend it. Finally, when America did enter the war, Chandra wrote in a jubilant mood on 11 December 1941: "America has entered the war with unity and determination and though the final round may not come for a long time, no one doubts that the eventual outcome will be [a total] victory for the United States and Great Britain."

Communication with home was of course delayed and disrupted. It took several weeks and sometimes months for letters to cross over, and this made it impossible to maintain any coherent correspondence. India, which was very little in the news prior to Pearl Harbor, gained a prominent place as Japan swept through Singapore, parts of Burma, and subjected Rangoon to daily air raids. Soon Calcutta and Madras were also bombarded. Sir Stafford Cripps came with his famous mission to enlist India's full support for the war effort. His mission, however, failed to convince the Indian leaders that India could attain its much cherished freedom, and Mahatma Gandhi saw no way out except to initiate the "Quit India" movement. "India has been in the news," Chandra wrote to his father on 15 August 1942. "I am afraid that many harsh things have been said about her both here and in England, and India continues to be the most maligned country. All this has been very painful. But Lalitha and I have tried to build a mental wall of self-defense but not with success." Chandra was referring to newspaper depictions of Gandhi as the fakir, a snake charmer with the face of Tōjō, performing rope tricks with a snake.

The Congress Party in India was immediately outlawed, and all its popular leaders, including Gandhi and Nehru, were imprisoned. Strikes and riots spread from Bombay and Karachi on the Arabian Sea in the west, to Calcutta on the Bay of Bengal in the east, and to Cawnpore, Lucknow, and Delhi in the north. Troops shooting down demonstrators became the prevalent scene. When Gandhi saw his life's dream of a non-violent political struggle begin to dissolve into violence and brutal counterviolence, he went on a fast until death. Chandra, upon hearing the news, wrote to his father on 26 February 1943: "Lalitha and I have been anxiously following the bulletins on Gandhi's health. . . . The fact that we have no place except the prison, no life except one of hardship, and no end except that of starvation for the greatest man of our times is an ironic comment on the 'ideals' of this war, which is supposedly the inauguration of the 'Century of the Common man': fine phrases and meantime the greatest apostle of the common man is allowed to perish. . . . And all the while India is of course maligned from every possible direction. . . . We have only to hope that history will eventually set things right." *

"Everything appears irrelevant in the face of our present political

---

* Chandra recalls, for instance, an interview on the CBS network with Lord Halifax, Great Britain's ambassador to the United States. He was asked, "Well, Lord Halifax, in this country

conditions," wrote Balakrishnan to Chandra on 6 June 1943, "your science, my art, everything. . . . India fills my eyes first. The rest of the world I can see only across our Indian prospect—with our greatest men and women in prison, humiliated and insulted, the nation lying abjectly prostrate and helpless, the imperialist vampire at its throat, even the tremendous sacrifices and suffering of Russia serving to help the imperialistic reactionary forces—and meanwhile, autonomous cancer-cells like [Muhammad Ali] Jinnah growing fast inside our vitals. It is enough to make one's blood boil, but this results in nothing more than, say, an epileptic fit or a psychic equivalent of it. I can do nothing." Chandra shared these feelings of the utter helplessness of an individual in the face of a chaotic world.

In spite of these disturbing, unhappy, and painful developments in India, Chandra did not waver from his strong sympathy for Britain and his conviction that the world, including India, faced the worst danger if the Axis powers were victorious. Lalitha and Chandra participated in the civilian effort to help the British. They also planted victory gardens every year and produced enormous quantities of vegetables; "We learned canning," says Lalitha. "For weeks we had to eat tomatoes for all three meals."

Lalitha and Chandra spent October to December 1941 at the Institute for Advanced Studies in Princeton. Chandra had maintained contact with John von Neumann, his friend from Cambridge days. Von Neumann had often expressed the hope that Chandra would make an extended visit to Princeton. After some effort, and with the help of Henry Norris Russell and H. D. Smyth (heads of the astronomy and physics departments), he was able to gather a sum of $1,000 for Chandra's visit. It was the first time since their arrival in America that Lalitha and Chandra had left Yerkes for such an extended visit to another city. Chandra wrote to his father on their arrival at Princeton on 3 October, "To be in the same Institute as Einstein, Weyl, Pauli, and others, it is a privilege! But terrifying all the same. I have an office at the Institute with a marvelous view. I have to give two lectures a week and am planning to attend a fair

---

most people think of Gandhi as being misguided and pro-Nazi and pro-Japanese: What do you say?" Halifax responded almost angrily, "I do not share this view in the least. Anyone who has the slightest knowledge of Gandhi knows that he is the most moral person and completely opposed to anything that the Nazis stand for. It is ridiculous to think that he is pro-Nazi." This was pleasing to hear for Lalitha and Chandra, but the vast majority of Americans thought of Gandhi and India as against America and against American interests.

number of lectures by others. Three months will pass before we even notice them."

On 7 December 1941, while Chandra and Lalitha were still at Princeton, the Japanese struck at Pearl Harbor and America entered the war. Chandra decided to take the initiative and join the rest of the scientific community, which was unanimously behind the war effort. However, because he was not a citizen of the United States, he could not enter into classified work, although he was made aware that he could be drafted as a private in the military services of the country any time.

Von Neumann was engaged in war-related work. He was a consultant for the Ballistic Research Laboratory at the Aberdeen Proving Grounds (APG) in Maryland, and he thought Chandra's services could be useful there. He had written to Chandra earlier, on 18 October 1940, "There is great interest at the Ballistic Laboratory at Aberdeen and among men working there on interior ballistics in general, in the theory of gases, and especially in questions of the equations of state of very dense gases. In spite of the dissimilarity of the interior of a white dwarf and of the explosion chamber of a gun, it is probable that your past experience in this field would be very valuable. If you feel like following up such a subject, will you let me know, or write directly to Mr. R. H. Kent."[14]

Robert H. Kent was the leader of a group of mathematicians, astronomers, and physicists assembled at the Aberdeen Proving Grounds in Maryland to study and solve the ballistics problems. "APG was a very large encampment with thousands of troops in various stages of training in the handling and maintenance of weapons," says Robert G. Sachs.[15] It had isolated test areas sticking out at the northern end of Chesapeake Bay, known as Abbey Point. Apparently, according to Sachs, the Ballistic Research Laboratory was established during the First World War with a small group of scientists whose task was to study the ballistics of artillery missiles and prepare firing tables to be used on the battlefields to set the elevation of the artillery piece so that the missile could be fired at a certain predetermined range and then hit the target. Kent had been drafted into the army when he was still a graduate student in physics and was put to work making such ballistic calculations. He continued the work after the war ended and between the two wars he was the only professional ballistician in the United States. Talented scientist that he was, he had studied basic ballistics and developed many new ideas. When the Second World War began to seem inevitable, he was naturally in a position to put his ideas into practice. Since the calculation of firing tables was a tedious business, he got people to develop automated calculators. He got the second differential analyzer built for Aberdeen and with von Neumann developed an electronic computer for ballistic purposes.

Chandra wrote to Kent after returning to Yerkes in the early part of 1942, but it took more than a year before he got the necessary clearance to work at the APG. He was a British subject; the clearance had to go through British intelligence channels. The appointment, after considerable bureaucratic delay, was finally authorized by the Secretary of War on 27 January 1943, and Chandra started his work at the APG in early February of that year. The work carried a remuneration of $30 per day with travel allowance. From February 1943 until the end of the war, Chandra commuted between Yerkes and APG—three weeks at Yerkes, three weeks at APG. "It was pretty strenuous," recalls Chandra. "But the entire scientific community was behind the war effort. No two opinions as in the case of Vietnam. I was part of that effort. I didn't mind the strain." In some ways it was a greater strain on Lalitha, who recalls, "I was always somewhat lonely because Chandra's preoccupation with his science left very little time for anything else. I had come to accept that. But those war years were especially hard since Chandra used to be totally away for three weeks at a time. And when he returned to Yerkes, he was busier than ever. He would double his teaching to make up for the three weeks, plus his research. . . . I used to marvel at him though, how he could switch from one type of work to another."

At APG, Chandra became a part of an outstanding group of scientists which included John von Neumann, Ronald Gurney, Joseph Myer, L. H. Thomas, Martin Schwarzschild, Edwin Hubble, Robert Sachs, and many others. Chandra shared an office with Sachs and worked on ballistic tests, the theory of shock waves, the so-called Mach effect, and transport problems related to neutron diffusion. He felt stimulated by the novelty of the applied, war-related work, the companionship of scientists from diverse fields, and the feeling of service and sacrifice in the cause of humanity.[16]

At the same time he was exposed to ugly incidents of harassment and humiliation because of the color of his skin. Scarred by earlier incidents of discrimination, Chandra totally avoided bringing Lalitha to Aberdeen, a rural community that, although not in the South, shared southern attitudes when it came to segregation. On the proving grounds he was given accommodations in the officer's annex. After an embarrassing first incident, when he was refused service at the officers' mess, things were straightened out. "Kent went to the commanding general," recalls Sachs, "and laid it on the line." Yet Chandra had to be extremely careful to adhere to a strict routine. "I had no problems once I was on the grounds, began my stay at the assigned quarters, and registered at the mess," recalls Chandra. "But every three weeks when I returned from Yerkes, I had to register at the mess again, and if the man there was new and not informed from the higher-ups who I was, I had trouble. Any time I was not on routine, I encountered trouble. Once, for instance, by mistake I got to a different gate than the usual one for entrance. The man there shouted, 'Eh blackie there, you just wait until I come!' I had to wait there for an hour before he let me in. It required considerable effort to persuade him to allow me to use his phone to call my office and Kent. So I had to be extra careful that all the arrangements had been made beforehand."

Chandra's friend, Schwarzschild, on the other hand, faced a problem of a different sort, which, although somewhat of a digression, is worth recounting. He came to this country to escape the persecution of Jews in Germany. "But as an enemy alien, a noncitizen, still young, and not as advanced in my science as Chandra, I couldn't do a thing," Schwarzschild recalls. "I could stay and teach or get myself enlisted in the army as a private, which I did when America entered the war. . . . By a strange coincidence, I found myself with Chandra at Aberdeen for half a year. We enjoyed each other's company enormously, and the fact that I was in uniform gave Chandra the most marvelous time for teasing me and pulling my leg." Schwarzschild was then sent to the Italian front, attached to the Allied High Command of the Air Force. His duties were to follow the infantry and get as fast as possible to places where the Allied forces had won a victory to see what actually had happened. "That put me in a position of having maps, taking notes, asking everybody questions, taking photographs, which made me the perfect picture of a German spy. And I had the record of being taken in as a German spy on the average of once every two weeks, at gunpoint! It's extremely serious for the first ten seconds and then somewhat serious for the next minute, and if you survive that, everything else is all right. In retrospect, of course, it becomes terribly funny."

There was no simple way to assure that he was not a spy. He could carry any number of identification papers, but of course a spy has all the right papers too. "The only way was to behave absolutely quietly," says Schwarzschild. "Keep quiet, do absolutely nothing once the gun is pointed at you that may make the finger pull the trigger. For me that was not hard to learn because I am a slow reactor. My nature is to do exactly nothing. For me it was much harder to see the person with such hate in his face." Schwarzschild mastered the technique of keeping quiet and getting himself locked up. Once he was behind the bars, the other person relaxed, and Schwarzschild could slowly convince him that, in spite of his German accent and all, he was not a spy. After a few incidents of this nature, the commanding officer got so concerned that he appointed a sergeant to always accompany Lieutenant Schwarzschild on such trips. "The sergeant was a truck driver from New York, and extremely strong, tall, and heavily built," Schwarzschild recalls. "Just a magnificent person, unbelievably courageous, but uneducated as the dickens. The two of us were on these trips together for weeks, and when I was with him, he could say I was all right in his New York accent. Even then we could not always be sure. Quite a number of occasions the gun would be pointed at us."

In the United States, meanwhile, von Neumann attempted to enlist Chandra for the Manhattan project in 1944. Chandra knew about the A-bomb project vaguely, since he used to talk to Eugene Wigner, Edward Teller, Enrico Fermi, and others at Chicago. Their indirect questions regarding his work on radiative transfer and neutron diffusion gave him some hints. In any case, on 9 March 1944, in a letter marked confidential, von Neumann wrote:

They are convinced, and I am in full agreement with them, that the project would gain very much if it could acquire your collaboration. A number of their problems are such, that you are the one logical man to deal with them considering your astrophysical work—others [some other problems] would be quite logical continuation of your present work on shocks. Your joining the project would certainly further several essential phases in a way in which nothing else could. I have no doubt that it will be successfully completed in a very finite time, and that the possession of its results by one country or the other and the manner of its use will greatly influence and, with a not inconsiderable probability, decide the future of the world. . . . The scientific problems of the project are very interesting, the group which deals with it is first rate. I would only say that I have never lived before and never expect to live hereafter in a better intellectual atmosphere.

Hans Bethe, the man in charge of negotiations, followed up von Neumann's letter and wrote on 20 March 1944 how happy Teller, Weisskopf, Oppenheimer, and he would be if Chandra joined them. "We are in great need of your help," Bethe urged, "and we believe that you would be the best man to ask to take charge of certain calculations which have some loose connection with work you have been doing in

Aberdeen. We have no other person who understands this type of problem with the exception of Johnny [von Neumann], who is here only a fraction of the time, and is then very busy with certain other problems. Apart from taking charge of this work, we would also like to have you here in general because there never seem to be enough intelligent people to do all the theoretical work that is necessary."[17]

Chandra agreed to join them, although he had considerable reservations regarding leaving Yerkes and living at Los Alamos. Clearance procedures were set in motion and in September 1944 Chandra was informed that he was cleared; an apartment had been reserved for him, and Oppenheimer would do everything to make the work and surrounding circumstances as attractive as possible. However, by then Chandra had changed his mind. He felt that the war was coming to an end. He had strong obligations to Yerkes. Moving to Los Alamos would also mean a total disruption of his work. "You are certainly right," wrote Bethe on 27 September 1944, "that it looks as if the European War would be over in a short time and that is quite a decision to go full time into war work just at this time." Nonetheless, von Neumann continued to exhort Chandra to go to Los Alamos. But new difficulties cropped up in his efforts. "A very regrettable red tape 'hitch' has developed," he wrote on 25 December 1944. "Your U.S. clearance would have sufficed for your employment by the project at the time of H. A. Bethe's negotiations with you. In the meantime there have been some new regulations which necessitate some extra steps (through British channels) for British subjects." In the end the idea was dropped and Chandra did not go to Los Alamos.

In spite of the disruptions caused by the war and commuting between Yerkes and Aberdeen, Chandra's research continued, and he produced a steady stream of papers and books. He was promoted to an associate professorship in 1942, followed by a promotion to full professorship in 1943. A letter from Milne in 1941 gave indirect hints that he was going to be nominated as a Fellow of the Royal Society of England, with a warning, however, not to be unduly disappointed if the honor was delayed. "I have known people to eat their hearts out in disappointment at delay in election, yet accidents may prevent the most likely candidates from getting in," Milne said. "It is unlikely that more than one, if any, astronomical or astrophysical candidate will be taken any year, so some will have to wait a longer time."[18]

To be named a Fellow of the Royal Society, to be able to put "F.R.S." after one's name, meant a great deal to British scientists, especially to scientists in India and other British colonies. Since only a handful of scientists held this distinctive honor in India, next to the Nobel Prize, to be elected to the Royal Society had become a symbol of supreme achievement in science and it made the person a celebrity in India. Scientists recognized by such an honor were national heroes whose triumphs were triumphs for India in her struggle for independence. Chandra was no exception. From the beginning of his brilliant career, he was expected to join the ranks of Saha, Raman, Bose, and other luminaries. "I am looking forward to the day when you will be elected to the Fellowship of the Royal Society, an honor much coveted, at least on this side of Suez," wrote his friend K. S. Krishnan on 16 July 1934.[19] His teachers and, of course, his father had similar expectations. Hence, although he personally had gone beyond such feelings, the good wishes and well-meaning reminders kept Chandra in a state of some anxiety as two years passed without news of his possible election to the Royal Society.

Milne wrote again on 20 January 1944: "I hope you will not be discouraged by the inevitable waiting period that happens to nearly everyone who is put up for election to the Royal Society. Go ahead with your best work, publish only what you yourself think first-class, and I am sure that, humanly speaking, all will be well in the end." Milne was now a member of the mathematics committee that tendered advice to the council on mathematics candidates. But the proceedings of such committees were strictly confidential, and Milne was not at liberty "even to hint what happens" at such meetings.

I cannot tell when or how soon, or indeed whether, you will be elected to the Royal Society. I will only say that in my personal opinion your work strongly deserved this recognition. But the number of mathematicians who can possibly be elected every year is strictly limited and there is severe competition. This letter will have achieved its object if it prevents you being discouraged at whatever delay in fact occurs. Please treat this letter as strictly confidential and private. I do not think it an improper letter to write, for I have known of people who *have* been discouraged by deferred election, and who have gone sick at heart. I therefore beg you to remain your happy self; snap your fingers at fortune, good or bad! It is right to have ambitions; life would lose its savor without them; but that man really achieves who can continue to work whether his ambitions show signs of realisations or not.

Milne, perhaps, was projecting his own innermost fears; how he would have felt under similar circumstances. In any case, the suspense came to a delightful end. On 16 March 1944, Chandra was elected to the Royal Society. He received cables from his friend Harold Davenport and

Milne. The day before the election Milne had sent a long letter in which he wrote,

. . . now I want to say how much I regret having sent you my gloomy letter of January 20th. It was written immediately on my return from the meeting . . . and I had come away from the meeting not expecting you to be chosen this year. I had been disappointed with my own advancing of your claims—I was the one at the meeting who knew most about your work—and in the hurly burly of competitive discussions of the rival claims of the various candidates I did not succeed in placing you as high in the list of names forwarded to council as I thought your work merited. . . . I cannot divulge my own confidential information about what happened at council. But in the distant future, when we meet again, and some of your competitors are safely in the society, I will tell you the inner history of your election this year. [That occasion did not materialize, since Milne died in 1949 and they never met again.] I will only say that my letter of 20 January was intended to prepare you against a possible disappointment, and I am extremely delighted at the turn events have taken. . . . I may (also) say that Eddington with whom I had a long conversation about you in Trinity in early January this year, was very definite that you ought to be elected to the Royal Society this year, largely because the way you have encouraged and stimulated theoretical astrophysics in America. He was not on the Mathematics Committee or Council this year, I may say, but I thought you would be interested in this tribute to you from your and my ancient enemy.

The controversy between Chandra and Eddington was a thing of the past. Chandra had successfully tucked it away and gone on to other things. Warm friendship between the two had prevailed through the war years, and the news of Eddington's death in 1944, sudden and unexpected, distressed Chandra deeply. Despite Milne's "enemy" role in that unhappy episode, Chandra and he had become personal friends in a group that included David Shoenberg, Harold Davenport, Harold and Freye Gray, and, of course, Leon Rosenfeld. Lalitha and Chandra did what they could to brighten their European friends' dreary lives during the war; they sent gift parcels of chocolates, crackers, fruit cakes, and Wisconsin cheese regularly at Christmas and other times.

As the war ended, the process of rebuilding the universities began everywhere. Scientists involved with war work began to return slowly to their home institutions. Hutchins was on the move to rebuild the University of Chicago. "He made several new 'distinguished service professor' appointments," Chandra recalls, "though the convention at the university had always been that the title of a distinguished service professor was given to someone who has in fact served the university for a number of years. But people like Enrico Fermi, Harold Urey, Carl

Rosby, Marshall Stone, and James Franck, who did not fit this category, were appointed as distinguished service professors." On the other hand, Struve, who *had* served the university for a number of years, was not offered one. Chandra was responsible for reminding Hutchins of the oversight, and Hutchins went on to give Struve the appointment. At about the same time, Chandra received an offer of a research professorship from Princeton to succeed Henry Norris Russell. To the delight of Russell, Chandra accepted the offer. "Your letter of the 26th August gives me very great pleasure," wrote Russell on 29 August 1946. "Both from the professional and more personal standpoint I am delighted to know that you have definitely accepted the offer of the Research Professorship at Princeton. I am honored to have you as my successor."[20] Struve, who knew of the offer, had not made a serious attempt to retain Chandra. An increase in salary from $5,000 to $7,000 was not a sufficient incentive when compared to the Princeton offer of $10,000 and Russell's chair.

When Hutchins learned of the Princeton offer, however, he called Chandra to his office in the presence of Struve and Dean Bartky and offered him a distinguished service professorship and matched the Princeton salary. Chandra was in an awkward position. Hutchins approached him again and asked to see him alone. Chandra recalls that meeting with Hutchins vividly. "If you feel that your work is going to be enhanced by going to Princeton, I wouldn't stop you," Hutchins said. "Because what is in your interest and in the interest of your future scientific work, is also my interest; and I do not gain anything by pressuring you, by asking you to be here, offering you higher salary, and so on. I do not want to do that. You have to make up your own mind whether there is something scientifically which we lack in the way of support. If there is nothing lacking then you ought to stay." Hutchins continued, "One thing we cannot offer you is the honor of succeeding Russell because we don't have a Russell. However, this honor of succeeding to positions is ill-advised. It is far more honorable to leave a professorship to which it is honorable to succeed than to succeed to an honorable position." He drove home his final point by asking Chandra whether he knew the person who succeeded Lord Kelvin, who was a professor in Glasgow for fifty years.

After this meeting, Chandra decided against going to Princeton and wrote tactfully to decline. "I am very sorry to inform you," wrote Chandra to Russell on 15 October 1946, "that in consequence of my discussions [with Hutchins], I have reversed my earlier decision and [have decided] to continue my association with the Yerkes Observatory and the University of Chicago. . . . I would, however, like to say that the two factors which have weighed most heavily are first my increasing conviction that I could not do justice to the great tradition you have built at Princeton and second, that having already built to some extent theoretical astrophysics at the Yerkes Observatory during the past ten years, I did not want to start all over again."

Russell accepted Chandra's decision graciously. In his reply of 4 November, Russell said that he had received

a very pleasant letter from Chancellor Hutchins, written from the standpoint of the administration and closing with, "we put terrific pressure on Chandrasekhar. Perhaps I should ask your forgiveness too." Let me take this opportunity to say I don't feel that any one need ask my forgiveness. Chancellor Hutchins was playing and playing fairly, the usual intervarsity game, and you were wholly justified in accepting the offer which gave the best prospects for your scientific work. Also it would be very far from me to fail to recognize the affectionate claim Dr. Struve and the Yerkes Observatory have upon you. So let me assure you that this episode ends as it began with all good will, and does not in the least disturb our personal cordial relations.

He also wrote to Hutchins, quoting President Dodds of Princeton University, "You know this is all 'hoss-tradin.' I am old enough by this time to know that 'hoss-tradin' can be done fairly, courteously, and yet strenuously and can result in complete good will."

That was the last time Chandra ever considered leaving the University of Chicago, even though offers began to come from other institutions in the States and Cambridge, England.

This particularly happy phase in Chandra's life ended in the early 1950s. In 1952 he became the managing editor of the *Astrophysical Journal;* while the editorship constituted a distinctive service of the highest merit to the scientific community at large, it tempered the pure joy of doing science, since the position imposed serious demands on his time and made him vulnerable to criticism by others on whose work he had to make judgment.

Other events also intervened. Acting as chairman of the department in 1952, in place of Bengt Strömgren, who was spending the summer months in Copenhagen, Chandra came into an administrative conflict with his colleague, who had been a friend since Chandra's stay in Copenhagen in 1932–33. Strömgren, though an able astronomer, proved to be an inefficient administrator. And Chandra thought that it was "a frank, straightforward matter" to say to Strömgren, whom he considered a close friend, "Bengt, things are so disorganized. How can you continue to do that? In the best interests of the observatory, you should resign. Let Kuiper be the director." Apparently, Strömgren did not take the criticism well. He talked to other members of the faculty. A deputation went to President Kimpton, who reappointed Strömgren as the director with a letter to the department expressing his and Kuiper's full confidence in him. Chandra recalls, "People made statements, Sigrid Strömgren [Mrs. Strömgren] for instance, that I had been jealous of Bengt's superior position."

Furthermore, during the same year (1952), Strömgren named a committee entrusted with the task of revising the curriculum that Chandra had designed and taught for the preceding fifteen years. It became apparent that in the committee's revised curriculum Chandra had no place. At a faculty meeting where the new curriculum was on the agenda, Chandra had to leave before this item came up for discussion. As he was about to leave, Strömgren asked him for comments on the revised curriculum. Chandra recalls having said, "If you have a committee and you want to revise the curriculum, and if the department votes positively on it, of course you have the right to go ahead, and I shall have no objection. But there ought to be one thing about which there is no misunderstanding. To the extent that I have had no role in revising the curriculum nor been consulted, I retain for myself the right to find a place in the university outside the astronomy department if I so choose." Notwithstanding the clear warning from Chandra, the faculty went ahead and voted to adopt the new curriculum. "It became a matter of self-respect for me. I had no choice," Chandra says. "If the people at Yerkes did not want the kind of science I was doing at the time, I had to find a place in the university where such science could be done."

As Peter Vandervoort[21] says, "If you look at the graduate catalog of the astronomy department during the years 1938–52, you will see that Chandra did an enormous amount of teaching during those years, giving six or more courses a year. Basic courses: a three quarter sequence on stellar interiors, a three quarter sequence on stellar dynamics and galactic structure. He followed that up the following year with a three quarter sequence on stellar atmospheres, a three quarter sequence on interstellar matter, and so on. After 1952, you see an abrupt change in the program. You see courses called basic astrophysics, spectroscopy, photometry I, II, and III, and so on. It is quite clear that there was a major, abrupt change in direction which was very different from the one

in previous years. Since Chandra was responsible for the previous curriculum, one could say that the department had repudiated the things
that Chandra had done."

By coincidence, not long after this incident, Enrico Fermi invited him
to become a member of the Research Institute (now the Enrico Fermi
Institute) and the physics department of the University of Chicago. And
fortuitously, Marvin Goldberger in the physics department wanted to
take a leave of absence but could not do so unless he found a substitute
to teach his course on mathematical physics. Chandra offered to teach
this course. To no one's surprise, students liked Chandra's lectures. When
the chairman of the department (Andrew Lawson) asked Chandra
whether he would like to continue teaching in the physics department,
Chandra readily agreed. Since 1952, although he did not officially resign, Chandra has rarely taught a course in the astronomy department.

Thus, for the second time in his career, Chandra faced an uncalled
for, unceremonious humiliation (though of an entirely different kind),
and he felt he had to make a decision. Should he indulge in wasteful
confrontation or should he go on to other more important things? He
chose the latter course and remarks, "I think, on the whole, this experience in the early 1950s did as much good for my science if not more
than my earlier episode with Eddington, because it made me associate
with people like Fermi and Gregor Wentzel, whom I would not have
had close contact with if I had stayed at Yerkes. I set up an experimental
laboratory in hydromagnetism with Sam Allison. I taught all the standard
courses in physics, quantum mechanics, electrodynamics, etc. I was the
first one to teach relativity at the University of Chicago, which of course
led me to research in relativity."

This turn of events, however, was not without an effect on his attitude towards his colleagues. It ruffled his inner self. It marked an end to
a period of blissful naivete, a period in which he was totally oblivious to
all matters other than his science. He had continued his research, which
brought steadily increasing recognition and stature in the scientific
world. Thus, he was elected to the Royal Society in 1944, was awarded
the Bruce Medal in 1952, and the Gold Medal of the Royal Astronomical
Society in 1953. With the exception of Eddington, no other astronomer
had received both of these medals by the age of 42. And up to this time
Chandra had lived in America without thinking in the least or even
being aware of the pettier, the darker side of human nature, at least with
respect to his colleagues. Now things changed. Forgotten incidents that

had been disturbing but which he had ignored and buried in his mind began to emerge with great clarity.

Chandra, for instance, had considered Strömgren to be one of the nicest, most wonderful people, in spite of the fact that in Copenhagen in 1932–33, when Chandra was working on rotating polytropes, doing all his calculations long hand, Strömgren used to boast of his hand computer, but never once volunteered to share it. "I still considered him as a close friend," Chandra says. "I appreciated his brilliance, and to a certain degree even his carelessness, his not completing things. For example, in 1946 he was a visitor at Yerkes. He did an important piece of work on the state of ionization in stellar matter and left behind a half-finished paper. I did the clerical work of writing it all up; it was published and became one of his famous papers." Strömgren, however, did not reciprocate such gestures of friendship. Although Chandra was greatly instrumental in getting him the chairmanship, Strömgren had, without Chandra's knowledge, come to an agreement with the university administration to spend four months of the year at Copenhagen so that he could retain his position there as well. Chandra began to recall other such incidents over the years, unfriendly rebuffs and manipulative exploitation of his friendship.

Chandra's thoughts also wandered to his early years at Yerkes, to the discrepancies and discrimination in Struve's treatment of him relative to his colleagues Strömgren, Kuiper, and Morgan. They had all been appointed as assistant professors in the first year, whereas Chandra was designated a "research associate." The following year, they were promoted with tenure. Chandra was reappointed as an assistant professor with no change in salary and remained an assistant professor for four years, until 1942. When they joined Yerkes, Kuiper and Strömgren were reimbursed for their travel expenses—Kuiper from Boston, and Strömgren and his family (including his maid) from Europe. Struve made no such offer to Chandra, who had in fact borrowed money from his father to pay for Lalitha's passage from India. They all had secretaries of their own, while Chandra had no secretary until 1944, after he was elected to the Royal Society; even then he had only a part-time secretary, Frances Herman. Manuscripts for his scientific papers, books, and his correspondence were all handwritten during his first seven years at Yerkes.

However, remembering these incidents did not in the least interfere with his work. "I became aware," he said, "but not affected. They had no real impact on my daily life. The incredible fact is that in earlier years I was not even aware that something impolite, something improper had been done to me." The rosy colors in which he had viewed his relations with his colleagues faded. He became disillusioned; cynicism colored his own vision. He consciously developed a frame of mind which made him immune both to criticism and approbation. "But I am afraid," he goes on to say, "up to a point, I was largely responsible, because people began to take me for granted, to treat me any way they liked, and I let them."

The years that followed as managing editor of the *Astrophysical Journal* only worsened matters. No longer wrapped in pure science, his mood of optimism and innocence gave way to one of pessimism and frustration.

10

The Autocrat of the Editor's Desk:
The *Astrophysical Journal*
Chicago, 1952–1971

. . . there is no such thing as the Chandrasekhar limit.

Jean Sacks, Journals Department, University of Chicago Press

Chandra was the managing editor of the *Astrophysical Journal* (*ApJ*) from 1952 to 1971. When he began, *ApJ* was essentially a private journal of the University of Chicago. Chandra played a decisive role in transforming it into the national journal of the American Astronomical Society (AAS). When circumstances thrust upon him the editorship, he became, by his own admission, "autocratic, the complete master, and totally responsible for the journal."

During the first twelve years of his tenure, the journal staff consisted only of Chandra and a part-time secretary with an office in Yerkes Observatory. "Between us we took care of all the routine work," says Chandra. "We took care of the scientific correspondence. We prepared the budget, advertisements, and page charges. We made out the reprint orders and sent out the bills." In the final stages of the publication of a paper, Chandra personally transcribed the author's corrections on the galleys to the printer's copy. "This task would appear to be needless," observes Chandra, "but I am sorry to say that the corrections made by the authors are often in the form of hieroglyphics, which only a person who has some understanding of what is being written can understand."[1]

In his first year as editor, six issues of the journal totaling 950 pages were published. Every year the journal grew. In 1968, it became twelve annual issues, and starting in July 1970 it became twenty-four issues, totaling over 12,000 pages a year. In addition to the regular *ApJ*, Chandra was also responsible for the *ApJ* Supplement series, and, as though this was not enough, in 1967 he started a separate *ApJ* Letters section* to be able to publish short accounts of important discoveries faster.[2] At the end of his tenure, he left behind a reserve fund of $500,000 for the journal. He saw to it that the journal became financially independent of the University of Chicago, with its own production office and production manager, paying the entire cost of production plus overhead to its publisher, the University of Chicago Press.

The journal, according to all those I talked with, improved in quality under Chandra's leadership to become the leading astrophysics journal in the world. "It was a kind of Golden Age for all of us," says Eugene Parker.[3] "Many of us, at times, had difficulties publishing papers in the *ApJ,* but it was the Golden Age compared to other times and other journals." Chandra selected the referees, sought their advice if needed, and communicated the relevant, edited version of the referees' remarks to the authors. He handled controversial issues deftly and diplomatically, but never relaxed his high standards regarding scientific substance and presentation. The final decision to publish a paper or not was entirely his. He was not immodest, but only truthful, when he wrote to Herman Bondi, "The policies of the *Astrophysical Journal* are my policies."[4] This authoritarian rule did not go unopposed. Threats of impeachment ensued, but by and large there was universal acclaim of his stewardship, to which he responds:

A journal is what the authors write. The editor doesn't solicit articles; the articles come to him. If the editor has in some way encouraged publication of good papers, promptly, efficiently, and fairly, he has done a little service, but the credit for the quality is not the editor's. It belongs to the astronomical community.[5]

This may be true; however, few would disagree that the journal required an enormous investment of time and energy on Chandra's part. For nineteen years he wore two hats: that of editor and that of teacher-researcher. With strict self-discipline, apportioning his time meticulously between the two roles, he maintained his usual level of research activity. He sought no reduction in his teaching load or other responsibilities. For nineteen years he remained tied to the job without sab-

---

* In 1989, *ApJ* published a total of over 16,000 pages, including the Letters section and the Supplement. Starting in 1990, there will be thirty-six annual issues of both *ApJ* and the Letters and twelve annual issues of the Supplement.

baticals or periods of leisure. He imposed upon himself an isolation from the astronomical community in order to be fair and without prejudice for or against particular individuals. He thus rejected invitations to conferences and symposia—opportunities to travel and socialize.

"Why? Why do you do this?" asked Enrico Fermi once, seeing Chandra carrying a stack of manuscripts to the Press. Chandra had no answer then nor does he have one now. "In retrospect, it was a mistake," he says, "a distortion of my personal life. I had no idea I would keep it for so long when I took it. I had no choice then."

The story of how Chandra became the managing editor and how he finally disengaged himself is a significant component of the *ApJ* history. More important, it is an account of his extraordinary contribution to the intellectual community and an unswerving commitment to his discipline. An overview of the journal's prior history will clarify the circumstances that forced Chandra to become its editor.

During the nineteenth century, the majority of astronomers were concerned mostly with the positions, motions, and distances of stars. Barring a few exceptions,[6] the majority evinced little interest in the physical nature or the internal constitution of the stars. Things began to change towards the end of the century when the telescope together with the spectroscope and the photographic plate formed an important new tool for astronomical research. The conjunction of these three instruments made precise measurements of stellar and nebular distances and motions possible and also provided a powerful means for the chemical analysis of near and distant luminous bodies. Physicists' laboratory atomic spectra and astronomers' stellar spectra found a common ground in their attempts to unravel the physical nature of the universe. The new science of astrophysics was born.

George Ellery Hale, who had a tremendous influence on the development of astrophysics in its embryonic stage, noticed that the spectroscopic literature, which had become so crucial to astronomers, physicists, and to a certain extent, chemists, was widely scattered among several journals, publications of learned societies, and the annals of various observatories. Few investigators could lay their hands on all the published information. In his extensive tour of Europe in 1891, Hale took it upon himself to discuss the need of a new journal for the new science with the leading astronomers and physicists on the continent. Encouraged by their endorsement of the idea and their promise of cooperation, he founded a journal called *Astronomy and Astrophysics* in 1892 which replaced the pioneer American astronomical journal *Side-*

real Messenger. The former editor and owner of *Sidereal Messenger,* W. W. Payne, became the coeditor of the new journal along with Hale. The astronomy section, edited by Payne, continued the work of the *Messenger,* while Hale edited the section treating the new science of astrophysics.

This arrangement continued for three years. The rapid growth of astrophysics, however, compelled Hale to return to his original idea of an independent journal exclusively for astrophysics. In 1895, he founded the *Astrophysical Journal,* with the subtitle, *International Review of Spectroscopy and Astronomical Physics.* Hale and James E. Keeler, another preeminent American astronomer, became the joint editors. It had a board of editors which included astronomers and physicists from England, France, Germany, Italy, and the United States. The University of Chicago purchased the journal, and the first issue of the present *Astrophysical Journal* was published in January 1895.

Hale and Keeler continued to be the editors until 1905, when Hale left Yerkes to become the director of Mount Wilson Observatory. The editorship of *ApJ* and the directorship of Yerkes were passed on to Edwin B. Frost, who had been an assistant editor since the inception of the journal. Frost was in charge until 1932, when Otto Struve became the Yerkes director and with that position inherited the editorship of *ApJ.*

Because of Hale's efforts and the traditions he established, there were frequent review articles on spectroscopy, cosmology, stellar motions, and other astrophysical topics by internationally renowned scientists. The astronomical publications in the journal, however, primarily consisted of the results from the Yerkes and the Mount Wilson Observatories. Other observatories maintained and took pride in their own observatory bulletins and publications. The situation changed slightly in the mid-1940s, mainly due to the efforts of Struve, who envisioned that the *ApJ* would one day become a national journal, the equivalent of the *Physical Review* of the American Physical Society. Struve persuaded the Lick and Harvard Observatories to publish in the *ApJ.* With the participation of these four major observatories, a large bulk of the astronomical work done in the United States began to be published in the *ApJ.* Theoretical papers were also being submitted, and Struve, who needed help in deciding about them, took on Chandra as the associate editor in 1944.

Still, at this stage, the *ApJ* was far from being a national journal. There was no refereeing as such; the editors acted only in an advisory capacity. The financial and editorial control rested completely with one institution, namely the University of Chicago. The participating observatories could publish up to a maximum of 100 pages free of charge, but they had to pay fees to the University of Chicago for publication costs of all additional pages published. These costs seemed to grow astronomically for participating institutions and individual authors in the late 1940s. Poor management of the journal by the University of Chicago was largely to blame,[7] and criticism against the *ApJ* started mounting among members of the AAS.

The need for a national journal sponsored by the AAS was evident, but small membership made it virtually impossible for the society to have a separate journal of its own in competition with the *ApJ*. For the time being at least, the only practical way to acquire some measure of control over the *ApJ* was to become a cosponsor of the journal with the university. That was not as easy a matter as one would think. The two institutions, the AAS and the University of Chicago, had their separate bureaucratic hurdles, rules, and regulations.

The right opportunity came when Struve became the president of the AAS in 1947. He passed the position of managing editor of *ApJ* to his senior colleague, William W. Morgan, with Chandra continuing as the associate editor. As one who had always hoped to see the *ApJ* become a national journal, Struve instituted in 1949, through the AAS executive council, a Committee for Publications to carry out a comprehensive study of the journal situation and consider the matter of joint sponsorship of *ApJ* by the AAS and the University of Chicago.

After a year of intense work on all aspects of the journal, this committee, headed by Lyman Spitzer,[8] came up with its report and recommendations in the summer of 1950. Based on this report, Alfred H. Joy, who had succeeded Struve as president of the AAS, prepared a proposal in July 1950 and sent it to the president of the University of Chicago, Ernest C. Colwell.

Joy noted that the society had a small membership; it was neither interested in nor capable of entering into a large business or assuming editorial responsibilities. The council had voted to continue its support for the *ApJ*. But, "while the older members of the society, especially those who had worked with Hale and Frost," wrote Joy, "were exceedingly jealous of the good standing and success of the *ApJ*, and proud of its fine record, its basic contributions to astrophysics, and its unequalled standing among scientific journals, there was mounting severe and sometimes bitter criticism of the journal and the University of Chicago among newer members, authors, and subscribers. Submissions were being shunted in increasing numbers to other journals."[9] Therefore, there was an urgent need to take immediate steps. He endorsed the publication committee's recommendations, namely, that the journal be sponsored jointly by the AAS and the University of Chicago under the supervision of an editorial board of seven members nominated by the AAS and approved by the university; the managing and associate editors be two members of the board serving a rotating five-year term; every regular member of the AAS be required to subscribe either to the *Astronomical Journal*[10] or the *ApJ*. The regular membership would be increased to include the subscription to one of the two journals. Joy requested that an agreement based on his proposal be signed between the AAS and the University of Chicago as soon as possible.

Colwell sent the draft to Morgan, the managing editor, for study and advice regarding a response. Morgan entrusted the responsibility to Chandra, as Chandra was then the acting chairman of the astronomy department at Yerkes.

Chandra recognized that the society was asking too much and paying too little to the University of Chicago in return. The draft, if accepted, would mean the virtual editorial control of the journal by the society without it in any way sharing the financial responsibility. The members of the board of editors would be nominated by the society, and although, to begin with, the managing and associate editorships would remain with the University of Chicago faculty (if the existing arrangement continued), there was no guarantee that, in the long run, the editorship would stay under the university's control. This drastic change in the control of the journal clearly would not be acceptable to the university. Taking this into consideration, Chandra prepared an eight-page draft letter for Colwell. He described in detail the extent of support the university was providing to the successful operation of the journal and argued that as long as the university continued to provide that support, both the managing editor and the associate editor should belong to the University of Chicago faculty. He also argued that the board members should be nominated by the university and approved by the AAS council.

Joy's initial reaction to the modified proposal was quite favorable. He recognized the need for the continuing support of the University of Chicago, and it did not seem to matter to him who nominated or who approved the board members, as long as there was agreement between the two parties. This attitude on his part, however, did not prevail in the next round, when Chandra formulated the draft proposal in the form of an eight-point formal agreement to be signed by the society and the university. Now Joy had second thoughts. It appeared to him and to the council members as though the society had virtually no role at all in the journal. Joy wrote to Colwell on 11 December 1950:

. . . your stipulations do not meet the hopes that we had in mind. . . . I had hoped that some practical way could be found by which the members of the Society could feel that they had a definite interest in the journal without jeopardizing the fine reputation of the journal or infringing upon the unquestioned rights of ownership by the University.

Chandra felt that there was still room for negotiation and accordingly drafted a response seeking clarification in view of the earlier favorable response. Colwell sent the letter, but added his own prefatory remarks,

I am sorry that the terms proposed in my letter of November 27 do not come up to your expectations. We do feel that as long as the University has the total financial responsibility for the *ApJ,* it must retain the editorial responsibility.

These remarks infuriated the council members, who interpreted them to mean the foreclosure of further negotiations. The AAS council promptly rejected the University of Chicago's proposal at the annual 1950 Christmas meeting in Haverford, Connecticut, and informed Colwell of the decision on 5 January 1951. The council considered the negotiations closed and voted to reactivate the Committee on Publications to come up with alternate proposals.

This unexpected turn of events occurred at a most inopportune time for Chandra. Lalitha and he were due to depart in less than two weeks for England and India for a three-month sojourn. They had waited for this trip for a long time. Fourteen years had elapsed since they had left home, and Lalitha was extremely anxious to visit with her mother, who was quite elderly and in poor health. After World War II ended, Chandra had tried and failed to elicit an official invitation to visit India.[11] Finally, they had saved enough to make the trip on their own.

Recalling the events of some thirty-odd years ago, Chandra says,

I was to leave from New York on Monday [5 February 1951]. I wrote to Lyman Spitzer and Martin Schwarzschild: "I am arriving on Sunday [4 February] at 8 AM. Could we get together and discuss this matter?" On Sunday morning we got together. I sat on one side of the table. Lyman and Martin sat opposite to me on the other side. I said, "Lyman, you are on the council. You rejected the draft because of Colwell's remarks. Tell me what is wrong with the draft itself which I had prepared. You make the changes. I will take it upon myself to get it through the university. You get it through the council." That evening there was a party at the Schwarzschilds' home where we were staying. Lyman brought a handwritten draft to the party. I took a look at it and said fine. Next morning, I sent a copy of the draft to Morgan asking him to write to me in England if there were any points which were unsatisfactory to him. And we left New York that day.

There were indeed only a few changes made to the original draft of the agreement as stipulated by Chandra. The editorial control would remain with the university, as the managing editor and the associate editor would be from the University of Chicago faculty. The editorial board, however, would be selected and approved by the council of the AAS from a list of names (at least two names for each candidacy) provided by the managing editor. The editorial board would formulate the general policies, decide on matters of subscriptions, page charges, and so on. The execution of the policies rested with the managing editor. The new draft met with the approval of the reactivated publications committee and Joy expected its approval by the council as well. After Chandra returned from India, he met with the university's vice-president, R. Wendell Harrison, because Colwell had resigned and left the university. Harrison agreed to sign the document on behalf of the university and send it to Joy for his signature on behalf of the AAS. This was in June, and Chandra thought the matter was settled. In October, however, he learned that Joy was still waiting to hear from the university.[12] Chandra says,

I went to Harrison and asked him why he had not done what he said he would do. He said, "Well, you know the University Press is disinclined," and so on. I told him, "I do not know if it is fully realized how much diplomacy and tact were needed behind the scenes to reopen the negotiations which the AAS had broken off last December. I had a long conference with Dr. Lyman Spitzer, chairman of the Committee on Publications of the AAS, in February this year, and it was only after explaining to him in great detail the special responsibility of the University of Chicago in the *ApJ* that he could be persuaded. And on the strength of what you said to me in June, I informed the council it was alright, and the council was persuaded to accept the contract with enormous difficulty." Harrison again said, "Well, the Press has some objections." I said, "Why don't you call the director of the Press to your office right now. You are the vice-president. You can do that." "Well, you *are* in a hurry," Harrison said. "Mr. Harrison, I have put all my personal prestige on the line with the society in saying you are going to sign. Now you tell me you won't. I have no choice but to resign from the university." Harrison said, "Oh, don't be rash." I said, "I mean it. Either I leave this office with your signature on the contract or with my resignation in your hand." Well, the director was brought in. We argued, and in the end it was signed and sent to the president of the AAS.

Chandra had done everything he could do. It was now up to the AAS to accept the contract. It had to pass through the council and the general body meeting of the AAS at its 1951 Christmas meeting. Since

Chandra was a council member, he could "play the strings from both sides." He recalls, "In the council, I was spokesman for the university, and in the university, I was spokesman for the society." With Spitzer, Schwarzschild, and President Joy on his side, the contract went through the council smoothly. But it was a different matter with the general body of members. The requirement that every member of the AAS subscribe to the journal and the editorial control of a national journal by one institution were touchy issues. The old stalwarts, like Harlow Shapley, resented the domination of the journal by a group of young Chicago astronomers who were all in their late thirties or early forties.

As Chandra remembers, President Joy handled the situation quite diplomatically. "He was simply magnificent. He told me, 'Chandra, you must be completely in the background.' I said, 'O.K., I will go and sit in the last row and I won't say a word.' He had it all arranged, who would speak and who would not speak." Still, Chandra noticed that Joy was quite nervous throughout the meeting, the reason for which became clear only after all the statutes were approved by the general body. When the meeting was over, Joy came over to Chandra and showed him a five-page telegram from Shapley addressed to the assembled members. The telegram expressed Shapley's disapproval of the joint sponsorship. If it had been read, the strong support he enjoyed among the general membership would have created quite a stir against accepting the contract. "The telegram came too late, didn't it?" said Joy to Chandra. Both smiled and agreed.

With this bit of conspiracy, finally the path was clear for all the formalities to be completed in the early part of 1952. The *ApJ* became a national journal sponsored by the AAS, published by the University of Chicago Press. The editorial board[13] was formed. It held its first meeting in May 1952 on the University of Chicago campus.

Two major policy matters concerned the board in its first meeting. One was the publication of the Supplement to the *ApJ* in addition to the regular numbers. The other was to develop guidelines for dividing the publication material between the *ApJ* and the *Astronomical Journal*.

At the insistence of Paul Merrill, a member of the Committee on Publications, Chandra had agreed, although it was not part of the signed contract between the university and the AAS, to the publication of the Supplement, which would contain bulky material, such as numerical tables and tables of wavelengths. As *ApJ* was to become the national journal of astrophysics and replace the observatory bulletins which normally published such material, Chandra thought it was a reasonable stipulation and, accordingly, had informed Morgan about his decision. But, while Merrill was content to publish the bulky material in an inexpensive way, using inferior quality paper, Chandra proposed that the Supplement be published with the same care and quality as the regular journal. Furthermore, the Supplement would be, according to Chandra, a natural place for review articles and articles which could not be accommodated in the regular journal because they were too long. The board went along with Chandra's proposal.

In the meantime, however, Morgan felt that it was all Chandra's idea to publish the Supplement in the first place. Morgan felt he was overworked. His own research was suffering, because the regular journal was taking up too much of his time. He did not want this additional responsibility. Chandra, according to Morgan, had sold the journal down the river by yielding to the AAS council. But Chandra had kept Morgan informed of every detail while the negotiations were going on. "Bill, you have seen everything," Chandra explained. "You never objected to anything earlier. You told me I had the ball. I've done what I thought was best. It's your job now."

Morgan, who was not feeling well in those days, would not be appeased, and remained uncooperative. At the first board meeting, for instance, in order to develop guidelines for the division of the material between the *Astronomical Journal* and *ApJ,* the editor was advised to bring a list of papers published in the *ApJ* during the previous five years which did not belong in the *ApJ.* Morgan was not present at the meeting, but he submitted a list of some thirty items, twenty of which were Chandra's papers. Embarrassed, the board ignored his list in making its recommendations.

After the meeting Morgan remained unsupportive. Chandra had to carry out all communication with him through the mailbox, although their offices were in the same building. The only recourse left for Chandra was to go to Morgan and tell him directly, "It is quite clear that you are dissatisfied with the way I negotiated the contract. I thought that I had done it fairly. You had copies of all correspondence. Apparently the end product is not to your taste. There's nothing I can do about it. On the other hand, I can see very well that being the managing editor with me as an associate editor is not satisfactory. Since you have the responsibility to carry on, I will resign my associate editorship."

To which Morgan responded, "Will you put that in writing?"

"Certainly," Chandra said, and he tendered his resignation to Strömgren, the director of Yerkes, that very day, Monday, 17 June 1952. But within two days, on Wednesday, Morgan resigned, allegedly for health reasons.

It was clearly a crisis situation. There was no one on the University of Chicago faculty other than Chandra who was qualified to assume the responsibility. After the protracted struggle to keep the editorship with the university, how could the university now say that there was no one at the university willing to be the editor? "I had no choice," Chandra says. "I was essentially forced to take the journal when Walter Bartky, the Dean of Physical Sciences, asked me. I felt a personal obligation. I had been responsible for the negotiations which brought about the change and the transfer of the journal, and the change required that the editor be from the University of Chicago. I resigned my associate editorship on Monday. As it turned out, I became the managing editor on Friday."

It was thus not choice but peculiar circumstances that forced Chandra to become the managing editor of *ApJ*. He had a difficult task ahead. The astronomical community was not accustomed to the ideas of refereed papers and page charges as requisites for publication. A substantial fraction of the society felt that the *ApJ* would go to shreds. "He is a theoretician. He doesn't understand astronomy," some said and continued to say.[14] The journal faced financial difficulties. According to the terms of the appointment of the editor by the chancellor of the university, the editor was responsible for the budget of the journal and running it as a self-sufficient, nonprofit, scholarly enterprise of the university. But, at the end of the first year of his tenure, Chandra found himself with a deficit of $7,000.[15] Soon all this changed. With his efficient handling, prompt correspondence, and the general feeling that the *ApJ* was to attain new high standards, papers started streaming in. The journal grew in size, and so did its income. Within five years, the journal was operating in the black. As Carrol G. Bowen, managing editor for the Journals Department of the University of Chicago Press, wrote on 30 July 1957:

> You are the splendid steward of intellectual assets and your responsible exercise of these duties is demonstrated in every way; greater income, greater circulation, greater volume in pages, and all with increasing surplus.
> You must run a school for other editors, when you retire!

He set the tone of the journal. The editorial and refereeing policies were his. He alone was responsible for content and style. "I was very autocratic," confesses Chandra. "I used to reject papers, rewrite them sometimes. I simply did not consider the editor's office as a transmittal of referees' reports to the author and back." The following excerpts from a letter to Hermann Bondi[16] illustrate his ideas and procedures:

When I receive a paper, I first examine it to see if it can be accepted without any reservations. I should say that about 10 percent of the papers are in this class; a paper from Merrill on the spectroscopic characteristics of a variable star or a paper by Schwarzschild on a new model for the sun clearly requires no refereeing. I consider myself as an adequate referee for about 10–15 percent. The others are sent to external referees who are selected by me. I need not add that the obviously crank papers are summarily rejected.

You have asked me if refereeing delays the publication of papers. I see to it that it does not! I ensure this by a number of devices: since I select the referees, I know who can be relied upon and who cannot. If a referee is dilatory, I send a polite but firm letter of reminder; such letters go to the referees if they have not returned the papers in three weeks. Any referee who keeps a paper for more than a month is asked (with due apologies!) to return the paper without his report. Seriously, I should say that I have obtained from my referees the utmost cooperation. This stems apparently from their recognizing that I am doing a job of service and that I am entitled to their cooperation.

I might add one general comment: A Managing Editor with responsibility and strong policy can serve his profession far more effectively than one with constraints and interference. But to be successful he must at all times consciously try to be fair, not inject his own personal prejudices and consider that his primary responsibility is to his science and to his profession. At least this is how I look upon my own duties as the Managing Editor of the *Astrophysical Journal.*

Indeed, he adhered to the avowed principles as strictly as was humanly possible. Generally speaking, this was well appreciated by the community; and with such a high personal stature to back him up, he had no problem in maintaining absolute control. However, as is to be expected during such a long span of editorship, there were instances of tempers flaring between the authors and the referees, accusations of bias and partiality, priority fights, and threats and abuses. Authors' tirades against referees often took the following forms:

. . . I consider that all the referees' comments are unimportant or sniping. I do not feel that the MSS needs further revision . . .

. . . You have selected a referee who is evidently not at all a disinterested person. If I am to judge by his comments, which I have shown to be replete with trivia, confusions, and downright errors, he is also not a competent person in this matter. What is more, the whole process of refereeing this paper has proceeded extremely slowly.

In my opinion you have grossly misused your editorial authority in this instance. In due course I propose to place a full account of the matter before all the members of your Editorial Board.

. . . the referees have not only demonstrated an incredible ignorance of the literature basic to the development of the field, but have also attempted to pad out an incompetent review with well-known material developed by the authors themselves, with irrelevant comments, and fatuous personal attacks.

Chandra took them all in stride, handling them promptly, forthrightly, and often with a flair of his own. His responses varied, but in general they tended to defend the referees, as he wrote in one instance:

In selecting referees for papers, this office is very anxious that they are competent persons in their fields. Moreover, only those parts of the reports are transmitted which have the confidence and support of the editor. You can therefore appreciate that it is incompatible with the policies of this office to accept charges which impugn the integrity of our referees: it is equivalent to a similar charge against the editor.

On another occasion, he wrote:

With very large numbers of papers passing an Editor's desk it is of course impossible that occasional errors of judgment are not made. The best an editor can try is to treat all papers which are received for the journal on an equal footing by the same procedures; I am not aware that in this instance I have followed any other.

A particularly difficult and noteworthy case was that of an author whose uncompromising attitude, and his insistence on his own brand of theoretical interpretation of his observational data, created a recurring source of trouble for Chandra and the referees. After receiving the first of a series of three papers, 108 pages long (accompanied by 17 zincs and 27 halftone illustrations), Chandra wrote in his rejection letter to the author,

. . . you stated very correctly "the publication and refereeing represent severe problems." Indeed since they are, may I, transgressing editorial protocol, state quite frankly what my own scientific position has been with respect to your work, an attitude which I cannot pretend has not influenced the manner in which I have proceeded editorially.

It was clear to me that you have made some discoveries which will in the long run profoundly influence developments in cosmogony and cosmology; but in your presentation the relevant and the not so relevant tended to be mixed. And it is possible that in my selection of two independent referees, I have been unduly influenced by my own skepticism with respect to your more directly theoretical deductions. But I did explain to both referees that vague and general reports will not be useful; and that in order to be useful they must be concrete and detailed.

The referees did give a concrete and detailed report. The paper was rejected unless it was extensively revised by removing the theoretical speculation and interpretation of the observed data. The author refused to make the revisions. The next two papers met with a similar fate, with Chandra's response:

I have examined this paper together with a referee; and his attached report (modified as per my judgment) is self-explanatory.* You will notice that this paper shares with its companion, the same difficulties for persuasion and understanding. However, the paper is somewhat different in the sense that it includes, as the referee states, observational material of an inspiring quality. I wonder if I could ask you to rewrite this paper including only the observational material and excluding all interpretations. Perhaps it is unfair to ask you, but I am afraid that, in the present form, it is unacceptable for the *ApJ* for the same reasons I have explained in connection with your other paper.

I hope you will forgive me if I say quite frankly that it is very difficult for me to read your paper consecutively: the language and the arguments you use are outside the range of my experience. I could of course be specially impervious to the compelling nature of certain types of observations; but I am sorry to say that many others (including the referee I consulted) seem to share my difficulty. To use a phrase much in vogue, I am afraid you have created a credibility gap.

*Contrary to my decision in the case of the other paper, I am enclosing the nearly full report. While it is too much to expect that you will agree with any of it, I hope that it will at least convince you of the magnitude of the task you have if you are to convince any conventional scientist.

The author, after the rejection of all three of his papers, responded:

Two of the most knowledgeable people in that field (which is not the Editor's field of special knowledge) had read the paper and unequivocally recommended its publication before I submitted it to the *ApJ*.

The technique for rejecting papers in the *ApJ* is well-known to astronomers. In each of the three papers under discussion, a single hostile referee was selected to evaluate the paper. The rude and unobjective report of the referee, transmitted over the signature of the editor, is *prima facie* evidence, in my opinion, of improper editorial procedure.

My experience is by no means unique. I know a number of authors who experienced so much more difficulty than other authors that they are effectively excluded from publishing in the *ApJ*.

He sent copies of the letters to all the editorial board members, urging them to make an inquiry and impeach the editor. Chandra's response was:

May I thank you for the frank way in which you have brought to the attention of the Editorial Board of the *ApJ* your disappointment concerning my editorial decision with respect to the two papers you submitted in the spring. . . . The policy with respect to refereeing I have followed with respect to papers submitted for publication in *ApJ* is, apart from minor variants due to special circumstances, the same as that of many other periodicals with which I am acquainted. I try to be fair and impartial; but, of course, misjudgments are possible and indeed inevitable in view of the volume and range of the material submitted. However, from your letter I gather that displeasure with the Editor of the *ApJ* is widespread. It is better for me to know this fact than to live in a fool's paradise or should I say, Editor's paradise.

The author's impeachment plea did not go far. Chandra let the other board members decide. Edwin E. Salpeter, the chairman of the board, wrote to the author, staunchly defending Chandra, saying, "he [Chandra] does as outstanding a job as is humanly possible under the present ground rules (an editor of the highest scientific stature who exercises his own judgment in the selection of the referees and who uses the reports of the referees merely as advice and makes his own final decisions). There are, to be sure, occasional complaints, but we feel that these are more than balanced by the overall competence and integrity of Chandrasekhar's editorship."

There are several other instances on record[17] which, in retrospect, are hilarious. Being refereed and asked to make revisions did not always sit well, especially with prestigious institutions. "We at the Lick Observatory know what papers to publish," wrote the director in defending a paper of one of his scientists. "I suggest that you publish it as we sent it." Chandra, of course, said he would not and rejected the paper.[18] The irony was that the referee (anonymous of course) was Henry Norris Russell, who, having detected a serious error, had recalculated and worked out the entire paper for the author. Chandra recalls that this referee's report was the last piece of scientific work that Russell did. In another instance, a good friend, well-known astronomer, and the chairman of the department, in submitting a manuscript of a visiting Russian astrophysicist for publication, wrote:

We are not specially proud of this paper. Nevertheless in view of the fact that this is the first communication by a Soviet astronomer while residing in the United States, I would recommend a sympathetic attitude toward the manuscript. I am sure that you will realize that rejection will cause embarrassment all around. Therefore, I strongly urge its publication.

"I am sorry that the standards of the *Astrophysical Journal* cannot be relaxed to make your political problems easier," was Chandra's response in rejecting the paper for publication.

Chandra became known to be particularly harsh toward new Ph.D.'s who wanted to publish their thesis work as papers. "First burn your thesis," he was known to say, "and then write the paper." However, this harshness was generally directed toward the thesis advisor:

. . . since it appears that the paper is the student's thesis carried out under your supervision; and the reports I have received on the paper are so adverse that I have hesitated to send them to the author since it might discourage the person unduly; and I hope you will forgive my saying that it appears that the poor shape of the manuscript is in some measure due to inadequate supervision.

It was Chandra's policy not to let the pages of the *ApJ* be used as a debating ground or to carry on prolonged controversy. He would step in and stop such controversies, as the following excerpts from two different instances show:

. . . My own tendency, when I have been involved in similar situations, is to let an error bury itself. Do you feel that in this instance the burial should be more ceremonious?

. . . I can see that X may deserve a reprimand for his handling of the available data. While the publication of X's original paper might have been unwise, it does not seem to me that authors of your distinction and experience need to undertake this reprimand. It seems to me that the best thing one can do about a misguided paper is to ignore it. I greatly hope that I can persuade you not to give X's paper any additional publicity it clearly does not deserve. I should be unhappy if you insist on publishing your letter; in that event I would of course have no choice.

Perhaps the most touching incident was that of John Waddell III, who used to submit semicrankish papers which were generally rejected by the referees. "He sent one paper which was very, very long," Chandra recalls. "It went back and forth, and finally was rejected. One morning I got a call . . . from John Waddell the Second. I didn't realize that Waddell the Second was different from Waddell the Third, so I said, 'I'm sorry, your paper has been rejected.'

"The caller said, 'I'm not John Waddell the Third. I'm John Waddell the Second, the father of John Waddell the Third. Are you the editor?'

"'Yes,' I said.

"'Do you know you killed my son? My son came from Europe this morning and there was a letter from you rejecting his paper. He was so angry that he took his car and drove away. He was involved in a collision and was killed. Therefore, you are responsible.'"

Before Chandra could respond, he continued, "You damned foreigners! Why do you have anything to do with American astronomy?"

Then Chandra said, "Mr. Waddell, your son got killed this morning in an accident. I'm terribly, terribly sorry it happened that way. I can understand your sorrow and feeling of tragedy . . ."

John Waddell the Second said, "I insist that the paper be published as it stands."

Chandra responded that he could not promise him that, but if he sent his son's manuscript, he would see what he could do.

When the manuscript arrived, Chandra enlisted one of his referees to retrieve and condense from the sixty-page paper five or six pages of publishable material. The referee prepared a short paper, which was published with an editorial note, and Chandra sent a reprint to the father, saying, "You called me at a very tragic moment and asked me to publish your son's paper as it was. I could not publish all of it with the integrity of an editor, but I have succeeded in getting it reduced to an amount which I thought was reasonable to publish. This is all I can do." Subsequently Chandra received a letter from John Waddell II, apologizing for his behavior.

The personal involvement, care, and attention with which Chandra carried out his editorial responsibilities, as well as the enormous expenditure of time and the drain on his energies they required, should be evident from the few instances encapsulated here. Nonetheless, it is amazing that Chandra did not let this interfere in any way with his own scientific work. When he took on the editorship, he had made up his mind to that effect, and indeed, when one looks at the record of his scientific work, which continued essentially undiminished in quantity and untarnished in quality, it is difficult for anyone to believe that he was also the editor of *ApJ* for nineteen years.

Furthermore, he did not let the journal affect his teaching responsibilities, nor did he seek in any way to reduce them. "No, the idea never occurred to me," Chandra says. "In those years, I integrated my research with my teaching, which I have always felt to be an essential component of my scientific efforts." How did he manage? "By being extremely strict about apportioning the time." The journal office, although located in the same building, was separate from his usual office and had strictly prescribed hours. Outside those hours his mind was firmly closed to journal business. If the phone rang and the call was about the journal, he quickly responded, "The journal office is closed."

"Am I not talking to the editor?"

"Yes."

"May I speak to you for a minute?"

"I told you, the journal office is closed. Please call the office tomorrow at ten o'clock in the morning."

This kind of encounter was a novel experience for the astronomy community and did not go well at the beginning. But Chandra was strict and consistent about it with everyone. As he says,

When you do this constantly, people get to know and understand it. Even the people who resented these remarks at an earlier time began to respect it. For example, take my friends Martin Schwarzschild and Lyman Spitzer. They used to call me on personal matters and then they would say, "I have a small question about the journal." And I would say, "Why don't you call me tomorrow morning about it?" Even to my friends . . . they never resented it. They perfectly understood that if you want to be away from the journal for certain periods of time, you simply had to be away. It doesn't matter whether the problems come from your friends or your colleagues, you can't be distracted.

Thus, with strict allotment of his time, being methodical to the point where all human aspects of the problem were eliminated, Chandra could keep the often irritating, disturbing editorial affairs from affecting his science. He had to pay a heavy price for his rigidity, however, as he says:

. . . it took an incredible amount of my time, which meant I had to forgo many, many things. For example, I don't think I left the university for longer than a period of three months during those nineteen years, and even that probably only on three occasions. If I was gone for a week inside the country, the office would call at 2:00 PM to tell me what was going on, and I used to leave instructions. My editorship essentially required that I remain in Chicago. Normally a scientist after he is forty, when he is moderately lucky and has sufficiently established himself, travels around, goes and spends time at other places, and so on. But I never experienced any of these advantages. The first time I left the university for any period longer than three months was in 1971.

Those nineteen years also took a heavy toll on his professional relationships. The demands he made on himself, his uncompromising standards, even his wish to be absolutely impartial and fair led to an isolation from the rest of the astronomical community. As Chandra reflects,

. . . in 1952, just before I became the editor, I spent a month at Caltech. During that period, I was entertained, met everyone on the staff (Edwin Hubble, Harold D. Babcock, Walter Baade, Ira S. Bowen, and Rudolph Minkowski were there). I was at Caltech again in 1971 after having served the astronomical community for the intervening period. And even though several astronomers at Caltech (must have been some twenty or more of them) had called me on the telephone hundreds of times, talked to me about their papers, not a single one of them called or talked to me. I don't think they did it out of ill-feeling, but I think it was due to this isolation I created for myself . . . the paper as the paper, never the man behind it. . . .

In contrast to the situation with astronomers, his relations with the staff of the University of Chicago Press were full of warmth and affection. Over the years, he had excellent rapport with the copyeditors, editorial assistants, the typesetters, the printers, and the business managers.

"Everyone at the Press really thought the world of him," according to Jeanne Hopkins, who was the chief technical and copy editor from 1967 until her death in 1984. "He was willing to do anything. He made everybody feel that the *ApJ* was really the most important thing that they were doing." Jeanette Burnett, who worked as his editorial assistant, shares the same feeling: "He had a great deal of respect for the people who actually did the typesetting work, for their craftmanship. He liked to go over there and actually work with the compositors; he would talk to them about what made a good-looking formula. He had a sort of aesthetic sense about that. He really cared. You can tell that when you look at his handwritten formulas." Speaking about the work routine and how closely he used to work with the staff, Burnett says:

He would come in two or three days of the week. We'd just fly through the materials, the stuff that had accumulated during the time. We had procedures all worked out, everything was mapped out so that it could operate pretty much without him. We would set everything up so that when he came in he could just look at the material and make decisions very quickly. Very efficient. He was very much a person who saved time. He didn't do a lot of extraneous stuff, and he didn't believe too much in real formality. He was more interested in getting things done and getting them done quickly with a minimum of fuss. . . . He also did practically everything on the journal. That is, he looked at all the proofs and transferred all the authors' corrections from the proofs that they sent in to a master set of proofs. The editorial assistant would sit there and kind of turn pages, make sure that everything was caught. We would just go down the paper together, but he would actually do the corrections.

Chandra himself likes to tell a story about his "extremely good relations" with the Press. In February of 1963 Maarten Schmidt called him from Pasadena and told him that he had found an enormous redshift—a redshift of 0.2, the biggest at that time—showing that the quasars were at cosmological distances. Would Chandra publish his findings in the journal?

Chandra said to him, "Well, today is Wednesday. If you send the manuscript by airmail, I will get it on Friday morning. The next issue of the journal is going to press on Friday. I will hold it, and I will have your paper typeset. And I will read the proofs over the weekend. Your article will come in the next issue." He told the Press about it, and they agreed not to print the journal until Monday.

The six-page paper arrived on Friday, not by the morning mail, however, but by the afternoon mail. When Chandra took it to the typesetters at two in the afternoon instead of ten in the morning, as he had anticipated, the foreman said, "Mr. Chandra, we can't do anything about it because we are not supposed to work on Saturdays. So how are we to do that?" When Chandra looked sad and disappointed, the foreman said, "Wait, let me see what my men tell me." He returned after talking to them for a few minutes and said to Chandra, "Well, Mr. Chandra, nobody need know that we worked on Saturday. Right? But one condition: on Monday, at lunch time, you will have to tell us all what this is all about." The paper was typeset and proofread on Saturday, and the journal rolled off the presses on Monday. And on Monday, at lunch time, the compositors and the proofreaders of the Press were the first in the whole world to learn about quasars! "I don't know whether they understood it," says Chandra, "but a few months later, when Maarten Schmidt appeared on the cover of *Time,* copies of it were plastered all over the place. From that time on, anything I said went. They were extraordinarily nice."

In 1967 Chandra started the "Letters to the Editor" part of the *ApJ.* When he announced its formation at the annual AAS meeting in Madison in 1966, the conversations that Chandra overheard at lunch tables went, "Well, Chandra just wants to imitate the physicists. That is his weakness; he wants to do everything the physicists do," alluding to the *Physical Review Letters,* which was started some years earlier (July 1958) by the editors of the *Physical Review.* Chandra, however, foresaw the need for the Letters section as the number of important discoveries in astronomy (such as quasars, X-ray sources) increased in frequency.[19] Rapid dissemination of the new knowledge was desirable, and Chandra felt that it was his responsibility to provide this service within the astronomical community. He was indeed proved right. The momentum of discoveries continued to increase. If the Letters section had not existed, there would have been a demand for one. And according to Chandra, "It was better for the journal to foresee the need than to be forced to improvise some solution when it was being pressed from the outside instead of from the inside."

The Letters, however, placed heavier demands on the editor than the regular journal. The need for rapid publication allowed less time for the customary author-referee interaction. It introduced more subjective elements, such as whether a contribution was urgent and important enough to warrant rapid publication, and whether it was written clearly enough to be useful outside a highly specialized area. Chandra took up himself the task of refereeing *all* the submitted letters and deciding their fate. He made *all* the corrections, and if necessary to save time, he consulted authors on the telephone. It became an immense chore, especially since publication in the Letters soon began to acquire special weight and prestige. Astronomers attempted to use the Letters as a vehicle for quick publication to claim priority, for unwarranted speculations, and ill-thought-out ideas. Chandra was forced to reject a large percentage of the submitted letters. That did not contribute to his popularity. To one author, for instance, he wrote:

> I hope you will permit me to say, in my personal capacity, that you appear to be diluting your unquestionable superior abilities in too many small publications, for all of which you seek the urgency of a letter. The fact that an idea is novel or that it has never been stated before does not necessarily justify its prompt publication. Please forgive me for transgressing editorial protocol and expressing these views.

Around 1967, Chandra began to think seriously of disengaging himself from the journal. Lalitha had been ready for him to resign as editor ever since 1964. He had decided to enter into a new area of research in general relativity; in addition, his work on ellipsoidal figures of equilibrium had reached a stage of completion such that Chandra could say, "I have a complete perspective now; I have done all I can. The subject needs to be rounded up, integrated by tying up all the loose ends. It merits a book." In fact, Chandra says he felt compelled to write a book based on the Silliman Memorial Lectures he gave at Yale in 1963. Perhaps most important of all, Lalitha was urging Chandra to unburden himself so that they could have a little more time together.

Furthermore, in spite of the journal's growth in size and stature and in spite of its sound financial basis, Chandra felt, and rightly so, that the administrative structure supporting the journal was highly unstable. The *ApJ* was a national journal owned by one private institution; essentially one man was responsible for all phases of its operation. What if something happened to that one man? "In fact, as late as 1969, if I had died accidentally or otherwise," Chandra says, "no one would have known what to do with the journal, because I did everything. My office took care of the page charges, the proofs, the refereeing, advertisements, budget. . . . It was simply not fair that the journal, which had acquired the national prestige it had, should be so fragile in its structure. A national journal should be a national responsibility."

His personal disengagement and the transfer of the journal to the care of the AAS were not simple matters. Having shouldered the responsibilities for so long, he would not even dream of just relinquishing it by tendering his resignation, as he could have done. He felt obligated, he says, to bring about a smooth transition, which meant getting it through

the society, through the university, each with its own rules and traditions. It became a complicated process which lasted four years.

First, it took two years to find someone who was willing to take on
the responsibilities of managing editor. Informal inquiries led nowhere.
A typical reaction was that of Donald Osterbrock, one of his former students, then a professor of astronomy at the University of Wisconsin,
Madison. In response to Chandra's invitation to become the managing
editor of the *ApJ*, Osterbrock wrote:

> I am terribly pleased that you would like to see me succeed you, and I only
> wish I could do it. But I feel it is far too much for any one person to do and
> continue with his scientific and teaching work also, unless he is prepared to
> make very great personal sacrifices. I have tremendous admiration for you for
> making these sacrifices, but I'm afraid I can't make them myself.

In December 1969, Chandra finally found Helmut A. Abt. Abt wanted
a new organizational structure for the journal to be established at Kitt
Peak National Observatory, where he worked. It would include, besides
Abt, an assistant managing editor, a separate editor for the Letters, a
number of scientific editors, and, of course, the editorial board. It would
no longer be a one-man operation. Kitt Peak would provide a "pleasant,
11′ × 15′ office" and the necessary space to house the editorial staff.
The journal would not be completely under the control of the AAS; the
University of Chicago Press would continue to be its publisher. Martin
Schwarzschild was then the president of the AAS. With his help, the
transfer and a new agreement took shape, and as of 1 April 1971,
Helmut A. Abt became the new managing editor.

Chandra also worried about the production end at the University of
Chicago Press. Someone had to be responsible for all the jobs he used to
do or oversee, like making up the budget, the page charges, proofread-
ing, and copyediting.* A year before he left *ApJ*, he created the position
of production manager. The obvious candidate for the job was Jeanette
Burnett, who had worked with Chandra for a number of years, but the
Press establishment would not have her in such a high position. They
could not, however, refuse Chandra when he offered himself for the
position. Burnett then became the assistant production manager for a
year and took over the job from Chandra when he resigned.

In addition, there was the problem of the reserve fund of $500,000,
which, strictly speaking, did not belong to the university but was offi-

---

*The current editorial and production staff for *ApJ* at the University of Chicago Press includes twelve full-time employees and one part-time employee.

cially a part of the university's general funds. The board of trustees had to approve the transfer of the fund to the AAS. Edward Levi, who was then the provost, was at first disinclined to the transfer. "Chandra, the *ApJ* is one of the goodies of the university. Why do you want to give it away?" Chandra convinced him that there was no alternative. If the university could not run it the way it should be run, it should give the journal away rather than see it destroyed. Then Levi said, "It seems to me that in your loyalties to the university and to the journal, the journal always wins." "Would you have it any other way?" Chandra responded instantly. "No," Levi said, and together they got the approval of the board of trustees for the transfer of the reserve fund.

Thus an era ended without much fanfare in the astronomical community. To some it came as a surprise. They could not imagine Chandra relinquishing the journal on his own initiative. They thought that he must have been forced by the university to resign. "I received a number of job offers," says Chandra.

How does he feel now about the journal? In typical Chandra fashion, he says,

> You know, I have developed complete, total neutrality. . . . The journal has left no residue of feeling in me. I don't feel any sense of accomplishment and I don't resent the fact that I did it. It was a job I had to do and I did it.

One would have thought his singular dedication and service would have resulted in some kind of a personal attachment to the astronomical community. That was not the case. In retrospect, Chandra feels that the opposite happened. It produced a distortion of his personal life and isolated him from his colleagues.

In contrast were his relations with the people he worked with at the Press: "The part of my experience with the journal which has left any residue with me," Chandra says, "is the personal friendships and the personal loyalty which all these people showed me." Indeed, when he resigned as editor, the Press gave him a farewell party. Jean Sacks, the head of the Journals Department, made a brief speech: "I have often come across, in the papers submitted to the *ApJ,* the term 'Chandrasekhar limit,'" she said. "I do not think there is such a thing as the Chandrasekhar limit."

11

In the Lonely Byways of Science
Chicago, 1972–1989

Why did not Sir Joshua—or could not—or would not Sir Joshua—paint Madonnas?

John Ruskin

Beginning in 1952, Chandra became more closely associated with the physics department on the main campus of the University of Chicago. He began to teach regular physics courses instead of only the astronomy and astrophysics courses he had taught in the years before and after the war. Furthermore, physics began to dominate Chandra's research, beginning in the early 1950s with his work in magnetohydrodynamics, stability of rotating fluids, plasma physics, and ellipsoidal figures of equilibrium. Eventually, beginning in the early 1960s, he became more and more interested in Einstein's general theory of relativity and the mathematical study of black holes and of colliding waves, which he has continued to explore.

Since the curriculum change introduced in the astronomy department at Yerkes had oriented it towards observational astronomy rather than the theoretical problems in astrophysics, Chandra's research students after 1952 came almost exclusively from the physics department.[1] The formal relations between Chandra and his colleagues at Yerkes after this had little or no effect on him or on the students and visitors at the observatory. Life went on as before. Chandra continued to be in charge of the weekly colloquium he had instituted. Every time these serially numbered colloquia reached a new hundredth mark, Chandra would give a special colloquium. Lalitha would provide a "birthday" cake. If there was a popular film in one of the neighboring larger towns, Lalitha and Chandra were always ready to take a car full of visitors or students. Their home was frequently open for afternoon teas. "I remember Lalitha's homemade crumpets," says Margaret Burbidge.[2] "They were baked in a special ring on a griddle in front of your eyes and then served with butter. Since my husband and I were from England, we perhaps had a better chance than some people of meeting Chandra on a more informal basis. Chandra loved to gossip about what was going on in Cambridge and in England in general. Another special thing I remember is the annual autumnal leaf-raking parties to which all his graduate students, research associates, and visitors were invited—those wonderful teas to follow an afternoon spent raking the leaves on Chandra's lawn."

Chandra's warm and affectionate relations with the children of his colleagues were another noted feature in the Yerkes community. He always had time for children, who never felt intimidated by him like their elders. "He liked children and we liked him," recalls Agnes Herzberg.[3] "We used to say he was 'magic,' because once on a picnic, everyone had rolled up their shirt-sleeves in the hot afternoon. In the cool evening hours, when they rolled down their sleeves, everybody's sleeves had a crease, but not Chandra's. On Halloween night, he would don his Cambridge black gown and try to trick us."

As Chandra often says, he had more informal and warmer relations with the children of his colleagues than with his colleagues themselves. Children say and do what their parents dare not say or do. As an amusing instance, he tells the story of Paul Kuiper, Gerard Kuiper's son. On one occasion, during the early fifties, when Paul was twelve or thirteen and his parents were away, Paul was invited to the Chandras' for tea. During the course of the conversation, Paul asked Chandra, "What do you think of my father's theory of Pluto as an escaped satellite of Neptune?" Chandra replied, "Paul, I haven't been following the subject closely. You know, I'm not really an astronomer." To which Paul instantly responded, "Yes, yes, I know, my father has told me that," to the great embarrassment of his father when the story got around to him.

Peter Vandervoort[4] recalls another story told to him by Chandra. Chandra had maintained a friendly relationship with Morgan's daughter Emily from her childhood. When she was a student at the University of Wisconsin, Madison, she used to call on him during the weekends while visiting her parents. On one such occasion, she said to him, "Chandra, I'm worried about you; this man Code, who is coming to Madison as the director of the observatory, is one of your former students, isn't he?" Chandra said, "Yes." "And T. D. Lee, who recently got the Nobel Prize, isn't he also one of your former students?" Chandra acknowledged that it was a fact. Then she said, "Chandra, I am worried. Here are all your former students going to great places. Why are you here, stuck in the mud in Williams Bay?" Recollections of such stories greatly amuse Chandra. He is proud of his extremely fond relations and the contacts he has maintained with the children of his friends and associates such as Milne, Davenport, and Shoenberg.

The year 1952 held yet another turning point in the lives of Chandra and Lalitha. After years of deliberating about what course to take, they decided to become naturalized citizens of the United States. In the beginning, when Chandra and Lalitha first came to America, they had no intention whatsoever of making America their permanent home. They were in the United States for a purpose, and that was Chandra's science. The world war had intervened. Years passed; they were unable to afford even a short visit to India until 1951. They had become more and more settled in their ways and had begun to contemplate acquiring U.S. citizenship. After the war, Robert Hutchins had often urged Chandra to become an American citizen so that Chandra could be elected to the National Academy. Chandra had not paid much attention to that, attributing it to Hutchins's role as the university president who encouraged many of his faculty members towards the honor and distinction of membership in the National Academy. However, Chandra, as a noncitizen, had often faced difficulties of a bureaucratic nature in fulfilling his scientific obligations. He could not formally invite foreign scientists as visiting lecturers and take care of the necessary procedures on his own. Papers had to be signed by someone else. Every time he left the country, he had to secure an income tax clearance certificate and a reentry permit. Yet, to relinquish the nationality of the country of one's birth and to adopt the nationality of another was an emotionally forbidding step for Indians of Chandra's generation and even for the following two generations. Their trip to India in 1951 had been a moving experience for both of them. It was their first visit home since their marriage. Many changes had come about in their families. Adults had aged, children had grown into adults, and some had their own children. Some loved ones had passed away. In his younger sisters, Chandra had found such a "complete human understanding, sympathy, and love," that it was a wholly new experience for him. "It has had the effect of a moral purification," he wrote to Vidya, Savitri, and Sundari immediately after his return to America. "Life here, in spite of its wholesome climate for my intellectual work," he wrote, "has the quality of distilled water, and I feel curiously desiccated."

Nonetheless, the extended trip after so many years convinced both

Chandra and Lalitha that the chances of their return to India permanently had become slim. They felt they were too set in their ways, and neither the Indian government nor any institution in India had initiated a dedicated effort to secure their return.

Moreover, the appearance of Adlai Stevenson on the political horizon in 1952 brought a resurgence of hope that an enlightened leadership might drive away the ugliness of McCarthyism which had plagued the United States for several years. Liberal spirits could envision a new America which would be in harmony with its professed values, including a genuine human concern for all people. There would be no shame or stigma in becoming American citizens or in declaring allegiance to the country since it would represent an allegiance to basic principles cherished by all peoples. Chandra and Lalitha, like many others, were swept away with enthusiasm and hope for a better America and, indeed, a better world. They joined the local Democratic party, with Lalitha participating in its deliberations and fundraising activities. However, since they were not citizens, this activity was not always looked upon graciously by citizens opposed to Stevenson. They decided then to become citizens, but that could not happen overnight. The then-prevailing immigration law, the so-called Oriental Exclusion Law, prohibited peoples of Asiatic origin from becoming citizens. They had to wait until December 1952, when the Immigration and Naturalization Act came into force, which in spite of its many objectionable features granted a small quota for Asians and made it possible for a small number of them to become citizens. During the spring semester of 1953, Lalitha and Chandra began to take a citizenship correspondence course offered by the University of Wisconsin on American history and the American Constitution. At the end of the course, both Lalitha and Chandra secured, predictably, an A grade (which probably spared them the literacy test in the customary citizenship examination). On 13 October 1953 Chandra and Lalitha became American citizens.

While their American friends were delighted by this event, Chandra's father and their circle of family and friends in India were not. It was looked upon as a disgrace, a contempt for the land of their birth, and an unpatriotic act. On 14 October 1953, the day after they took the oath of allegiance to the United States, Lalitha wrote to Chandra's father, explaining the important step they had taken and their reasons for doing so. "Over the years, something else has happened," she wrote. "Whereas before, the roots which gave us nourishment were in India, we noticed we had started growing roots here also. And so another thing happened:

we found ourselves slowly but steadily being sucked in from the fringes into the heart of things—our acquaintances had become friends; a knowledge of the history and aspirations of the people made us develop a bond of understanding; earlier misunderstandings gave place to the feeling that deep down all peoples are alike in their likes and dislikes, in their want for security, in their desire to help and receive, in their desire for leadership and progress, etc." She went on to say,

During the last elections, we began to feel that it was not right that we stand idly by not doing our little part to indicate what representatives we would like to govern the country. Apart from the need of an outstanding President for leadership in these times of crisis, there was another urgent factor which faced us—and that was to get rid of the menace that was McCarthyism. And there was more in all this. The United States had become the leader of all democratic peoples, and what it did, deeply affected every corner of the world. Also the world was becoming more and more interdependent: the United Nations had come into being and what each citizen said in his or her country had its bearing on the international level. In these days of increasing frustration, it is comforting to look at this happy side of things. If I may stretch the point, the world is becoming a community of nations, has in fact become a community of its two billion individuals.

Chandra, in his letter of 25 October to his father, echoed similar thoughts and sentiments.

But Chandra's father was not pleased. After strongly admonishing both Chandra and Lalitha for presenting him with an accomplished deed without prior consultation, he wrote scathingly,

Well, I said: India, Independent India, has used you "ILL"—in bold capital letters. Its enlightened citizens or those in power, except Mr. Bhabha,[5] have given no thought, nor cared for your work or your personality. So, "Take vengence; kick India out of mind," may have been the spirit which prompted you, though subconscious. "I have eaten the salt of U.S.A. for 17 years i.e. from Xmas 1936 to date. I owe them my service," you said to yourself—ignoring the right or wrong of things. . . . Domicile depends on the country of the birth of the father. If the citizens of that country have not shown you any consideration for your living during adult life, kick the country. And so father too gets the kick. . . . Consider for yourself your upbringing, growth and training. Born of the traditions of the Cauveri delta of Tanjore and Trichy districts with high intellectual aspirations for four generations, counting my maternal grandfather who walked to Navadweepa in Bengal for his education! Next the atmosphere of Madras with its sandy beach where you said to yourself as Balakrishnan wrote in Triveni,[6] "Oh! Let me become a Newton!" You were under my ostensible care from 19th October, 1910 to August, 1930 for nearly 20 years. You ate also the salt of the Indian tax-payer, though under the British regime for four additional years. During the next two years you received the Trinity Fellowship.

The six years you spent at Cambridge contributed in no mean way to your mental growth and later achievement, that is, you were bound with the British tradition. You will realise therefore that your life has borne the fruit of the best of the Hindu tradition combined with the British tradition. I say purposely British for as you are perhaps aware that while in Britain, I traveled to pay homage to the graves of Sir Walter Scott, Ruskin, and to the lake country, and the residence of Wordsworth, went to Marlowe Abbey and to Abbotsford, obtained a copy later by post after my return to India, of the picture postcard of Sir Walter Scott in Princess Street, Edinburgh for he had instilled into me besides the spirit of the few lines below, the spirit of the chivalry towards womanhood, to say only a few things.

> Breathes there the man, with soul so dead
> who never to himself hath said
> This is my own, my native land!
> whose heart hath ne'er within him burn'd
> as home his footsteps he hath turn'd.

Your choice of America as your future home does at least kill *my* spirit of home loved in childhood and adolescence.

These words made Lalitha and Chandra extremely unhappy. There was no simple way to deal with the naturally strongly rooted feeling against acquiring another nationality, especially since free India did not allow one to maintain dual citizenship. Chandra did his best to explain that he was not relinquishing or denying his heritage derived from "the traditions of the Cauveri delta of Tanjore and Trichy districts with high intellectual aspirations for four generations." Nor was he disclaiming his "loyalty to men and institutions who had helped and encouraged him" during his six years in England. The legal proceedings did not require any such denials; on the contrary, they demanded reaffirmation of one's heritage. The only thing that was necessary to state under oath was that it was their "honest intention to make the United States" their home. And however much they may regret this, it was a fact. "We have been here for seventeen years," wrote Chandra in reply, "and there seems every prospect that we shall continue to do so the rest of our active life (i.e., to my retirement age). The retirement age in this country is 65; and if we stay in this country up to that age, we shall have lived and worked here for 40 years; and would not this duration tend to justify our calling this 'our home'?" Chandra then, after reiterating the practical considerations that induced them to become U.S. citizens, pointed out that they were also influenced by their close associations with people from other countries—such as the Fermis (Italians), the Kuipers (Dutch), the Schwarzschilds (Germans), and many others—who had taken U.S.

citizenship and who had maintained cordial relationships with their homelands.

It is difficult to know to what extent Chandra's father reconciled himself to what had taken place, to what was beyond his control. There were mixed reactions among the other members of both their families. Some defended the step as inevitable and blamed the Indian government for its indifference, for not providing the right opportunities to make Chandra's return possible. Others were strongly critical of them and blamed it on a lack of patriotic spirit. It was 1961 before Chandra and Lalitha were able to return to India for a four-month visit. Chandra's father had been the victim of a sudden heart attack and had died on 6 February 1960. After his retirement, he had devoted his life to a scientific study of Karnatic music and had developed, together with his daughter Vidya, a notation for transcribing the music.[7] Dynamic yet eternally discontented, controversial yet loving, he was missed greatly by his family and friends.

Citizens or not, life in their adopted country was not without unpleasant incidents for Lalitha and Chandra. Segregation and color discrimination were facts of life which affected them both personally and politically. Aside from Chandra's experiences at the Aberdeen Proving Grounds during the war, the beginning of his career at the University of Chicago itself was shrouded in somewhat of a mystery which was not dispelled until the 1960s. Chandra's career, as we recall, began as a research associate with the rank of an assistant professor in 1936. What was mysterious at the time was that the appointment was offered directly by the president of the university, Robert Hutchins. On his first visit to Yerkes in 1936, Otto Struve had taken Chandra to meet Hutchins and they had had lunch together. After lunch Struve and Chandra stopped by the dean's office, but only Struve went inside. Later when Chandra was on his return trip to Cambridge, he received a cable on board ship, directly from Hutchins, offering him the position. "I felt there was something curious about all this," recalls Chandra. "I was being appointed as a research associate. Now such an appointment does not normally need the intervention of the president of the university." Finally, in the early 1960s, long after he had ceased to be the university's president, in a lecture widely reprinted in the newspapers, Hutchins contended that the members of the academic world were no better than other ordinary people when it came to the question of public morality. He had pointed out how the medical school of the university had "violently resisted admitting Negro students" during his tenure in office. He also asserted in the same lecture that the chairman of a science department had opposed the appointment of a leading theoretical astronomer to its faculty "because he was an Indian, and black." When this was published in the *New York Times* and the local Chicago papers picked it up, it created quite a bit of commotion in the administrative circles of the university. As Chandra recalls, George Beadle, who was then the president of the university, called to say that he was not aware of anything that could corroborate Hutchins's statement and that he was quite innocent of such prejudice. It then became clear to Chandra what had transpired in 1936, and he described the incident to Beadle. And when the *New York Times* called him to ask what he thought of Hutchins's remark, he said, "The University of Chicago was thirty years ahead of the times."

Chandra also had not known that it had required Hutchins's intervention for him to be able to lecture on the main campus. During the spring quarter of 1938, Struve had proposed the idea of an elementary course in astronomy as part of the curriculum of the University College. The Yerkes astronomers, including Chandra, would take turns teaching the course. But apparently Dean Henry G. Gale[8] did not approve of the participation of Chandra in any of the work on the campus and wanted him to be excluded. Faced with a potentially embarrassing situation among his colleagues, Struve had approached Hutchins, whose one-sentence answer to Struve was, "By all means have Mr. Chandrasekhar lecture." From all accounts, Hutchins was a remarkable man, universally acknowledged as the most forward-looking president of the University of Chicago. He was known to say often that perhaps the best thing he did for the university was to appoint Chandra. Chandra's brilliance and subsequent accomplishments aside, Chandra was the first nonwhite person appointed to the faculty of the University of Chicago.

Racial discrimination in American society was something that Chandra was quite familiar with even before he had accepted the appointment at Yerkes against the advice of his father and friends in India. His uncle, Sir C. V. Raman, had faced humiliating experiences during his visit to America in the late 1920s. In Boston, Raman had gone from hotel to hotel in search of overnight accommodation, until the taxi driver finally took him to a hotel run by a Japanese couple outside the city limits where he could spend the night. Raman had also narrated to Chandra an incident at Cornell University when Sir William H. Bragg,[9] who was a visiting professor at Cornell, had invited him to be his guest in his apartment in the guest house. Raman had arrived late in the evening; the next morning, he had gone down for breakfast with Bragg. When the maid who was serving saw Raman, she had dropped the tray as though in shock and left the room.

Chandra was therefore prepared for such confrontations and, indeed, Chandra and Lalitha have had their share of experiences of a similar nature. "I remember going to New York to attend a Beethoven concert by the New York Philharmonic in 1938," says Chandra. "Lalitha was with me. We had made reservations in Barbizon Plaza, but when we went there from Grand Central station, the desk clerk saw us and said there was no place and that there was no record of any reservation. There was no place for us in another hotel nearby either. Then I remembered Raman's experience. We were almost ready to go back to the station and take the train back home, but at the last minute I remembered Jan Schilt, the chairman of the astronomy department of Columbia University. I called him. He arranged accommodations for us at a hotel known as King Crown Hotel, which was, I believe, a university hotel."

Recalling other similar instances, Lalitha says, "The first reaction is one of shock that such things could happen to us. Then you say shocks are good for you in order to know the truth. Otherwise how will you ever know? Once you know, and know them as facts of life, you begin to look at the whole thing in a different light. Of course, I had read American history, about slavery, and the emancipation declaration. I used to feel sad that change was so slow in coming. Then I used to say to myself, 'Look honestly; look at the way the brahmans in India treat the untouchables. Why should I think that we are any superior to the Americans in the way they treat the black people?' Thinking this way, you learn a much deeper lesson than you would have otherwise. You become tolerant towards other people's weaknesses."

Williams Bay continued to be their home until 1964. During the years 1946–52, Chandra used to drive to Chicago to teach every Thursday, occasionally staying overnight at the International House at 59th and Dorchester. From 1952 onwards, however, overnight stay became a regular feature because of editing *ApJ* and teaching on the main campus. "I used to give two lectures," Chandra recalls, "one on Thursday and the other on Friday. I used to come early on Thursday morning to the campus. During the time I worked with Fermi, I spent Thursday mornings largely with him, and whatever time I had left over I used to spend at the journal office and/or attending colloquia. Fridays I would mostly attend to the journal work. I preferred direct meetings to correspondence with the people with whom I worked for the journal. The stay at the International House used to cost $3.00 a night, which was all that I could afford at the time." It also turned out that for many of his students it was a splendid opportunity to talk to him while waiting in the cafeteria line.

Students, research associates, and visitors who so desired got rides from Chandra on these trips from Yerkes and back. "He would take as many passengers as the space in his car would allow," recalls Peter Vandervoort, who began his graduate work with Chandra in the middle 1950s. "There was a strict pattern to it," he adds. "One had to be at a specific corner or place at the exact hour, to the minute. No matter what the weather conditions, we would start at 6 AM, and generally reach the campus by 8:30 AM. He would take the same route every time, stop at the same gas station, eat at the same Howard Johnson restaurant on the way; Chandra used to order for himself Welsh rarebit which would always come with three strips of bacon, and Chandra would then transfer the bacon to one of the nonvegetarian students." "He had a reputation for being a fast driver and was occasionally stopped for speeding," says Margaret Burbidge. Chandra himself loves to tell the story of how one winter, while driving back from Chicago with his student Esther Conwell[10] in the passenger seat, he encountered foul weather and icy roads, and skidded into a ditch on the outskirts of Williams Bay. Fortunately neither he nor his student were hurt. They managed to ring somebody's doorbell and find help. According to Chandra, Esther Conwell then said, "I can just imagine the headlines tomorrow—if the accident had been serious—COED FREEZES TO DEATH IN ARMS OF PROFESSOR!"

During those years of commuting between Williams Bay and Chicago, especially after Chandra started teaching regularly on the main campus in 1952, Chandra and Lalitha had often considered moving to Chicago. As an initial step in this direction, in 1959 they had rented a one-bedroom apartment near the university (5550 Dorchester Avenue) so that they could stay overnight during their weekly Thursday–Friday visits to Chicago. "But we had some misgivings about moving permanently to Chicago," they now say.

In Williams Bay, people were accustomed from the beginning to our way of entertaining friends and guests with only vegetarian food and nonalcoholic drinks. In Chicago, especially after the war, people had taken to giving cocktail parties. And during our stays in Chicago over the weekends and other visits, we were often invited to such parties and also to dinner parties with the Mayers,

Ureys, Fermis, and Andersons. We used to be invited at 7 PM, but sometimes the drinks went on for hours. We thought we would of course have to reciprocate such hospitality, but we couldn't imagine ourselves being able to reciprocate in the same manner, in the proper style. Gradually we realized that we didn't have to. And one day in 1964, as we were driving past the newly constructed high-rise building at 4800 Lake Shore Drive on our way back to Williams Bay, Lalitha suggested that we go and look at a model apartment and find out if any apartments were available. A two-bedroom apartment on the twenty-fifth floor was available, and we signed a lease. After staying there for three years, we moved to our present location on Dorchester Avenue.

Chandra's closer association with the main campus also brought about his closer association with the university administrators and his participation in university affairs. Besides Hutchins, who played such a significant role in getting Chandra to the University of Chicago and retaining him, Chandra had extremely friendly relations with two other presidents of the university, namely, Edward Levi and John Wilson. "Chandra always symbolized, to me at least," says Levi, "a kind of humanistic scientist, a person of enormously high standards, determined, idealistic, brilliant, and one who never seemed to deviate from his own ideals, ideals for a great scholar. I always thought of him as the kind of a person for whom and through whom the university existed."[11] Wilson expresses similar sentiments and says, "A genial, modest, and self-effacing individual, Chandra always stood for the very highest level in academic affairs. That is the essence of his being."[12]

On occasion, Chandra's word (or call and discussion) about some event or individual has carried a great deal of weight. The case of Noel Swerdlow is an example. Swerdlow, who is now recognized as an eminent Copernican scholar and historian of ancient astronomy, was denied tenure by the history department of the University of Chicago in the mid-1970s. The Dean of the Social Sciences had gone along with the recommendation of the department. "Chandra intervened, saved my job and my career," Swerdlow recalls.[13] "I had met Chandra in 1973 when we both gave talks on the occasion of the 500th anniversary of the birth of Copernicus. Since then I had occasionally met with Chandra and discussed my work with him. Chandra would ask me about ancient astronomy. It was clear to me that no one in the history department understood my work on Copernicus, my explanation of how he derived his heliocentric theory. It was all mathematics and not history to most members of the department. Apparently there was one member of the department who was supporting my candidacy on the basis that I was a true historian because I used manuscripts for my research! So you can imagine the situation. And the head of the committee in charge of my case apparently had written to people whose work I had reviewed unfavorably and had gotten letters that were unfavorable to me. He read those letters extensively to the department, I suppose in order to undermine letters from those people who had any knowledge of my work. So I heard that the vote of the department was split down the middle and the decision went, needless to say, against me. In that situation, I went to Chandra, because he was the only one on the campus who had any conception at all of what my work meant, and he was one member of the faculty for whom I had the greatest respect, more than anybody else, both for his work and what he is as a person. He intervened. I believe he went directly to the provost."

Chandra indeed had spoken to John Wilson, the provost, on behalf of Swerdlow. He convinced him that the history department was making a monumental error in denying him tenure and suggested that Wilson consult people like David Pingree and Otto Neugebauer, experts in the history of astronomy. Wilson overruled the decision of the department and of the Dean of Social Sciences and granted tenure to Swerdlow. A few years later Chandra intervened again. Things still had not gone well for Swerdlow. "I was given ridiculously low raises," says Swerdlow. "I came into financial troubles, not having even enough money to mail a manuscript to England. The chairman said all he could do was to lend me ten dollars. Three years later, I got an offer from Stanford. Before accepting it I sought Chandra's advice. He asked me which place was better for my work. He recalled his own experience when he had an offer from Princeton and what President Hutchins had said to him. I told him from the point of view of my work, I preferred to stay at Chicago. Then I believe he again went to the provost and got me an appointment in the Division of Physical Sciences with a place in the astronomy department, where I have been since then and have been quite happy. I cannot imagine any other place where I would be happier."

Chandra is happy that in a few instances like this he has been successful in influencing the administration. "I have been able to do so," Chandra says, "only because the people know that I would not go and ask them for anything I was not absolutely and totally convinced was right for the university. I am glad that in the Swerdlow case history has proved that I was right; his book on Copernicus has earned him a solid reputation in his field. I will be surprised if the university does not name him a distinguished service professor within a few years." Swerd-low was in fact elected to the American Philosophical Society in 1987 and also awarded a MacArthur Fellowship.

In recent years Chandra and Lalitha have struck up a close friendship with James and Annette Cronin, who live in the same apartment build-ing. "In the low period that followed Fermi's death," says James Cronin, "Chandra's loyalty to the University of Chicago and his remaining in Chicago while many well-known scientists left has been an enormous contribution to this university. With his monumental stature, he has served as the conscience of the department, acting strongly whenever needed in preventing bad appointments, bad scholarship, etc." However, opinions differ in regard to his having exerted strong influence within the department. As Valentine Telegdi, a great admirer and friend of Chandra, says, "In the early fifties, there was this unbelievable constellation of people. There was Fermi, there was Edward Teller, Gregor Wentzel, Joe and Maria Mayer, Harold Urey, Willard Libby, and a good number of other people whom I may have overlooked. Then came Murray Gell-Mann, Richard Garwin . . . After Fermi's death, a sort of decay began, and Chandra and Wentzel, who had such unique positions and univer-sal respect in the university and in the department, did not do enough to stem the decay. Of course, Chandra's reserved nature and character, his resolve not to say anything unless he felt he was absolutely correct, were responsible for his reticence in speaking out."[14]

Chandra took voluntary retirement in 1980 and ceased to have any teaching or other obligations at the University of Chicago. The univer-sity, on the other hand, which would have been quite happy to keep him on a full-time appointment as long as he wished, offered a post-retire-ment appointment with essentially no obligation on his part. Chandra acquitted himself of that privilege as well in 1985, at which time the university converted the post-retirement compensation into a research grant and conferred upon him the emeritus status. "I like this arrange-ment," Chandra says. "It gives me the freedom to invite at any time young people like Basilis and Valeria,[15] who want to work with me with-out going through the rigmarole of applying for a grant and so on."

When Chandra speaks of his singular association of more than fifty years with the University of Chicago, he has only good things to say. "I have been through the administrations of six presidents," he says, "and I had extremely cordial relations with all of them; more than what a fac-ulty member is likely to have with the chief administrator of the university. I have never felt that any other university could have or would have done anything more than what the University of Chicago has done for me." Chandra illustrates this with a story.

In the fifties I got several offers from other institutions—MIT and Berkeley, for example. I just declined them without mentioning the offers to anybody. Yet somehow word got to President Kimpton. He called me to his office and said, "Chandra, I hear that other universities are trying to seduce you away."

I replied, "You don't seem to have a high opinion of my morality."

He sort of laughed and said, "Can we do anything for you?"

I said, "Nothing really. If there was something, I would ask in the normal way."

Still, when he insisted, I mentioned the matter of a separate office for myself, which I was going to take up with Herbert Anderson, director of the Enrico Fermi Institute. Since I used to come for only two days a week, I was sharing an office with Donna Elbert on the third floor of the Fermi Institute. But with Donna's typing and computing going on, it was rather difficult for me to carry on my work. So I said to Kimpton, "Since you press me, I have been meaning to talk to Anderson to find out whether I can have an office for myself. But, please don't worry. I will talk to Anderson about it."

When I came to the office the next Thursday, Donna said to me, "Mr. Anderson has been looking for you all week."

I hadn't been in my office for ten minutes when Anderson came charging in and said, "Chandra, if you wanted a new office, did you have to go and ask Kimpton about it?"

I told him what had happened. I don't think he ever believed it.

Age seems to have little or no impact on Chandra's fervor for science and the pursuit of the life of the mind. Since his classic work on the mathematical theory of black holes, published in 1983, he has pursued the study of colliding waves and the Newtonian two-center problem in the framework of the general theory of relativity. These studies are again, as one has come to expect from Chandra, not in the mainstream of contemporary activity in the study of general relativity. And it may take some time to appreciate some of the startling results that have emerged from his studies. But that does not perturb Chandra in the least.

The year 1987 marked the 300th anniversary of the publication of Newton's *Principia*. Responding to lecture invitations in Cambridge, England, and other places, Chandra began to delve deeply, not only into the origins and circumstances of the writing of the *Principia,* but also into the very heart of the *Principia* itself as few people have done. After three centuries, Newton comes to life again, as Chandra describes his experience in studying some of Newton's well-known propositions. "I first constructed proofs for myself," he writes. "Then I compared my proofs with those of Newton. The experience was a sobering one. Each time, I was left in sheer wonder at the elegance, the careful arrangement, the imperial style, the incredible originality, and above all the astonishing lightness of Newton's proofs; and each time I felt like a schoolboy admonished by his master."[16]

A distinguished astronomer apparently asked him once some years ago, "Chandra, when will you come to grips with the real problems in astronomy?"

Chandra, recalling the opening question in John Ruskin's essay on Sir Joshua Reynolds and Holbein, apparently responded, "I do not feel up to painting Madonnas."

If someone asked a similar question today, he would perhaps give the same answer. He is content with another cautious step up the ladder, embarking on another adventure, as Chandra likes to put it, as a lonely wanderer in the byways of science.

Epilogue: Conversations with Chandra

The rich and the poor are two locked caskets, of which each contains the key to the other.

Karen Blixen

This biography, based though it is on Chandra's papers, his correspondence, and extensive conversations with him, is an account of his life from my point of view. The excerpts of conversations with Chandra in this epilogue present his thoughts in a more direct manner and hence, in some instances, supplement and enrich the story I've told.—Kameshwar C. Wali

Motivations for Science in India

Chandra grew up in what was a golden age for science, art, and literature in India, spurred on partly by the struggle for independence. J. C. Bose, C. V. Raman, Meghnad Saha, Srinivasa Ramanujan, and Rabindranath Tagore, by their achievements in scientific and creative endeavors, became national heroes along with Jawaharlal Nehru, Mahatma Gandhi, and a host of others active in the political movement. Did their success produce an enduring atmosphere for creativity, or did the prevailing cultural and social conditions hamper a healthy growth of sustained activity? Over the years, Chandra and I have discussed various facets of science in India and the experiences of some of the eminent Indian scientists. The following excerpts from these conversations reflect Chandra's thoughts on these matters more completely than I have described them in earlier chapters. Of particular interest may be Chandra's views on some misunderstandings surrounding the discovery of the Raman effect.

KW: Can you recall exactly what motivated you to be a scientist when you were young, when you first started studying physics? What influences did you feel?

SC: Well, of course, when one is young, one has ideas and motives that become refined later. To the extent I can recall, I am certain that the primary motivation was the attraction in some way of becoming well known. Even as a student in school, I was familiar with the name of Ramanujan—how he had gone to England quite unknown and had returned quite famous. My uncle Raman is another example.

KW: You mean the discovery of the Raman effect[1] and your being in Calcutta during the summer of 1928?

SC: Yes, being in the midst of people who had made important discoveries did influence me. But retrospectively, it is not the kind of influence which a young man should have.

KW: Why?

SC: Because it gave a false picture that making discoveries was easy. After all, the Raman effect was a very simple discovery in some sense, and young people in India had a wrong impression of what science was. I certainly had a wrong picture. Ramanujan became famous in four years. Saha's second paper produced the ionization theory attached to his name. Satyendra Nath Bose became associated with Einstein in the second or third paper he ever wrote. And Raman made a discovery and got a Nobel Prize. It gave a very glamorous picture, which was all right for these people, people with considerable standing. Although comparing some of them with other great men of science we know, they may not measure equally. But we must remember that these men came from a surrounding and a background that was devoid of modern science.

KW: Yes, let's discuss that a little more, the void of science in India.

SC: I mean it is a remarkable thing that in the modern era before 1910, there were no [Indian] scientists of international reputation or standing. Between 1920 and 1925, we had suddenly five or six internationally well-known men. I myself have associated this remarkable phenomenon with the need for self-expression, which became a dominant motive among the young during the national movement. It was a part of the national movement to assert oneself. India was a subject country, but in the sciences, in the arts, particularly in science, we could show the West in their own realm that we were equal to them.

KW: Do you think this kind of feeling was a conscious one among the young?

SC: Yes, I think it was. Let me give you a rather crude illustration of that. The ambition of every scientist at that time was to get elected to the Royal Society. Ramanujan became famous in India, not because people understood his mathematics but because he was the first Indian to be elected to the Royal Society. Then of course J. C. Bose, Raman, and Saha followed. A Fellowship of the Royal Society meant the British recognized you.

KW: Recognition from the West, from the British: it was an important factor even in my days there, in the forties and the fifties.

SC: It had a distorting effect. I mean it was all right in the twenties, but extended into later periods, it wasn't.

KW: But do you think the glamor of making a scientific discovery, the desire for fame and recognition, is inherently bad? Don't you think this strongly motivates one, at least when one is young, to do science?

SC: No. I don't think it's inherently bad. I agree entirely with you that it is an understandable motive and to some extent nothing to be ashamed of. Fame and recognition are other ways of saying that you want to be successful in science. After all, what is wrong with the wish to be successful? I know there is a feeling that there is something degrading in it. I find that feeling incomprehensible, for to be successful in anything, you have to do it well.

KW: There is this classical image of a scientist as one who does not care for fame or success or anything worldly. He is only interested in the search for truth.

SC: I don't believe in that. In my own case, as I said before, I started with a totally glamorous view of science that persisted so long as I was in India. And the fact that I had published one or two papers while I was still in college made me feel that I was on the right track. But going to England was a shattering experience precisely from this point of view. In Cambridge there were people like Rutherford, Dirac, Fowler, Eddington, and a host of other great men of science. I immediately realized that it may be all right to wish to achieve and be famous by contributing to science. However, one has to put an enormous amount of labor and hard work into it. One has to persist in a positive way. If you are moderately lucky, or let us say you are not unduly unlucky, you are sure to do something in the long run. Therefore, the balance to some extent was provided for me when I went to England. Others who do not find this balance, I have noticed, go often in the wrong direction.

KW: Am I right in that very few scientists in India during that time, after making some significant discovery, continued to work in a sustained way?

SC: Yes.

KW: Most of them gave up active science for administrative roles. They rested on their past glory. Do you think the same thing would have happened to you if you had returned to India?

SC: It's difficult to say. My feeling is that by 1936 my attitude to science had changed sufficiently for that not to happen. Besides, in my case, the really serious incident, the controversy with Eddington, had influenced my attitude to science. I don't want to repeat here what you have already elaborated and quoted. But briefly, the result of the controversy was that I decided that the most important thing in science is to continue to be productive and active, not to worry about the controversies. If I was right, people would know it in time. I changed my area of research in a deliberate way into something different.

KW: Yes, that was the starting point of your unique way of periodically changing your area of research, wasn't it?

SC: Motives for which one does science change with time. They should, they must change with time. You see, if I am right in believing that the primary motivation for the flowering of Indian science between, say, 1900 and 1930 was a part of the national consciousness, and that people somehow wanted to show that they were equal to the British, if that was the motivation, it was all right at the beginning and for some length of time. However, when recognition, glamor, reputation continued to dominate as the primary motives in individual scientists' lives, deleterious effects took hold. Those who had made significant contributions were constantly aware of those successes. They wanted to be regarded as unique individuals, and therefore they turned around and discouraged younger people or attributed all kinds of motives to their contemporaries. Do you understand, for example, why people like Saha, Bose, and Raman—all three such distinguished men who made such significant contributions—could not maintain harmonious relations among themselves?

KW: No, I don't. But I know the disharmonies were open and a matter of public knowledge. Dirac once told me that, during his visit to India, he found Bose and Saha so openly "hostile" to each other that they wouldn't be in the same room together.

SC: And Saha, Bose, and others were openly critical of Raman. They thought he was an amateur physicist since he worked on problems in acoustics—theories of bowed and struck strings, and musical instruments. They were annoyed and jealous of his friendship with Sir Ashutosh Mukherjee, the vice-chancellor of the Calcutta University, and that he was offered the Palit Chair in Physics. They had a tendency to ridicule him. Parameswaran, my teacher in Presidency College, for example, used to repeat quite openly in his class that Rutherford had asked him once, "Does Raman expect to fiddle his way into the Royal Society?" But, of course, once Raman discovered the Raman effect, people realized that he had made good.

KW: Once you told me that the rift between Saha and Raman had only intensified after the discovery of the Raman effect.

SC: Yes. The rift between them became extremely acute when Sommerfeld visited India in 1928. Raman had arranged a grand function at the Indian Association for the Cultivation of Science in Calcutta, at which he was to give a talk on his discovery with Sommerfeld in the audience. But after Raman's talk, Saha stood up and made remarks to the effect that the discovery was no more than a confirmation of what Smekal[2] had predicted. He tried to belittle the whole discovery in the presence of Sommerfeld. Later he wrote letters that were published in *Nature* pointing out that Raman's explanation was incorrect. The correct explanation was to be found in the Kramers-Heisenberg dispersion theory. Things became worse later when Raman became the director of the Indian Institute of Science in Bangalore. Before he left Calcutta, he made sure that Krishnan was appointed as the Mahendra Lal Sircar Professor, and he had his own plans to ensure the future of the association. At the last minute, an unexpectedly large number of former members of the association showed up at the meeting and voted against Raman's plans.

KW: Must be due to Bengali provincialism.[3]

SC: Yes. Then Saha, who succeeded Raman as the Palit professor, wanted to convert the Allahabad Academy of Sciences into the National Academy of Sciences. Raman opposed that move and created a rival academy, the Indian Academy of Sciences In Bangalore. Consequently, there arose an "exclusion principle" between those who belonged to one and those that belonged to the other.

KW: You were in Cambridge when all this happened, is that right?

SC: Not at the time the Sommerfeld incident took place. As you have noted, Krishnan and I became very good friends during the summer of 1928. He used to come to Madras quite often and, until 1930, he used to tell me everything that happened in Calcutta. I came to know what transpired at the meetings from conversations with him. And when I went back to India in 1936, I came to know all about the effort to oust Raman from the directorship.

KW: This is about Raman's alleged mismanagement of the institute in inviting Max Born and several other scientists who were fleeing Germany.

SC: Yes. And the adverse report by the so-called Quinquinium Committee sent to the viceroy, Lord Linlithgow. Raman wanted me to persuade Rutherford to write a letter to the viceroy, which I did when I returned to Cambridge. Rutherford did write a letter supporting Raman, but it didn't work. Raman had to resign from the directorship. He remained as professor of physics at the institute till he built his own institute. I did not know that relations between Saha and Bose were not harmonious. You say Dirac told you.

KW: Yes.

SC: Satyan Bose was one person who, when Raman made the discovery, did not associate himself with the others by belittling it. I think you have noted my recollection of Raman telling me what Bose had said to him after seeing the spectra, "Professor Raman, you have made a great discovery. It will be called the Raman effect, and you will get the Nobel Prize." Bose in some ways, from the human point of view, was the best of them all. He was very generous, gentle, easygoing, and not particularly caring about the glamorous aspects of science.

KW: Later, relations between Krishnan and Raman went to pieces, didn't they? A long time ago, back in India, I used to hear that Krishnan did not get proper credit for the discovery. He had played an equally significant part in the discovery, and therefore the discovery should have been known as the Raman-Krishnan effect. What's your opinion?

SC: Yes. They did have a falling out, caused, perhaps, by the kind of rumors you have mentioned. But first let me summarize my own view regarding sharing the credit for the discovery. My own view is that, in a genuine sense, the discovery of the Raman effect was possible because two absolutely original scientists, complementing each other, worked together. It is not so much a sharing of the discovery between the two, as giving the whole credit to each. That was my assessment in 1928.

KW: You were there in Calcutta soon after the discovery.

SC: Yes, and even though I was only seventeen and a first-year honors student, I had read Sommerfeld's *Atomic Structure and Spectral Lines.* I had read Compton's *X-Rays and Electrons.* To understand the Raman effect one does not need more physics than that contained in these standard books. In fact, having studied these books, I was perhaps better off than most of the experimental people working in spectroscopy and involved in only measuring and tabulating wavelengths and so forth.

KW: When did you meet Kirshnan the last time?

SC: I met Krishnan in 1951 purely by accident at the Madras airport. We had dinner together a few days later and I had a long conversation with him.

KW: Did you ever discuss the controversy with him directly?

SC: Well, Krishnan did tell me some things during that long conversation. If you go back and read the literature, you will find that the first announcement of the discovery was made in a letter published in *Nature* [31 March 1928], signed jointly by Raman and Krishnan. Then the first spectrum of the Raman effect with its correct explanation appeared later in another joint letter in *Nature* [5 May 1928]. However, between these two letters, there is a letter signed by Raman alone, which appeared in the 21 April 1928 issue of *Nature.* Krishnan told me that Raman had sent this letter to *Nature* without Krishnan's knowledge and was apologetic about it to him later. Krishnan saw the letter for the first time in print and he could not understand why Raman had published the letter. Raman also gave a public lecture [16 March 1928] that was subsequently published. With the exception of this lecture and the one letter in *Nature,* all other subsequent letters and publications, including the first detailed account of the discovery published in the *Indian Journal of Physics* [2, no. 4 (1928): 399] are under the joint names of Raman and Krishnan. And indeed, Krishnan felt that the announcement in a lecture and the letter Raman wrote under his name were intended to exclude Krishnan's name as a joint author. Nonetheless, Krishnan, to the best of my knowledge, never said a word; but despite Krishnan's discretion, Raman had both on private and semiprivate occasions openly impugned Krishnan's integrity. After Krishnan's death, Raman made a shocking statement to a *Times of India* correspondent: "Krishnan was the greatest charlatan I have known, and all his life he masqueraded in the cloak of another man's discovery." Nothing I have known justified such a statement. Besides Raman's published Nobel lecture includes ample reference to Krishnan and is fair.

KW: Why, why did Raman begin such a tirade?

SC: I don't know for sure. Krishnan told me a complicated story. I have given a rather detailed account that is deposited in the Royal Society Archives. According to Krishnan, it was in November 1948 that he, along with others who had been Raman's students, had assembled in Bangalore to celebrate Raman's sixtieth birthday. Krishnan was then the director of the National Physical Laboratory in Delhi and had close connections with Prime Minister Jawaharlal Nehru and other high officials of the government. Raman had just started building his Raman Institute, and had apparently run into financial difficulties. Krishnan, who knew about Raman's financial worries, had talked to Nehru and had come to Bangalore with Nehru's authorization to offer Raman a substantial annual grant from the government with only the formal requirement that Raman submit a financial statement every year. The arrangement, so it seemed to Krishnan, was a very generous one, and Krishnan felt that this support from the government would please Raman and relieve him of his financial worries. It was Krishnan's intention to announce the offer of government support as a part of the address he was to give at the official celebration of Raman's birthday. But the entire plan was aborted.

Apparently, the day before the celebration there was a private dinner in Raman's house attended by some thirty of his friends and former students, including, of course, Krishnan and Ramanathan. Krishnan was sitting opposite Raman, and during the dinner Krishnan explained to Raman the government offer that he was going to announce the following day. But to Krishnan's amazement, Raman got extremely angry and expressed his outrage that the government would ask of him any kind of report. Krishnan told me that he gave the speech the following morning as he had planned without, however, announcing the government offer. And as Krishnan expressed himself to me, that was the beginning of the end of his relations with Raman.

KW: A strange man, Raman. Strange story.

SC: Krishnan also told me that during January to April 1928 he had kept a diary of all the events that took place during that period. In 1971 on my visit to Ahmedabad I found that K. R. Ramanathan [an early associate of Raman who was working in Raman's laboratory in 1928 when the effect was discovered] was in possession of Krishnan's diary. In an article which Ramanathan had prepared for the issue of *Current Science* devoted to Raman (Raman had died a few months earlier), he had quoted extensively from Krishnan's diary. The published version of Ramanathan's article contains none of it.

KW: You mean the editor excised it?

SC: Yes, I know that to be the case since the editor himself confirmed it. Krishnan told me in 1951 at that dinner meeting that he was going to deposit that diary with the Royal Society, but later on I found out that the Royal Society did not have it. In fact, Krishnan did not leave any record with the Royal Society. I got a copy of the diary made and deposited it with my own recollections in the Royal Society Archives because I thought there should be some record of it.

KW: Can you recount any conversation you had with Raman in later years?

SC: I remember one conversation I had with him in 1961 that throws a characteristic light on him. It was my second visit to India after going to Chicago in 1937. While visiting my sister in Bangalore on that occasion, I went to call on Raman. When I arrived at his office, he was unwrapping my *Hydrodynamic and Hydromagnetic Stability,* which had coincidentally arrived by mail just then. He took the book in his hand and turned to me and said, "The only book of this size I have seen before is a novel by Anthony Trollope—absolute trash!" And he went on, "How do you manage to write a book of this size? I could never find the time to write a book. I always found research far more interesting. In 1926 I wanted to write a book on the scattering of light. I heard that Henri Cabbannes was writing one; so I did not. And the result of it was that I discovered the Raman effect and got the Nobel Prize while Cabbannes wrote the book." I could not help responding, "My God, I have lost four Nobel Prizes!" But he had the last word; he said, "Getting a Nobel Prize isn't that easy."

KW: Didn't you once say that he resigned his Fellowship of the Royal Society?

SC: Yes, he did in 1968, I believe. I heard the story from P. M. S. Blackett. Apparently the London *Times* had an article on Fellows of the Royal Society who had received the Nobel Prize, and it did not mention Raman. Raman blamed the omission on the Society and wrote to Blackett saying that unless a satisfactory explanation was forthcoming, he would resign his fellowship. Blackett wrote back saying that it was a London *Times* article; the Society had no part in it. The handbook of the Royal Society was available in all libraries and it clearly had the correct record. The *Times* should have consulted it. Blackett's response, however, did not satisfy Raman. He resigned anyway; he was not satisfied with the explanation.

KW: "The supreme egotist," as Ramaseshan [Raman's biographer] says. But Ramaseshan also says that in private conversation Raman showed such an unbelievable scientific humility as to make one wonder which was his true self. Apparently, Raman had booked two tickets on a steamship for himself and his wife in July 1930, the year he got the Nobel Prize, to enable them to reach Stockholm in early December!

Prizes were not announced until November. Had he waited until they were announced, he would have been unable to attend the ceremonies.

SC: Let me tell you another story told to me by Rosseland. I have recorded this also in my records for the Royal Society Archives. Robert Millikan had invited Raman to give some lectures at the California Institute of Technology in the summer of 1924. Apparently Raman was not too happy that he was not generally known in the United States. He was even more disappointed that so very few came to his lectures. He used to visit Rosseland regularly, who was then a Visiting Fellow at Mount Wilson in Pasadena. Raman's conversations with Rosseland had apparently one refrain: "You see Rosseland, no one knows me now. You just wait, when I go back to India I will make a great discovery and receive the Nobel Prize. Then everyone will know me." And Rosseland said that at the time he just wrote Raman off: "But I changed my mind when Raman made good on his promise."

KW: These rifts and rivalries between senior scientists were sort of inherited by junior scientists. It became a sort of tradition, did it not? And it continued. I recall my own experience in Banaras Hindu University during the years 1950 to 1955, when the rivalries between two professors went to such extremes that the students and associates of these professors were sharply divided into two groups. You belonged to one group or the other.

SC: Yes.

KW: It must have influenced your decision not to return to India after the Trinity Fellowship. I have a quote here from Krishnan's letter of April 1936 in response to your letter informing him of the offer from Yerkes. He says, "On this side at any rate, I know the conditions, and with all my optimism and enthusiasm for science, I feel the present scientific atmosphere in India quite oppressive, and it may be an advantage if you can postpone your return to India by one or two years."

SC: Yes, I remember that letter. Nonetheless, if a suitable offer had come through, I would have stayed in India. But no such offer came. I was not interested in becoming the director of an observatory.[4]

KW: Perhaps we can attribute all this tension to the fact that there were so few brilliant scientists at the beginning. And also to the Indian situation: the age-old tradition of unquestioned obedience to authority, originating in the family; respect for fame; open flattery . . .

SC: I think that is a part of it, yes. I am certain. If you are surrounded by people who always praise you and if, at the same time, you do not see among your contemporaries people whom you respect yourself, then it has a demoralizing effect.

KW: You get used to the flattery and open praise. If you are not constantly on guard, you begin to depend on it.

SC: Yes. For example, you can see this difference between the United States and Canada. I happen to know Gerhard Herzberg extremely well. After he received the Nobel Prize, the situation became quite different. There are so many things called after him in Canada now—the Herzberg Institute of Astrophysics, the Herzberg Medal of the Royal Society of Canada, and so on. He sees only a few comparable to him in his reputation. Whereas in this country and in England, the situation is different since there are so many people. I told you about that marvelous example of a Physical Society meeting in the Cavendish Laboratory in 1933. The Nobel Prize for Dirac had been announced just two days before. When Dirac walked in a few minutes late, there was tremendous applause that went on for a while. Then Rutherford got up and said, "That's enough. This is not the first time the Nobel Prize is coming around this way." Dirac sort of smiled; everybody smiled and took it in good spirit.

KW: In India, of course, western science, as we know it, was an import of the British, who also imposed their hierarchical educational system.

SC: That's right. We took over the hierarchical system from the British and added our own characteristic abuses to that. Nehru once made precisely the same remark: "We learned the hierarchical system from the British and we exaggerated it and kept it even though they are no longer here." But the amusing thing is that people turned around and said Nehru himself maintained the hierarchical system.

KW: Respect for authority, a false respect quite often, growing out of fear perhaps, led to a great deal of hypocrisy in relationships which resulted in an atmosphere totally unconducive to scientific thinking.

SC: Yes. Another facet was intolerance and lack of understanding. I was in Bombay in 1961, for example, and spent ten days at the Tata Institute. Bhabha[5] was still the director. I was really disappointed by the uniform criticism of him—criticism from his senior colleagues, students, and everybody in general. I turned around and asked them, "Why don't you be a little tolerant? After all he has built this institute. And so far as I know, this is the only institute where admissions are made on the basis of merit without the considerations of region or caste or creed.

If he has certain weaknesses, why don't you overlook them? Well, they wouldn't show any tolerance.

KW: So true. Also, quite often, such criticisms are not based on first-hand knowledge. The uniformity of criticism, as you said, shows that it was not based on personal experience or one's own independent judgment.

SC: So often the same story is told by many people. It turns out that no one has firsthand knowledge of the incident being recalled. Actually, my own view is that Bhabha did an enormous amount of good for Indian science, even though he had his weaknesses.

KW: What were his weaknesses?

SC: You might say exaggerated forms of personal prestige. You must have heard this story. When Pauli and a number of foreign visitors were in India, they all had to travel by bus when they went somewhere while Bhabha followed them in a limousine just by himself. Apparently Pauli got so annoyed at this that he left India the next morning.

KW: Really?

SC: Yes. Victor Weisskopf told me this. Chaim Pekeris, a famous geophysicist who visited India, told me another story. Apparently there was a public lecture; Bhabha came in along with Pekeris. Every seat in the auditorium was occupied. But when Bhabha came in and saw that there was no vacant seat, he just waved his hand at the front row and everyone in the front row got up and just disappeared.

KW: I had heard that he used to make people wait for hours—Indian scientists who had appointments.

SC: That is a common disease. I tell you, I was shocked by a similar instance. When I was in India in 1961, I was to see M. S. Thacker, the director general for the Council of Scientific and Industrial Research. When I went to see him, I was led to his office by the back door. Then Thacker said to me, "I'm sorry I had to ask you to come by the back door. I had forgotten that I had an appointment with Satyan Bose at the same time. If you had come by the front door, he would have seen you since he is waiting outside." It seemed unbelievable to me that an old man of 75 and of the stature of Satyan Bose would be kept waiting outside. It would have been perfectly all right with me and normal if Thacker had said, "I am sorry. There has been a mix-up. I have to see Bose first." Somehow there is this idea that your higher official position entitles you to make these lower ones wait.

KW: This tendency filters down to lower and lower levels.

SC: Yes. Complete asymmetry between your relations with people infinitesimally higher than you and those infinitesimally lower than you.

KW: That's very beautifully put.

SC: Of course similar things exist in all countries. Forms of insensitive, rude behavior are not unique to India. In India, however, such behavior plays havoc with younger people. It breeds hypocrisy. You cannot admire one person without criticizing someone else. You cannot say, "I like both of them."

KW: Yes. It is quite common that people say one thing in the presence of a person and something else behind his back.

SC: Again, I am sure that similar things, perhaps in somewhat diluted form, exist in all places.

KW: Don't you think the British, in the educational system, also contributed to this kind of attitude to some extent? You were exposed to British professors and British administrators in Presidency College. We didn't have any British professors or British administrators in my part of the country.

SC: Perhaps. I'm not clear on that point. The transformation of the educational system took place quite rapidly in India between 1900 and 1930. For example, Raman and my father were students in Presidency College between the years 1905 and 1908. At that time, professors in all the departments and the principal of the college were all Englishmen. In physics there was a man called Jones of whom Raman and Parameswaran, my teacher, spoke very highly. Jones was responsible for encouraging them to do science. By the time I was a student in the Presidency College, there were still a few Englishmen, Professor Earlam Smith in chemistry, for example. The principal was P. F. Fyson, an Englishman. But there were also quite distinguished Indian professors who were trained in England—Parameswaran in physics, B. B. Day in chemistry; K. Ananda Rao was in mathematics and was the most distinguished man on the faculty at that time. Ananda Rao was in Cambridge during Ramanujan's time and knew Ramanujan quite well. Hardy knew his work and had a great deal of respect for him.

KW: Was your attitude towards English professors any different from that towards the Indian professors?

SC: No. I don't think so. The English professors had a very good rapport with the Indian students. The Indian students liked them, but I don't think we looked up to the English professors more than to the Indian professors. We thought they were all good or bad depending upon their merits. But with the English professors, there was a funny relationship because of the political movement.

KW: Did any of them participate in India's freedom movement?

SC: No. These were men who were in the government educational service; anyone in the government service had to go along with the government. This was true of Indians as well. My father, who was in the government railway service, was not particularly pro-National Congress during those years.

KW: I think that was a reasonably common attitude among parents who were "British government servants."

SC: Of course the primary center of interest for middle-class parents at that time was to see that their children became successful lawyers, doctors, or engineers and continued to prosper in the middle class. That required one to play along with the British government.

KW: In your case, it was the English principal Fyson who was responsible for getting you the Government of India scholarship.

SC: Yes. But I remember an interesting incident in connection with him. You know that the Indian national movement began to crest around 1928, about the last of my college years. Most of us students were pro-National Congress and the Congress met in Madras in 1928. Nehru came to Madras to become its president. The students were told not to skip classes to attend his presidential address, but all of us did and all of us were fined. Principal Fyson had us called to his office to admonish us. He knew me quite well from the work I was doing but did not know that I was among those who had skipped classes that day. When he saw me in that crowd, he said, "Well, you too?" And then, "I am sorry. I said that none of you should strike. You went against the rules. I'm afraid I must fine you all." We paid the fine all right, but didn't go into a full civil disobedience movement. Fyson's attitude towards students, however, was not at all affected by the incident. In particular, it did not prevent him from supporting me two years later for the scholarship with which I went to England.

KW: That's remarkable.

SC: In fact, one of the characteristics of the Englishmen in India was that they would support everything in the country if it was not politically motivated. In science for instance, if there was an Indian who was competent, they would definitely support and help him. After all, Ramanujan would have been unknown at that time except for Hardy and Littlewood. Saha would have been unknown except for Fowler. And I believe that Raman's career in the early years was encouraged and supported by people like Lord Rayleigh and Rutherford. Satyan Bose of course went to Germany. He had Einstein's support; but his case is slightly different. So I think the English scientists and English professors have no reason to be ashamed of their role in helping India in science.

KW: What was the reaction of the British scientists, say the Cambridge establishment, to the political situation in India during your Cambridge years?

SC: Well, I found that most of the professors whom I met were rather conservative. During my first year in Cambridge, Mahatma Gandhi came to England and he came to Cambridge. The Indian students went to see him and I was with the students who met him, in fact in a room of the size of this office. He talked to the Indian students. Of course, once in England, the news one got from India was all very one-sided. But it's also true that once I got to England, for the first few years anyway, I felt I had so much to learn to become a scientist. After all, the scholarship was only for three years and I had no guarantee that I would be able to stay longer. So I felt I had to spend all the time I had in studying. My interest in politics, at least following the events in India, was rather sporadic. After I became a fellow, however, I followed the Indian politics more closely. While I got along with my contemporaries, people of my own age, who had attitudes very similar to mine, I found it very difficult to talk politics with people like Milne or Fowler because they had opposite views.

KW: They did have opposite views?

SC: Oh, yes.

KW: As scientists I thought they would have different views, that they would sympathize with the cause of freedom.

SC: No, not particularly.

KW: Were there any specific discussions?

SC: Well, I remember talking to Milne on one occasion. I told him I thought the British government had made a big mistake in not granting dominion status to India, because they may have to concede much more later. He asked me what I thought of Nehru. I said, "I think he is a marvelous man." It was quite obvious from his reaction that he did not share my views. "Well, do you think India will be able to survive, if the British leave, what with all the princes and so on?" You know, all the standard questions. On the other hand, there were some who had very different views. For example, Hardy and Littlewood. I remember Hardy once saying, "Well, I'm sure I wouldn't get anywhere if I nominated

Nehru for an honorary fellowship to Trinity." There were some. There were those associated with the Labour party, but of course during my time Baldwin was the prime minister. Baldwin and Chamberlain. No, it was quite obvious that the establishment at Cambridge was quite against the Indian movement. I mean, I could feel it very strongly as a Fellow. On the other hand, if you didn't talk politics with them, they wouldn't interfere; they didn't treat you any differently even if they knew your views were different. They didn't take that into account in their relations to you.

KW: Did it bother you that even scientists did not see the Indian side?

SC: I was a little unhappy about that, but during the second part [of my stay] in Cambridge, 1933–36, my emotional center was science. I sort of felt, either consciously or subconsciously, that so long as I was in England I had better try to do my science as best I could, and so long as I was not in India, taking part in political matters was not helpful either to me or to India. I sort of took an attitude of neutrality. It wasn't that difficult for me to separate the two parts completely. Some of my friends of those days, even after India became free, thought it was all a big mistake. Yes, some of my English friends. But when you expressed your disagreement, the conversation sort of ended without pursuing the arguments. The English, you know, have a way of politely ending the conversation.

Srinivasa Ramanujan

Chandra was not quite ten years old when his mother told him about a famous Indian mathematician, Ramanujan by name. At that time, neither Chandra nor his mother had any idea what kind of mathematician Ramanujan was. For that matter very, very few in India ever knew or understood Ramanujan's genius. But, as Chandra says, "Ramanujan's role for the development of science in India did not depend upon his being understood! The fact that his early years were spent in a mathematically and a scientifically sterile atmosphere, that his life in India was not without hardships, that under circumstances that appeared to most Indians as nothing short of miraculous he had gone to Cambridge, supported by eminent mathematicians, and had returned to India with every assurance that he would be considered, in time, as one of the most original mathematicians of the century—these facts were enough, more than enough for aspiring young Indian students to break their bonds of intellectual confinement and perhaps soar the way Ramanujan had."

Indeed Ramanujan greatly influenced Chandra; he was his role model for a life dedicated to the pursuit of science. Chandra has written and talked about him, and has been instrumental in many ways in perpetuating the memory of the tormented genius that Ramanujan was. Over the years, I have had several conversations with Chandra regarding Ramanujan. What follows are excerpts from those conversations.

KW: Balakrishnan writes in his biographical article in the *Triveni Quarterly* that, during your visit to India in 1936, you took the trouble to seek out Ramanujan's widow, and found her living in one of the dark, dingy, bylanes of Triplicane.

SC: Yes. It was because of Hardy, who gave a series of lectures on Ramanujan's life and his work at the Harvard Tercentenary Conference of Arts and Sciences in the fall of 1936. In the spring of that year, after my return from Harvard, I had a fair amount of conversation with Hardy about Ramanujan. On one occasion, he said to me that the only photograph of Ramanujan that was available to him at that time was the one of him in cap and gown "which makes him look ridiculous." And he asked me whether I would try to secure, on my next trip to India, a better photograph that he might include with the published version of his lectures.

KW: So that's what led you to Mrs. Ramanujan.

SC: Yes. Balakrishnan and I found her living under extremely modest circumstances. I told her that one of the professors in England was writing a book on Ramanujan and that he would like a good photograph of him to include in his book.

KW: Balakrishnan also writes that you invited her to your house, introduced her to your sisters, and made her feel quite at home. You told her "how the greatest professors across the seas revered the memory of your late husband as that of a guru, a great master."

SC: Yes. At first she said she had no photograph. After a while, she recalled that she had Ramanujan's passport in her possession. The passport contained a photograph. I escorted her to her house in my father's car and took possession of the passport. To my great delight I found the photograph sufficiently good to make a negative and copies, even after seventeen years. I left the passport with Balakrishnan to get the copies and the negative made and send them to me at Williams Bay.

KW: You later sent the negative to Hardy.

SC: Yes. The photograph appears in Hardy's book, *Ramanujan, Twelve Lectures on Subjects Suggested by His Life and Work*. I recall Hardy's reaction to the photograph. He said, "He looks rather ill, but he looks all over the genius he was." You know, it is this negative which has served as the basis for all photographs, paintings, and etchings of Ramanujan, and the enlargements are copies of the picture in Hardy's book.

KW: Yes. I also read Richard Askey's comments on the occasion of dedicating the bust Lalitha and you presented to the Indian Academy of Sciences in Bangalore, that Paul Granlund's bust of Ramanujan was inspired by the passport photo. It's a beautiful piece of sculpture.

SC: I do take a certain amount of pride in getting that photograph. Just as Hardy says, "Ramanujan was, in a way, my discovery," I like to say that the photograph was one of my most important discoveries.

KW: You also mentioned to me once Ramanujan's attempted suicide.

SC: Yes. I learned of the incident from a conversation I had with Hardy at dinner in Trinity during the spring of 1936. Hardy arrived a little late but bandaged, and he sat at the table opposite me. The fact that he was bandaged naturally aroused the concern of those around him, and when questioned, he told a remarkable story. It's rather elaborate. You may want to reproduce it from the talk I gave to a private club at the University of Chicago. [The following quotation is from S. Chandrasekhar, *Notes and Records of the Royal Society* 30 (1976): 249.[6]]

It appeared that he [Hardy] was in London during the day and while crossing Piccadilly Circus a motorcycle hit him and dragged him along. He was bruised but only superficially. Nevertheless, Hardy was escorted to the Scotland Yard by the policeman who had arrested the cyclist in order that Hardy could report what had happened. After Hardy had given the appropriate evidence and was about to leave, a messenger came to him and told him that a senior officer of the Scotland Yard wanted to see him. Hardy was slightly surprised. But the officer treated him with great courtesy, asked him to be seated and said, "Professor Hardy, I have been wanting to see you for many years. In fact, I have waited for this occasion for seventeen years. Do you know that we have had evidence in our files here to arrest you for giving false evidence?" Hardy was a little surprised and the officer continued. "Do you remember, Professor Hardy, that in February 1918 an Indian mathematician had tried to commit suicide by falling before the train in an underground tube station? His intention was thwarted and he was arrested and brought to Scotland Yard. I was in charge of that case. And you arrived later to give evidence." At this point I should digress, even as Hardy did while narrating his encounter with the officer at the Scotland Yard.

Ramanujan had been quite ill during the winter of 1917 and apparently in a state of extreme depression. Ramanujan in fact had tried to commit suicide in the manner described by the officer; but by a series of miracles (like the switch being turned off by a guardsman, and the train coming to a stop just a few feet ahead of where Ramanujan had fallen) he had been saved. To continue with the story.

"When you arrived at Scotland Yard, Professor Hardy, you told us that Ramanujan was a Fellow of the Royal Society and as such could not be arrested. We released Ramanujan and you left apparently believing that you had bluffed us; but in fact you had not. You knew as well as we did that Fellows of the Royal Society are not immune from arrest. But you also told a lie. At the time of this incident Ramanujan was not in fact a Fellow of the Royal Society. But you knew that he would be elected a month later but that is not the same thing. Nevertheless, on inquiry we found that the man whom we had arrested was indeed reputed to be a great mathematician and we in Scotland Yard did not want to spoil that life. And so we let you believe that you had convinced us. But I have always hoped that an occasion would arise when I could tell you that we knew all along that you were telling a falsehood and had perjured yourself. The occasion has now come; but I am not going to arrest you."

I cannot quite recall whether it was on this same occasion, or on a later occasion, that Littlewood referred to Ramanujan's attempt to commit suicide in connection with his (Ramanujan's) election to a Fellowship at Trinity. There is apparently a rule that one who is medically insane cannot be elected to a Fellowship at Trinity. And Littlewood (who was one of the electors during the year that Ramanujan was a candidate) was afraid that Ramanujan's "insanity" in having attempted suicide might be brought up to disqualify his election; and on that account Littlewood said that he had gone to the electors meeting with a medical certificate to the effect that Ramanujan was not afflicted by insanity. And Littlewood added that he was extremely glad that no occasion arose which required him to produce the certificate.

[Our conversation continues.]

You know there is a sequel to that story. In 1968 I gave the Nehru Memorial Lecture and the Ramanujan Lecture at the Indian National Academy of Sciences. They now call it the Indian National Science Academy. And, since it was the Ramanujan Memorial Lecture, I told the stories which Hardy and Littlewood had told me about Ramanujan's election to Trinity Fellowship and the Fellowship of the Royal Society. In that context, I mentioned Ramanujan's attempted suicide. But it appeared in the newspaper the following day with big headlines, "Ramanujan tried to commit suicide," and told the story in a garbled way. I was astonished at the opposition it created. Everybody thought I was ill-advised in telling that story. I went to Bangalore to see Raman, and one of the first things he said was, "Why do you want to defame that man?" Then there was a letter in the *Times of India* saying that my motive in recounting the incident was to enhance my own personal reputation at the expense of Ramanujan's. I was very depressed by that reaction. A few days later when I went to Madras, I was met at the airport by a young man who introduced himself as a neighbor of Mrs. Ramanujan; and he told me that Mrs. Ramanujan was anxious to see me. I was naturally worried that she would be very upset about it. However, she told me that my account of Ramanujan's attempted suicide had cleared up certain things about her life, which was so tragic. I have written up the things she told me in the Royal Society notes, but not for the public, because I don't know whether it is the kind of thing one should write without giving proper account of the whole story. Just one remark, Mrs. Ramanujan told me, for instance, that Ramanujan's mother put every kind of obstacle in their married life.

KW: It's not uncommon in India. I have heard many stories of that nature.

SC: It is known (recorded, for example, by Hardy) that during Ramanujan's years in England, he received very few letters from his family in India. And Hardy attributed Ramanujan's depression during his later years in Cambridge to his "misunderstandings" with his family. Ramanujan used to write regularly to his wife, but apparently she never saw any of those letters since his mother apparently did not pass them on. Ramanujan's wife could not write to him because she did not have the money for postage.

KW: These facts are not widely known. To realize that Ramanujan's life in England was overshadowed by such personal frustration.

SC: It throws an entirely different light. Yet he made some of his great contributions to mathematics during his four years in England. So I did write of my visit with Mrs. Ramanujan and sent it to the Royal Society Archives in London with his photograph and other things. They asked me whether they should publish it in their notes and records. I have not given them permission, but I have a feeling that it should be published sometime.

Chandra has continued to take an interest in perpetuating the memory of Ramanujan. In the late 1940s, he was instrumental in founding the Ramanujan Institute of Mathematics with the financial help of a former classmate, Alagappa Chettiar, who prospered as an entrepreneur, became a well-known philanthropist, and was knighted by the British government. After Chettiar's death, when reports reached Chandra about the imminent death of the institute, he promptly wrote to Prime Minister Jawaharlal Nehru. With Nehru's personal intervention, the institute survived. The institute owes a great deal to Chandra and André Weil for its continued existence after Nehru. They served as advisors to Madras Government in the choice of the director and staff. They helped to secure retirement benefits for C. T. Rajgopal when he retired as the director in 1968. Mrs. Ramanujan had a meager pension of 150 rupees a month and was living in poverty. In 1962, due to Chandra's efforts, it was more than doubled. Now she still lives modestly but comfortably in Madras. She is respected and participates in most ceremonies honoring her husband.

Finally, busts of Ramanujan cast by Paul Granlund have come into being only very recently, in 1986, largely due to the efforts of Richard Askey and Chandra.

John von Neumann and Enrico Fermi

Among the numerous scientists with whom Chandra has been associated, two stand out: John von Neumann and Enrico Fermi, two of the most extraordinary minds of the twentieth century. Regarded as a genius, von Neumann is legendary among scientists for his contributions to various branches of mathematics, computer science, and logic. After the war, he became a controversial public figure because of his continued, enthusiastic participation in weapons research, a hard line towards Russia, and his opposition to international control of atomic energy. Fermi is equally legendary for his pioneering contributions in physics: Fermi statistics, Fermi theory of β-decay, Fermi's experiments with slow neutrons which led to the discovery of trans-uranic elements, Fermi's discovery of the first pi meson-nucleon resonance, and so on. Further, his wide interests went far beyond the fundamental questions of physics and led to his collaboration with Chandra on some researches in astrophysics. The following is an excerpt from what Chandra has written about Fermi in his introduction to their joint papers appearing in *Collected Papers of Enrico Fermi* (Chicago: University of Chicago Press, 1962–65).

During all my discussions with Fermi, I never failed to marvel at the ease and clarity with which he analyzed novel situations in fields in which, one might have supposed, he was not familiar and, indeed, was often not familiar prior to the discussion. In the manner in which he reacted to new problems, he always gave me the impression of a musician who, when presented with a new piece of music, at once plays it with a perception and a discernment which one would normally associate only with long practice and study. The fact, of course, was that Fermi was instantly able to bring to bear, on any physical problem with which he was confronted, his profound and deep feeling for physical laws: the result invariably was that the problem was illuminated and clarified. Thus, the motions of interstellar clouds with magnetic lines of force threading through them reminded him of the vibrations of a crystal lattice; and the gravitational instability of a spiral arm of a galaxy suggested to him the instability of a plasma and led him to consider its stabilization by an axial magnetic field.

KW: When did you first meet John von Neumann?

SC: I first met him in Cambridge. He was visiting Cambridge from January to July 1935 and giving some lectures in mathematics. He was living near Trinity in a hotel or an apartment. As I recall, he was rather lonely at that time. English people, as you know, are quite a reserved people. They often don't pay much attention to a visitor. I used to see him quite frequently as he used to come and see me in my rooms in Trinity.

KW: Wasn't he older than you? Some ten years or so?

SC: Yes. Perhaps less. Six or seven years. He was comparable in age to people like Dirac. But when I first met him, he was in his early thirties. We were both young and consequently we got along quite well. He was surprisingly free of ostentation, very generous with his time, and an extraordinarily nice person in a human way.

KW: January–July 1935. That was the time when you were embroiled in the controversy with Eddington.

SC: Yes. Johnny became involved with that to a certain extent. We started working together with the idea of bringing general relativity into the problem of degenerate matter. The work remained uncompleted, however, when he left Cambridge. There are still some unpublished manuscripts which we started writing at that time. In fact, we had started using the equations of state for degenerate matter, which Oppenheimer and Volkoff used later in their work on neutron stars. I got to know him rather well.

KW: You met him again later when you came to this country. I read some extremely friendly letters from him to you beginning in 1937.

SC: Yes. If my memory serves me right, the very first year we were in this country, Johnny was visiting Eugene Wigner in Madison and both of them drove down to Yerkes to spend an afternoon with us. Later von Neumann visited us at Williams Bay more than once and stayed with us. And as you have written, he arranged my visit to the institute in Princeton during the fall of 1941. He was chiefly responsible for my being at Aberdeen Proving Grounds.

KY: You wrote papers together as a result of the work during the fall of 1941.

SC: Yes. On the statistics of the gravitational field arising from a random distribution of stars. Besides, you know, he was responsible for the publication of my paper, "Stochastic Problems in Physics and Astronomy," published in *Reviews of Modern Physics* in 1943. It's one of my most widely quoted articles. I read somewhere that somebody had made a study, and that it was among the most widely quoted articles during the period 1961–71, although it was written in 1941 and published in 1943. The editor of that study wanted me to write a piece on it. But you see, I had written that article for myself with no intention of publication. When I showed it to von Neumann, he insisted that I publish it and suggested the *Reviews of Modern Physics* as the place to send it. Since I was hesitant, he took it upon himself to send it in for me.

KW: Von Neumann was so different in temperament and character from you. I have read that he loved parties, in fact that his house in Princeton was a party place with plenty of drinks, off-color jokes, and so on. How did you two get along outside your scientific work?

SC: We had no problems. We were mostly excluded from such parties, but at the same time, he loved intellectual conversations, was very witty and extremely lovable. I remember one time when he stayed with us; we took him out to dinner and Lalitha asked him whether he would like to have a drink. He said no. A few minutes later he smiled and said, "I think I'll have a drink; the idea has sunk in." Another time I met him in Washington and (to give you an example of a typical conversation with Neumann) I said, "Where are you living now?"

To which he responded, "I am living in a hotel."

"Don't you have to find an apartment?"

"Yes."

"Isn't it very difficult to find an apartment?"

"Yes, the only way to get an apartment is to await a miracle."

A few weeks later I met him again and asked him whether he found an apartment.

"Yes."

"How?"

"By a miracle."

He was extremely quick. I am sure you have heard many stories of his quickness, absentmindedness, and his humanness.

KW: Yes. Did you come to know Mrs. von Neumann?

SC: Yes, of course. He was married twice. I met his first wife in Cambridge. After the divorce he married again. His second wife's name was Klara. Both his wives were from Hungary.

KW: Before and during the war, was he as famous as he became later?

SC: Certainly; he was quite famous as a scientist, but he was not a public figure before. After the war, he became one—a member of the Atomic Energy Commission and all that—belonging to the group of people like Oppenheimer, Teller, and Wigner. All these people after the war became more remote than they were formerly, probably for understandable reasons. I also got to know Wigner quite well during the war. Whenever he used to be in Chicago, we used to have regular luncheons every Wednesday. My discussions with him played an important role in my work on the negative hydrogen ion.

KW: You also met Fermi during the war.

SC: Yes. I used to have lunch with Fermi and Teller occasionally during the war, whenever I visited the campus. After the war, Fermi joined the faculty. During the first year I saw him only from a distance. We corresponded about colloquia, etc. Then some time passed during which I used to meet him for lunch in the Quadrangle Club. On one occasion he said to me, "Since you are interested in hydromagnetics, and I don't know anything about it, perhaps it would be nice if we could meet together regularly once a week. We could discuss astrophysical problems." That's how my close association with him began. In the fall of 1952, and in the winter and spring of 1953, I met with him once a week on Thursday at 10 AM. We would discuss a variety of astrophysical problems till noon and then go to lunch.

KW: These discussions led to your joint papers.

SC: Yes. After a few months of discussions, we accumulated a fair amount of material. I wrote it all up as a joint paper. When the paper was all finished and approved, I said to Fermi, "I'm afraid I made so many mistakes all during this time." He sort of said, "Well, I too made many mistakes in earlier times." Indeed, I have rarely come across a person so generous as Fermi in physics discussions. For example, when I said something with which he did not agree, he would say, "Well, it doesn't seem to me that what you are saying is right. Of course, I may be wrong, but let me assume that you are wrong and proceed. Then a point must come where it will be clear where I have gone wrong." And then, as the discussion progressed, a point would come where it became clear that he was right. At that moment he would change the subject at once. He would never stop to say, "I told you so."

KW: That's indeed rare. With some people you can discuss physics, forgetting the person entirely. Who is right, who is wrong never comes into the picture. The joy in understanding and solving the problem together is what matters. With some others, the person never disappears. You are afraid of making a mistake; you worry about what the other person thinks of you. Or you take a great amount of joy in proving the other person wrong, which is equally bad.

SC: Fermi was never like that. I must say von Neumann was also similar. He was so extraordinarily quick to grasp, but he would never rub in the fact that he was better than you. He used to work with a sense of equality that was extremely nice. I liked him enormously for that.

KW: Both Fermi and von Neumann died of terminal cancer. I have read somewhere that von Neumann suffered a great deal; that he died a completely broken-down man. In contrast, as you have written, Fermi faced his inevitable death in good humor, so to speak. [This is what Wigner writes about his boyhood friend:

When von Neumann realized he was incurably ill, his logic forced him to realize also that he would cease to exist and hence cease to have thoughts. Yet this is the conclusion, the full content of which is incomprehensible to the human intellect and which, therefore, horrified him. It was heartbreaking to watch the frustration of his mind when all hope was gone in its struggle with the fate which appeared to him unavoidable, but unacceptable.[7]]

SC: Yes, the two cases are marked by contrast. Von Neumann could not accept the fact that his death was inevitable. He turned to Catholic religion, got himself baptized, was in constant panic and was very demanding of Klara.

Fermi was quite different. The day after the operation when it became clear that he would die, Herbert Anderson and I went to see Fermi in the hospital. Let me read to you what I have written:

It was of course very difficult to know what to say or how to open a conversation when all of us knew what the surgery had shown. Fermi resolved the gloom by turning to me and saying, "For a man past 50, nothing essentially new can happen and the loss is not as great as one might think. Now you tell me, will I be an elephant next time?[8]

KW: How long did he live after that?

SC: Just three months. The operation was in September and he died in November. He was in Italy the previous summer in the Varenna school, and during that time he was unable to eat well. And when he came back to Chicago everybody was horrified to see how thin and emaciated he was. One doctor said it was psychosomatic and so on. But the people at the university got very worried and saw to it that a good doctor saw him. And in fact I had a conversation with the surgeon. As I say here in my account of it:

At first the doctors could not diagnose what was wrong with Fermi. It finally became clear that the cause of the illness was either congestion of the esophagus or cancer of the stomach and the intestines. And the matter would be settled by surgery. Dr. L. R. Dragstedt, who performed the surgery, told me of his conversation with Fermi the night preceding the operation. Dragstedt told him that if the problem was congestion of the esophagus the surgery would be complicated and would take a long time. But if it was cancer of the stomach and intestines, then they could probably do very little about it and the operation would be of only short duration. And the next day on returning from surgery Fermi opened his eyes and noticed he had not been in surgery for very long. He turned to Dragstedt and asked him, "Has metastasis set in?" And the answer was yes. Then he asked him, "How many more months?" And Dragstedt replied, "Some six months." And Fermi went back to sleep.

Actually he didn't live the six months, he only lived three months, and in fact he died of a heart attack in his sleep. Which was fortunate because by November, I think he died in November, the pain began to be very, very severe and the prospect of three more months of this pain was very worrisome to Laura Fermi. But he just quietly died during the night.

KW: You probably saw him often in those three months.

SC: I didn't see him very often, but I did see him a few times, yes.

KW: Did he remain the same way?

SC: Oh, yes. Most of the time he was completely normal, but occasionally a sentence or two would come out. I remember one occasion when Laura was wondering who should go and pick up Emilio Segrè, when he was coming, etc. Fermi said in a slight fit of annoyance, "Do we have to discuss it here?" or something like that. Clearly irritated slightly. Except for one or two vague remarks like that, he showed no particular feeling or exhibition of his inner feeling that he was going to die very soon.

KW: Did he show any interest in physics?

SC: Yes. He actually thought he had six months to live, so he began writing out some of his lecture notes. He had a secretary come and take dictation, but after a few weeks he found he was under too much pain to do that, so he stopped. He tried to keep up his interest particularly with regard to writing his lecture notes and a book that was going to come out on pion physics, I think. But he could not because he was in too much pain.

KW: He didn't show any concern about his life? Frustrations?

SC: He didn't show it to any of his friends. I knew Laura very well. She behaved, at least during the time we were with them together, perfectly normally and naturally. Of course, she was a very brave woman.

KW: We have two examples of brilliant scientists who reacted so differently. It's very interesting.

SC: Yes. It's very interesting. Fermi remained an agnostic, whereas von Neumann somehow couldn't reconcile himself with the idea that he was going to die.

KW: But most of his life he was not religious, right?

SC: No, he was not.

KW: He suddenly became religious towards the end.

SC: You know, this affected Klara von Neumann very badly. We knew her quite well and, in fact, we saw her in 1958 or 1959 when we were in La Jolla. She married Carl Eckart, and we remember particularly one evening, sitting on the La Jolla shores with her describing the incredible mental strain of going through those times with Johnny. How he insisted on having her near the bedside all the time, and how it sapped her energy. And you know Klara von Neumann died a very unhappy death. She was drowned walking in the ocean on the La Jolla shores.

KW: No, I didn't know that.

SC: They had a party, in fact, for Maria Mayer when she got the Nobel Prize. It was one of those West Coast parties, with plenty of drinking, going on very late. Mrs. Eckart, that is the former Klara von Neumann, returned at 3:00 AM but the next morning Carl Eckart couldn't find her. They found her body on the shore of the ocean where the descent was deep. Even though she tried to change her life, she married Eckart and they seemed to be happy, I think the influence of the last months of Johnny must have been severe.

KW: Yes. To see the transformation of one who was such a genius to something so different, so helpless at the end, probably did affect her. Well, to change the subject a little, when you worked together were your discussions confined only to science or did they extend beyond?

SC: With Fermi, it was mostly science. Von Neumann used to talk quite a bit about politics. He was extremely concerned about how things were going at the beginning of the war. After the war, you know, he thought that the most rational thing to do was to strike at the Russians. His view was, "It's something which would be the logical thing to do, but of course we won't do it, and I don't recommend that we do it. That will solve the problem, but I don't expect it to be solved that way."

KW: Did you discuss your controversy with Eddington and express your personal feelings and frustration to von Neumann when he was in Cambridge?

SC: Well, he knew the story. But I had the feeling that although it was important to me personally, it was too insignificant for a person like him. It was a matter that did not require his attention. The fact is that I have always been doing things which are outside the mainstream of what people care about at a given moment. Hence, I don't think that at any particular time I have felt that the work I was doing was sufficiently important to speak about it in any exaggerated terms. I didn't have the kind of confidence which others express in their writings or talks.

KW: But it is there in your papers. When in 1932, for instance, you wrote, "Great progress in the analysis of stellar structure is not possible before we can answer the following fundamental question: Given an enclosure containing electrons and atomic nuclei, what happens if we go on compressing the material indefinitely." If you read those papers, it's quite clear that you had the full appreciation of their importance. You sound so prophetic.

SC: There is a peculiar dichotomy in my views. When I concentrate on my work, I am confident that what I say is right. But I am never sure whether my work has value in an area outside of my knowledge, whether the particular things which are my concern are also those of other people in the large. Hence I never . . .

KW: I suppose, to some extent, it is the Indian heritage and background. Maybe we should say, you are an example of modesty in science. I remember the story you told me of your conversation with Felix Bloch in Jerusalem some years ago. You were telling him that there were two kinds of modesty: social modesty and modesty in science. And Fermi was an example of the second kind. Then Bloch asked you to what class you belonged. When you said the latter, he thundered before the gathered crowd and said, "*That* you certainly are not!" I recall amusing myself when you told me this story with this logical thought: By his own admission he is not socially modest; Bloch does not think he is modest in science; therefore Chandra is very immodest.

Second Visit to Russia (October 1981)

Chandra visited Russia for the first time in 1934. He was a young researcher then in his first year of a Trinity Fellowship. After forty-seven years, in October 1981, Chandra made a second visit to Russia, this time accompanied by Lalitha. He was invited by Viktor A. Ambartsumian, president of the Armenian Academy of Sciences and director of the observatory at Erevan, to participate in a symposium to mark the fortieth anniversary of Ambartsumian's and Chandrasekhar's pioneering contributions to the field of radiative transfer and, more particularly, of Ambartsumian's discovery of the principles of invariance.

Chandra had met Ambartsumian in Leningrad during his first visit. He knew the latter's work on planetary nebulae for which he had become well known. Later, when Chandra began his work on radiative

· transfer in 1943–44, he became aware of Ambartsumian's earlier papers on invariance principles and had incorporated them in the body of his work to solve a large number of problems which up until then had not been considered possible to solve. "The fact that I gave prominence to Ambartsumian's work on radiative transfer," says Chandra, "and later to his work in which he had used the methods of invariance to study intensity fluctuations caused by interstellar matter, had created a sense of good feeling between the Russian astrophysicists and myself." Chandra had also been responsible for Ambartsumian's foreign membership in the National Academy of Sciences in the United States, for his foreign membership in the American Astronomical Society, and for his being awarded the Gold Medal of the Royal Astronomical Society of London. Principally because of these associations, Chandra eventually accepted the personal invitation from Ambartsumian to participate in the symposium, in spite of initial hesitations. At that time, Andrei Sakharov's house arrest and confinement to Gorky had created an uproar among western scientists. A strong sentiment against cooperation with the Russian scientists, including participation in conferences and symposia in Russia, was rampant. Members of the National Academy of Sciences were urged to boycott meetings in Russia.

The conversations excerpted here took place in November 1981 soon after Chandra's return from Russia.

KW: Forty-seven years is a long interval between the two visits. Weren't you invited earlier?

SC: I was. I was invited to Russia at least three times during the sixties and seventies, once in connection with some celebration in Leningrad, once in connection with Ambartsumian's seventieth birthday, and once also in some other connection with Ambartsumian. Each time I had declined, principally because of all the hassle involved in traveling to Russia and the short time they allowed to make preparations.

KW: What made it different this time?

SC: Well, the letter said that the symposium was to mark the fortieth anniversary of Ambartsumian's and my work. The opening addresses would be shared by the two of us. I thought it was rather unique for the Russians to make a celebration of the work of one of their great men and someone else outside the USSR.

KW: Did you correspond with Ambartsumian during the intervening years?

SC: I wrote to him, but he never replied. But after the war, he was in

England, from where he sent me, through K. S. Krishnan, a set of reprints of his work on radiative transfer. Reference to my work in those reprints was put in by hand.

KW: Not printed?

SC: No. Therefore, I had a feeling of annoyance and irritation about him. While I had given his work a great deal of prominence and did whatever I could to bring his contributions to light, he had not responded the same way. In his own writings on the subject of radiative transfer, he did not refer to my work, and even if he did, he did it in a rather tangential way.

KW: Had it something to do with Russian politics?

SC: Perhaps. That's what I thought and left it at that. When my book was translated into Russian, the translator's introduction says that the entire book is based on Ambartsumian's work with some minor additions. Quite an unfair statement. Nevertheless, over the years, during the sixties and seventies, I used to get Christmas cards regularly from two of his associates, Viktor V. Sobolev and V. V. Ivanov, both from Leningrad. Sobolev had been closely associated with Ambartsumian during the war. They always wrote rather nicely to me even though I had never written to them.

KW: You mentioned to me once that Lalitha played an important role in making up your mind about going.

SC: Yes, yes. My decision to go finally was based on her insisting that, since I always had admired Ambartsumian for his scientific contributions, the small feeling of irritation should not stand in the way; perhaps it could be cleared up in a personal meeting and remove the cloud of unpleasantness. She insisted rather strongly, purely for human reasons.

KW: So were you happy that you went?

SC: Certainly. It was a total experience—a most moving one, which Lalitha and I will never forget—because of the personal warmth shown to us by scientists at all levels, from young students to well-known established scientists like Ambartsumian and [Ya. B.] Zel'dovich. Of course, I have been abroad and given lectures at various places. I do get large audiences and very patient hearing, but the personal warmth which I experienced in Russia on this trip was exceptional. I simply had not experienced anything like that. I might have told you before that before we went to Russia, we went to Poland. There was only a ten-day interval between our return from Poland and going to Russia. Therefore, I wanted to make all the reservations, etc., before we went to Poland, but the visa for Russia hadn't come. I sent a telegram a week before our departure to Poland, saying that I might have to cancel my visit if the visas were not available by the end of the following week. And, you know, I got three or four telegrams from Russia and a telephone call from the Russian Embassy, the Consulate Division, saying that the visas were there. I thought it was extraordinary; it indicated that they were extremely keen on getting me there.

KW: Where did you go first in Russia?

SC: We went to Moscow. Mr. and Mrs. [Igor D.] Novikov and V. Lukash were waiting to receive us at the airport. They were all so nice. They took care of us and took us to the hotel in waiting cars. Then, somebody or other came to the hotel at 7:30 AM every morning to keep us company at breakfast and see that we had no difficulty.

KW: Were there other scientists from the West?

SC: There was one other person from the USA, F. Zweifel, who was in Europe at that time and came to the symposium. There were some scientists from eastern Europe and none from the western European countries. The symposium was organized by the Armenian Academy, not by the Soviet Academy. Hence, it was not an international meeting in the strict sense. Just a few were invited personally. I think, in a way, the academy was anxious to avoid inviting people and getting rejected because of the political situation. It didn't want to confront that.

KW: How long were you in Moscow?

SC: We were there for four days. Zel'dovich made me give a talk at Shternberg Astronomy Institute. I lectured on black holes to overflowing audiences. There was an enormous interest in the audience. Ginzburg was there and so were all the scientists whose names I had known. And of course a large number of students. I could tell they were students because, at the end of each lecture, there was always a long line of them with copies of my books to autograph. I never autographed so many copies of my books in my life. We had also arranged visits to the Bolshoi Theatre and to the Kremlin.

KW: You then flew to Erevan?

SC: Yes, it was a long flight. You go almost to the Turkish border. The plane was late, but L. Mirozoyan was waiting for us at the airport. He took us to the observatory guest house, where we were given a special double suite of rooms. When we arrived there, Ambartsumian and his wife were waiting to receive us with flowers and a basket of fruits and talk to us.

KW: How was the conference, or symposium rather?

SC: It was very nice. It was a well-attended meeting. Ambartsumian told me—and he also said this in his opening remarks—that because of my presence, the registration at the meeting had doubled from what it had been before. There were astrophysicists from all over the Soviet Union, from Vladivostok, from Moscow, Leningrad, Kiev, Kharkov, Tiflis, and Tashkent. They all wanted to talk to me in one way or another. There were some young people working on the subject—N. A. Mnatsakanian, an extremely nice young Armenian. I also met B. E. Markarian, who is very famous for having discovered the so-called Markarian galaxies. They took us to museums, had special permits to see special exhibits. I also remember pleasant walks with Ambartsumian in the gardens near the observatory that he had built, and talking to him with the understanding which comes from having worked with each other's ideas and each respecting the other. It was a very moving experience for me. He was extraordinarily nice to me. He remembered the time we met in 1934. In fact, before we left for Russia, I had received a very warm letter from him saying that he remembered the time forty-seven years before. At that time we had very similar views on science. He stated that the world had changed beyond comprehension, but he wondered whether our scientific views would still be the same.

KW: Did the question of the minor irritation ever come up?

SC: No. It did not occur to me. It was long forgotten. He made it clear from the way he talked, the way he introduced me, and the way I was given prominence. In his lecture he said that the radiative transfer is built on four legs, two of which are the principles of invariance [which he had formulated] and two are the extensions which I had done. I think that is a fair statement. Further, Sobolev, at the conclusion, made a very moving speech. He said, "It is so marvelous to have the two people, whose work has influenced the work of all of us, here in person. One of them [all these years] was represented only by his book. Now, he is here in person."

KW: Where did you go after the symposium?

SC: We flew to Leningrad and then home. Ivanov, who had come to Erevan, flew back with us to Leningrad. Ivanov was sitting next to us; he was surprisingly frank and open when Lalitha made some cautious inquiries. He talked about the terrible times during the Leningrad siege and the constraints in which they continue to live. He said how exciting it was for him to meet someone whose book he had read from cover to cover. He said, "You tell me a formula in your book and I will tell you on which side of the page it is. It's your book that taught me everything." And Sobolev said the same thing, that to them the introduction to the subject was through my book. They knew that my work was not simply a trivial extension of Ambartsumian's.

KW: Previously then, there must have been some political reasons for Ambartsumian's behavior.

SC: I suppose so. On the other hand, you know, as one grows older, one becomes less possessive and more objective. One is not contending for fame, etc., in the same way as in one's youth. But there was one incident that comes to my mind which caused me a small irritation that had nothing to do with Ambartsumian.

KW: What was that?

SC: Pulkova Observatory near Leningrad, about 20 kilometers away, is run by a man named Krat who had ascended the academic ladder. He is a member of the academy and the director of the observatory. There appeared to be a play of precedence between Krat on the one hand and Sobolev and Ivanov on the other. Sobolev wanted me to lecture at the university in Leningrad, but Krat overruled that. The lecture had to be at the observatory. Then after the lecture he led me into his office. Several people, nearly a dozen, were waiting on the way to and in front of his office. All well-known names in Russian science. They were greeting me and referring to the fact that they had met me here or met me there, they had read this or that, all referring to their association with me. The former director, a member of the academy, an old man past 80, for instance, was among the people waiting. I had met him in Leningrad the previous time forty-seven years ago. He remembered it and had come to the lecture because of it. Krat led me through all such people and, behind closed doors, he had tea with me along with only one other person, the vice-director. The tea went on and on. When we came out, these other people were still standing and waiting.

KW: Very strange. Did anyone comment afterwards?

SC: Yes, Ivanov said, "Well, you see the kind of society in which we have to live. I felt sad because I knew you wanted to talk to us, but you were taken away leaving us standing outside." That's the only part of the entire trip that really irritated me. I didn't encounter any of it in any other place. Ambartsumian is a powerful man, both scientifically and politically, but he did not treat his colleagues that way.

KW: Other than this one incident, how did you feel about conditions in Russia in general? Did any of the scientists invite you to their homes?

SC: Well, conditions in Russia were certainly much better and life much easier than in Poland, where we had been just a few weeks before we went to Russia. More cars; the shops more full. Scientists who are not necessarily at the highest level in society lived reasonably well, I thought. Yes, we went to several homes. Mnatsakarian, who is fairly young and also fairly high in the administration at the observatory at Erevan, took us to his home. His wife is a professor of philology and his father (or father-in-law?) is the dean of the university at Erevan. We went to Lifshitz's home, and Novikov took us to his apartment. Zel'dovich wanted to have us in his apartment but couldn't do so because his wife unfortunately had had a heart attack, and so did Ginzburg, whose wife was also ill.

KW: Did you discuss political matters on such occasions?

SC: Not really. The only discussions of a political nature we had were with Ivanov during the flight to Leningrad. He volunteered. The sum and substance of his statements were that there was no point in hiding the fact that they lived in a highly constrained society. The choice for them was to do science or bang their heads against a concrete wall, not do science, not do anything. We had a wonderful time with E. M. Lifshitz. He told us the story of Landau's arrest and release. He told us how it happened.

KW: In the thirties?

SC: Yes. Landau was arrested in 1938. Lifshitz, who worked with Landau closely, was in the habit of calling him every morning at his home before leaving for the institute. That particular morning, there was no response; no response from his office at the institute either. He rushed to the institute, but inquiries concerning Landau made to those whom he met in the halls and corridors were returned with glazed faces. He went to Kapitsa, who was the director of the institute. Kapitsa told him that Landau was arrested and that he [Kapitsa] was going to write a letter to Stalin to know the cause of the arrest. Lifshitz then stopped and asked, "Do you realize what that means?" and then answered his own question, "It is like jumping naked into a den of lions!" But Kapitsa did write to Stalin, saying that he was the director of the institute; he had been given a Stalin prize the previous year for having discovered superfluidity; that Landau was an invaluable member of the institute, and indeed, the only person in the Soviet Union who could interpret and understand his (Kapitsa's) experimental results on liquid helium. As the director who had daily contacts with Landau, he should like to know the cause of Landau's arrest. When the letter failed to elicit any response, he wrote to Molotov stating that he had written to Comrade Stalin and that he had gotten no response and that the work at the institute had come to a standstill. After some three months Kapitsa was called to the Kremlin; he was ushered into an office and some high official, a general perhaps, placed in front of Kapitsa a dossier six inches high and asked him to read it and find out for himself the reasons for Landau's arrest. Kapitsa's response was, "I don't want to look at it. I'm not a legal man. I understand neither legal matters nor legal language. You, who are familiar with what is in there, tell me in simple words and in a few sentences what it contains." At this point, Lifshitz explained to us that Kapitsa realized that he would get nowhere if he even so much as touched the file. And in spite of repeated insistence on the part of the official, he refused to touch the file, always asking the official to explain the reasons. Some vague responses with words such as "enemies of the people," "spying," etc., brought Kapitsa's request for more explicitness. "I know Landau. I know he is a physicist. You say he is a spy. Whom did he spy with? Did he spy with physicists? With ordinary people? I know most of the physicists in Russia and other countries. You tell me the names." The interview ended in a deadlock, each person refusing to do what the other wanted, or so it seemed. But two months later Landau was back at the institute, and Lifshitz concluded his dramatic account by saying, "The world's community of physicists will never know how much they owe to Kapitsa for that."

KW: It's a wonderful story. Shows how the courage of one man *can* do wonders. Did you talk about Sakharov?

SC: Yes. With Ginzburg. I had appointments with some scientists. He got me out of them. "Forget science," he told the other man, and he took me to see the art gallery in Moscow. He's really a marvelous person. He told me abruptly and in passing, "I was in Gorky the other day," and then continued his description of the painting we were seeing! That was a very courageous act. It would be very, very nice if people would show that kind of courage *everywhere*.

KW: I agree. Many times we find people quick to criticize the Russians, but if it . . .

SC: I don't see how we can *demand* of somebody else anything. I mean, when we are not experiencing the same thing. We should be glad if more people show courage, but that is not to say that we must *require* of them such qualities. Besides, you ask yourself, how many scientists, colleagues, and friends of yourself here stand up for what is right? Take, for example, Eugene Wigner. Everybody agrees that he has the most re-actionary political views. But I have seen him expound his views at lunch tables with several people sitting around, not one saying a word.

KW: I guess respect for his achievements in science provides an excuse.

SC: Well, really, does respect for his scientific achievements allow one to smother one's conscience to that extent? I have seen him sit at lunch table here in Chicago, praising Nixon, praising our involvement in the Vietnam War, accusing the allies, accusing India, accusing Pakistan, saying that defoliation in Vietnam did not do any harm, and so on, and so on. Everybody just listens. Once he turned to me and asked, "Chandra, isn't it terrible the way Indian soldiers are treating the Pakistanis in the Bangladesh War?" I told him, "I have no brief for them, and I don't see why you are asking *me* that question." But, the point is he went on in that way without anyone directly contradicting him, although I imagine Wigner sensed that he was not convincing his listeners. He certainly sensed that from me.

KW: Yes, it would be nice if we were all more objective, more frank and honest. Peter Freund told me of an instance when you were accused of being anti-Semitic.

SC: Oh, yes. Last year there were the so-called refusenik seminars here in Chicago in some people's houses. I guess there was one in Peter's house. Jim Cronin and Leo Kadanoff were there and apparently someone got up and made a terrible diatribe against me. He said, "Some astronomer in this university whose name I do not know is anti-Semitic." It became clear, apparently, that he was referring to me, and both Peter and Jim said, "We know the person you are referring to. You are absolutely wrong. We won't have you accusing him."

KW: Why? What was the origin of that?

SC: I don't know for sure. But my suspicion is the following. There was a Russian immigrant who came here to the institute and spoke very emotionally about the problem of Russian Jews. But he concluded rather generally that to emigrate from a country is a human right which must be allowed at all costs and at all times. I got up and said, "I'm not arguing with your particular case in question, but I don't think you can generalize. Look, from India this past year, 90 percent of the students who graduated in medicine left the country and are practicing medicine in the United States. Every three out of four graduates from India want to come to the States for advanced degrees, and most of them want to settle in this country. They are more driven by better living conditions, more money, etc., not necessarily by positions in par with their education and accomplishments. Many times they fill low positions. Do you think it's wrong if a developing country like India attempts to stop this 'brain drain'? China, for instance, has been able to develop its internal resources to a point. Human beings are also internal resources." So I said, "I don't think you can generalize."

KW: He took that to mean that you defended Soviet policy, and that you were anti-Semitic.

SC: I suppose so.

KW: Yes. Besides, it's silly to talk about free right of emigration when this country is so selective about who it wants to take. There are no open borders. The poor starving Haitian refugees, for instance. It is so inhuman to send them off.

SC: We use certain phrases with which nobody can disagree, but they are always applied selectively.

KW: Returning to your trip to Russia, how would you summarize it?

SC: I was enormously impressed. "Impressed" is perhaps the wrong word. I mean, I felt the trip, though it was tiring, though it interrupted the writing of my book on black holes, was well worth it. I had never felt that way about a visit, that it was worthwhile because it was worthwhile to others. Both Ivanov and Sobolev were in tears when they said goodbye. It's sort of strange, you know, to feel there are people who have known you through your books and through your papers. But because of political constraints they were not free to meet someone with whom scientifically they felt so close. Therefore, the sudden encounter was so unusual. It's a new kind of experience for a scientist. I had never experienced it. I would say it was a confluence of many things. First, the work—radiative transfer—which related me to these people was work which to me was also most pleasurable. Second, it was studied and cultivated by some Russian scientists. Some of my other books are also used extensively in Russian astrophysical studies. One of the students who came to take my autograph had all five books of mine!

Becoming U.S. Citizens; Connections with India

In Chandra's time, it was extremely rare for an Indian to settle abroad permanently. Chandra's successful career in science, however, prolonged his stay abroad, made him a permanent exile, and eventually led to his becoming a U.S. citizen. Did Indian institutions or the government of India attempt to secure his return? Having left India so early in his youth, how does he feel about India? What led to his relinquishing one nationality in favor of another? Do Chandra and Lalitha feel they are integral parts of the American society? What does Chandra think of pure versus applied science for developing countries? We often discussed these topics in our conversations.

KW: I would like to ask you about India, your connections with Indian government and institutions. Was there at any time a serious attempt made to get you back?

SC: Well, as you know, I went to England on a Government of India scholarship with the stipulation that on my completion of a Ph.D. degree, I should serve the Indian government for five years; otherwise return the entire money I had received. So when I returned to India in 1936, with the University of Chicago offer in hand, I searched for a position in India—not only because of the stipulation, but I sincerely wanted a position in India. I saw a number of officials, like the Director of Public Instruction, and asked them whether I could have a position of a readership in some university and do theoretical work. No, they said. There was no such position; no position at all. Actually they were rather rude. Then I asked them to absolve me of my commitment. They weren't willing to do that either. I said, "You can't enforce me to pay the money back when you say you don't have a position. I have an offer of a three-year appointment from Chicago; if I accept it, I don't know when I will return to India. I want to clear this matter." After a great deal of wrangling, they absolved me of any payments.

KW: You knew scientists in high circles, people like Krishnan, Saha, Bose, and, of course, Raman. Didn't they offer to make an effort?

SC: Well, it's difficult to say. I didn't want to approach Raman for personal reasons. I met Krishnan in 1939 in Cambridge and asked him if I could get a readership in Dacca where he was a professor. Krishnan said he would explore, but nothing came of it. Then the war intervened.

KW: In 1944, you were offered the directorship of the Kodaikanal Observatory. Was it on their own?

SC: Yes. That was on their own, but stimulated, I'm sure, by my election to the Royal Society that year. The war was still going on. Besides, the kind of position, the directorship of a solar observatory, was not suitable for me. It was not appreciated that I was a pure theoretician and just the fact that I was at an observatory did not qualify me for an administrative position. I wrote back saying, if I could have a comparable position without administrative responsibilities and without relation to observational work, I would be willing to consider it. Nothing came of it. The same offer was repeated again by S. K. Bannerjee, who became the director general and visited this country after the war. Again I told him the directorship of the Kodaikanal Observatory was not in my area of competence; a position in a university was what I preferred, or at least something that did not have a commitment to the observational facilities. He said he would explore it, but again nothing came of it.

KW: Well, after independence, did the situation change? It was a time of great change in India. National laboratories, the establishment of the Council of Scientific and Industrial Research, the Tata Institute of Fundamental Research . . .

SC: Yes, I remember Jawaharlal Nehru's visit to this country and Chicago. I had a short conversation with him privately. He said, "You ought to return to India. If there is anything I can do, I will do it." I didn't know how seriously I should take this, but it was a positive statement on his part. Then in 1951, when we went to India for the first time after we came to the States, I met Bhabha. He had just started building his institute. He offered me a position at the Tata Institute. But by then I was not sure to what extent I would fit into the Indian scene. Left to myself, I rather imagine I would have been tempted by Bhabha's offer to a greater extent than I was. But I was advised against it by several people, not that they influenced me. For example, Raman thought it would be very unwise for me to join Bhabha's institute. "I don't think you should play second fiddle to him," he said. I didn't particularly like his arguments. What does a first fiddle or a second fiddle mean? Anyhow, my visit at that time was too short, and I couldn't make up my mind, and I didn't pursue it further.

KW: Didn't Bhabha pursue it?

SC: He didn't until two years later when he was in this country. In fact, he was interested at that time in a problem on which Fermi was working. Hence he came to Chicago to see Fermi, and I met with him at that time. He renewed his invitation and just left it to me. I didn't take it seriously enough, I suppose. So I don't think it can be said that after 1944 genuine efforts were not made by people in India to get me to return. The first one, the directorship, was impractical from my point of view, but the second one, by Bhabha, was certainly one I should have considered seriously. Retrospectively, I'm not absolutely certain that I was right in not pursuing it. In 1953 we became U.S. citizens, and the picture changed completely.

KW: It must not have been an easy decision. What made you decide to become citizens?

SC: Strange as it may seem, both Lalitha and I were tremendously impressed by Adlai Stevenson. He brought a spark to an uneventful political horizon. Lalitha became an active member of the local Democratic party to find out what she could do to support Stevenson's candidacy, such as raising funds and getting petitions. (She was told that noncitizens could join political parties.) I joined her in some of her efforts. People were mostly appreciative of our efforts, but there were some odd characters who reported to the police about our soliciting funds. A policeman drew us aside and said, "Professor, I don't want to upset you, but there have been complaints. Maybe you should not be soliciting money." "We thought it was legal," we said to him, and it was. The outcome of all this activity was that at the State Jefferson-Jackson dinner in Milwaukee, people approached us as heroes for having participated in opposing the conservative wave. And we were introduced to Truman, who was the guest of honor. Also, Lalitha got some appreciative letters from Senators Proxmire and Nelson.

KW: Was Stevenson's campaign the only reason for your becoming citizens?

SC: No. It contributed. During our extended trip to India in 1951, after being away nearly fourteen years, it became clear to us that the prospect of our returning to India permanently was less probable than we had thought. We were too settled in our ways here. Once that was felt, we thought it advisable that we become citizens in order to participate fully in life here. There were a number of occasions in which I could not discharge my scientific responsibilities because I was not a citizen. Minor hindrances, but they made matters cumbersome. I also recall, as you have written, President Hutchins, long before 1952, saying to me, "Why don't you become a citizen, because then you can become a member of the National Academy and participate in the intellectual environment with greater freedom than you can now?" Well, then I had thought that as the president of the university, he just wanted more members in the National Academy. But as years went by, perhaps having decided to stay here, I began to think that I could discharge my obligations as a scientist and my obligations to the scientific community better by becoming a citizen. But the convincing argument was Lalitha's: "So long as we intend to stay in this country, why should we stand on the sidelines and not participate in the political process?"

KW: It's interesting. My own experiences parallel so much your experiences in this respect. In 1975 we were together in India for the first time as a family, my wife and three daughters who had all graduated from college. We traveled. We realized that all five of us would not be able to return to India and pursue our careers. In 1976, we became citizens.

SC: Yes, we have quite parallel experiences.

KW: I was also tired of getting the so-called sailing permit everytime I went abroad, visas to European countries, reentry permits, tax clearance, etc.

SC: Yes. Those were the types of annoyances one avoided by becoming a citizen.

KW: After you became citizens, did you begin to feel you were participating fully in the political processes of this country? That you became a part of it?

SC: Stevenson struck a very responsive chord in everybody. His nomination sent euphoria among the intellectuals, and we became very active in politics. In that framework we worked very actively for the elections in 1952 and 1956, making collections, contributing to advertisements, signing ads, knocking at doors, and so on. Also in 1960 for John Kennedy's election. So I should say we actively participated more than what one would normally expect during the years 1952 to 1964. But starting with 1964, the Vietnam War and the Nixon era caused so much frustration that we began to feel it was fruitless to be active in politics. Since 1964, we have participated only to the extent any normal person does; we have contributed to political candidates, signed national petitions, "Scientists for Johnson," "Scientists for Carter," and so on.

KW: So in the Stevenson-Kennedy era you were wholly a part of this society.

SC: Yes. In fact, I remember going to teach classes with a Stevenson button. It was not supposed to be done. It was not very common in those days. In any case it was not something you did in a university. But when I wore this Stevenson button, (Robert S.) Mulliken, who was staunchly on the other side, became quite furious; he came the next day with an "I like Ike" button. Then everybody followed and started wearing buttons.

KW: Returning to your Indian connection, what happened after you became a U.S. citizen?

SC: Well, we went to India again in 1961, this time under the auspices of the government of India. I went all around India giving lectures and spent nearly six months. I met Nehru at his residence and had dinner with him. He reiterated his desire to see me in India and told me that Thacker [the director general of the Council of Scientific and

Industrial Research set up soon after India became free] would talk to me, and Thacker did. He told me that he would give me a national professorship provided I would relinquish my American citizenship. Then he said, "I will see that you have every facility, you see, I am so close to Nehru. And as long as that is so, anything can be accomplished." I asked Thacker, to what extent the national professorship was permanent. He said it's given for five years and is renewed every five years. He could guarantee it so long as he was the director and Nehru was the prime minister. I had to reply, "All men are mortal." I don't know how he took it, but I didn't see how, after spending so many years thinking about it and finally becoming a U.S. citizen in 1953, I could just relinquish it. My last official connection with India was in 1968. I got a telegram from Vikram Sarabhai saying that the Indian government wanted to give me *Padmavibhusan* [the highest honor the Indian government bestows on a noncitizen] and would I be willing to accept it; I said I would. I later received the medal and the citation from Morarji Desai at a ceremony in the Indian Embassy in Washington.

KW: Morarji Desai?

SC: He was the deputy prime minister. Since the award had to be made directly, the arrangement was made that he would award it when he was visiting Washington. That was in October 1968, and in November of the same year I gave the Nehru Memorial Lecture. On that occasion I saw Indira Gandhi three or four times. She presided at my talk. I was actually enormously impressed by her. I was to give the lecture at 7:00, and according to the protocol, I was to be in the anteroom at 6:30, the prime minister, Indira Gandhi, at 6:40, and at 6:50 the president of India was supposed to come. At 7:00 we were to go into the lecture hall together. Mrs. Gandhi came in at 6:40 exactly; she came directly from the Lok Sabha, and it was obvious that she was terribly tired, without having had the opportunity even to freshen up after an oppressive day at the Lok Sabha (where, as I learned from the papers the next morning, she had been shouted at and heckled). That evening, the first thing she said to me after shaking hands was, "You can have no idea as to how impossible it is to do something positive in this country; everybody wants to criticize and to find fault; and yet there is so much work to be done." She was obviously in quite a rage. Of course, it was all protocol from that point on. We entered the lecture hall and Mrs. Gandhi introduced me with a speech for about ten minutes without notes and without preparation; it was a marvelous speech which was faultless in its form and its content, without the florid language which people generally use in India.[9] I was quite impressed by that. After the lecture, she left. Later that evening there was a dinner at her home.

KW: What was your talk about?

SC: I talked about astronomy in science and human culture. I took great care because I didn't know exactly what to say. I had a crash course in ancient Hindu astronomy in which Otto Neugebauer helped me. I told him I wanted to read about 200 pages in Hindu astronomy. And he gave me manuscripts and books, so I acquired some degree of knowledge in astronomy in ancient India. I talked about that a little, about half the lecture, and for the other half I talked about the role of astronomy in science. I talked about the role of astronomy in human culture, essentially with respect to Hindu culture, and then I talked about the problems of astronomy and how they reflect on the larger attitudes of life. Well, it's not the kind of lecture one normally wants to give.

KW: A general popular lecture.

SC: Yes. It was the second Nehru Memorial Lecture. The first was given by Blackett the year before, and I thought it was a recognition by the Indian government of the highest sort. I felt I had to do something reasonable about it. After the ceremony, we went to Indira Gandhi's house for dinner. I recall there were two circular tables. I was sitting at one table, and there was one seat in the middle which was left vacant. Everyone was seated when Mrs. Gandhi arrived a few minutes late, and obviously everyone got up. It was apparent that she had had time to change; but she still appeared to be in a state of tension. When I got up to help her with the chair, she said, "Don't help me. I'm not used to being helped." That remark dampened everyone at the table. I wasn't exactly sure how to start a conversation. There was a dead silence. Then I remembered a statement from Schlesinger's book about John F. Kennedy. When Indira Gandhi was in New York, President Kennedy had called her on the telephone.

KW: Was she accompanying her father on that trip?

SC: No, she was by herself. It was after Nehru's death. She was the Minister of Information in Shastri's government. In fact, she visited the University of Chicago at that time and I had a brief conversation with her. But the conversation which is reported in the Schlesinger book is that Indira Gandhi told JFK that she was to appear on *Meet the Press* on NBC the following day. Apparently JFK said to her, "Appearing before Mr. Spivak and his panel is like being thrown to the lions." And I asked her if that was true. She suddenly relaxed and said, "Oh, yes. Next morning when I appeared at the NBC studios, some ten minutes before we were to go on the air, Mr. Spivak asked me, 'Madame Gandhi, can you recite the Gettysburg Address?'

"I asked, 'What for?'

"He answered, 'To see how well you know American history.'"

And she said, "My response to it was, 'Your president talked to me yesterday and told me that appearing before you and your panel was like being thrown to the lions.'

"To which Mr. Spivak said, 'Oh, no Madame Gandhi. We are all very kind and polite.'"

She responded, "Oh yes, Mr. Spivak, I know that. I've just come from Africa. I saw a lot of lions there. They were also very kind and polite."

This recollection by Mrs. Gandhi eased her tension and broke the ice; and during the rest of the dinner she was an exceptionally gracious and vivacious hostess. Next morning we went to see her in her office in the Parliament building. The reason we wanted to see her was that, when Bhabha died, her office called me and wanted me to accept the chairmanship of the Atomic Energy Commission. These people were clearly unprepared, because I asked her aide who called me, "But you know I am an American citizen. You will have security problems." He said, "Oh, yes, we can take care of that." He hadn't even thought of the problem. Again it was clear that to be in a position like that required familiarity with Indian politics and required administrative capacities. I wanted to decline it, so I wanted to explain to her the reasons. That was why I asked for a special interview. "Well, all this was not made clear to me," that's what she said. Then she turned to me and said, "You do not know how difficult it is to administer this country. Yesterday you talked to us about astronomy. Suppose someone had told me that you went to the Ganges this morning and had a bath to prevent Raghu from swallowing the sun, what can I make of you? That is the kind of people I have to deal with." It did seem to me from these few remarks that she was a very sensible person.

KW: What did you think of the Emergency she declared in 1975?

SC: Well, I could understand that she saw no way in which things could change. People tried to get her out; they wanted her to resign from the government before the official date of the election, even though she had a plurality. They tried to accomplish that by mass meetings, asking the army to rebel, which essentially meant subversion of the constitution. It did seem to me that the Emergency Proclamation, even though it had many undesirable features, was by and large justified. And certainly the sudden absence of strikes from schools, universities, and colleges, the fact that the Indian heavy industry for the first time made profits, all that showed to me it did a lot of good. I've always, even though my knowledge is probably not very adequate, been in sympathy with the kind of government which Nehru and Indira Gandhi pursued. Of course, you see, the fact that the individuals of the government were nice to me personally is something which might subconsciously contribute to this feeling. But to the extent that I can be objective about it, it does seem to me that there was much one could be proud of in what was accomplished in the first twenty years after independence. And it was not at all clear to me that the later witch-hunting after her, without any coherent policy in national matters, was going to be beneficial to the country in the long run.

KW: You and I are among the small minority which thinks that the Emergency did some good. Was 1968 the last occasion when you met Mrs. Gandhi?

SC: No, the last occasion was in February 1982. Lalitha and I had the privilege of meeting her at her home in No. 1 Safdarjung Road. She was gracious enough to give me an appointment for the asking of it and for no particular reason. I began by asking her whether I was wrong in thinking that the pride and the hope which characterized India of the twenties and thirties were no longer the characteristics of the India of 1982.

She responded with a certain grimness, "I am afraid that is so. National pride is no longer a national commodity; instead criticism and distrust seem to be the present national pastimes."

My last question to Mrs. Gandhi was, "Are you optimistic about the future of India?"

Her instant reply was, "I cannot afford not to be optimistic."

At that, her face, which had till then been grave, broadened into a most charming smile. . . . She was a great lady if ever there was one.

KW: Yes. I agree with you wholeheartedly. Have you maintained contacts with Indian scientists?

SC: After 1961, I'm afraid my contacts with Indian science have not been very close. Largely because I have been away so long. People I knew were past their prime and no longer active in Indian science. But in recent years I have resumed contacts with young men like Radhakrishnan, who is the director of the Raman Institute; Ramaseshan, the director of the Indian Institute of Science; Vishveshwara; my own students and post-docs like Bimla Buti, Surindar K. Trehan, and Trilochan Pradhan, who have all done good work and risen in academic ranks.

KW: Do you perceive any changes in the new generation?

SC: Oh, yes. Some of the younger people have been trained in this country and returned to India. Since I know these people, I have come into closer contact with Indian science again, and my general impression is that the personal rivalries and politics which engulfed the generation prior to mine in Indian science and also my own contemporaries have largely been eliminated, and on the whole these younger people seem to have a fresh attitude. They do not crave the British recognitions and so on. They are more concerned with what's going on in the country and anxious to develop high standards. For instance, the new scientific journals in India that have come up follow a strong refereeing policy. The few aspects I have seen is a measure of the general tendencies. I think it's fair to say that Indian science is coming into maturity.

KW: Should basic or pure science receive a high priority in a developing country like India? In view of the widespread poverty, hunger, lack of basic education, etc., shouldn't basic research, the luxury of a few, be relegated to a lower priority? Shouldn't applied science come first?

SC: It's a subject on which I don't think I can make too deep or any original remarks. I haven't thought too much about it. Sure. Significance of applied science for developing countries is obvious. That one should have considerable support for that is also clear. And, therefore, to develop pure science at the same time, especially experimental science such as high energy physics or observational astronomy, becomes difficult because it's so expensive. On the other hand, at least for some, life has value only if they pursue humane activities—literature, science, music, and so on. I don't subscribe to the view that one should take a totally materialistic view, that is, one should be concerned 100 percent with only eliminating poverty, increasing living standards, etc. I don't see how one can have a human society which does not support the pursuit of pure science as part of human endeavor.

KW: It's a complex question, how you balance the various priorities in a competing world. You cannot wait till you reach a certain standard of living to start thinking of higher pursuits of mind and so on.

SC: Exactly. Roger Penrose was in Pakistan some years ago giving lectures on some of the very new developments in relativity. I overheard a conversation between him and Ray Sachs.

"Well, Roger," he said, "You were in Pakistan trying to distort the people's views. Why do you want those poor people to do useless things?"

Somehow this remark of Ray Sachs annoyed me. I turned around and said to him, "Do you mean, relativity is good for us, you and me, but the Pakistanis are not good enough for it?"

That was a rude statement, but I don't know how else one could react to such remarks. Sachs was visibly upset. So I tried to apologize and asked him, "What do you think of this: In 1920, two young Bengali students,[10] as poor as anybody you can think of, translated Einstein's papers. Do you think it was a useless effort and that they were stupid and foolish to do that?"

It's a terrible dilemma, I agree, that for the future of India, the problems of population, poverty, and the misery in which the majority live, have of course to be solved. But does that need absolute stifling of all other things? I do not think so. I have the feeling that, in every country, pure science has a place that is not very different from the cultivation of literature, arts, etc. No one claims that the latter should be eliminated or that the priorities require one to suppress them. Questions arise in the case of science because of the money involved. But there are reasonable choices one can make; there are branches of science which are not expensive. One should not just blindly imitate the rich countries.

KW: You have had students from all over the world. What do you think of this dilemma from the viewpoint of other countries?

SC: For example, I had a student from Nigeria.[11] He is back in Nigeria. I find a great deal of satisfaction in this fact, not only because I introduced him to certain parts of current physics, but also because there is someone in Nigeria who writes papers on relativity. I think it does, for the world at large in any event, have a corrective effect. People think we cannot understand the Nigerians. We cannot understand the Indians. The point is the human mind works in the same way. It's reassuring that things we find pleasurable are pleasurable to other people in every part of the world. And the fact that there is a common interest emphasizes that there is a common heritage.

Nobel Prize

The announcement of the 1983 Nobel Prize for physics, which Chandra shared with William A. Fowler, was greeted with great joy and appreciation throughout the scientific world. Soon Chandra was inundated with telephone calls, telegrams, and letters of congratulations and good wishes from his former students, associates, scientists, heads of scientific institutions, and governments. They came from many parts of the world, reflecting his multifaceted associations.

. . . To the hundreds of messages that are pouring in, may I add sincere congratulations from one who has admired you since he was 15, and counts it as one of the great privileges of his life that he has actually met you and gotten to know you. This award is long overdue and no one has done more to deserve it.—Werner Israel

I am overjoyed, Professor Chandrasekhar. It is forty years late.—James Kemp

. . . at last! Congratulations!
    The Swedish academy has managed to avoid what would have been one of the greatest injustices in history.
    . . . My heartfelt congratulations upon the recognition bestowed on you. I feel immensely proud to have learned from you what I know and to have the privilege of calling myself a student of yours.—Guido Münch

. . . still I cannot close without stating the simple fact that I and everyone of my colleagues who share the title of astrophysicist is personally in your debt. It is a very happy event to see a portion of that debt recognized in this fashion.—A. A. Penzias

Please accept my warmest congratulations. It is hard to imagine a Nobel Award that would be as widely and enthusiastically approved by physicists and astronomers.—E. M. Purcell

Sometimes in the course of human interests things happen that truly lift the spirit. For me your winning the Nobel Prize is such an occasion.
    I have always regarded my relationship with you as one of special inspiration for me. Your kindness, graciousness, absolutely uncompromising dedication to science, culture and integrity have really had a profound impact on me.
    Every now and then the Nobel Committee does something truly great. This is one of those times. I cannot adequately express how happy I am for you and this is a feeling shared by all those who have been privileged to know you.—Murph Goldberger

It is always good to see "Virtue Rewarded," so to speak, and I congratulate you most warmly not so much for this prize as for the career of such original and distinguished research of which one small segment now receives this recognition.—Lyman Spitzer

The occasion also brought responses from people not directly associated with Chandra through his science. The assistant master of Hindu High School at Triplicane, Madras, India, wrote to Chandra:

One of our Hindu School old boys, won the Nobel Prize, who by this, brought a name and fame to our school, through which you have studded the crown with a "KOHINOOR DIAMOND."

And S. Krishnaswami, a musician by profession, remembered the days when Chandra and he were students together. He wrote:

I am the happiest man that a classmate of mine is the Nobel Prize winner. May Lord Venkateswara, the Lord of the Seven Hills, bless you with long life for the service in the cause of science and the choicest gift of life.

But from the letters I read, I found the following one of the most touching:

Paris le 20 Octobre 83

Dear Sir

I cannot allow myself to congratulate you, but allow me to tell you that when I heard the news, I was turned upside down. For some reason, I feel so happy and proud that your long work has found such a world recognition. Not that you need—if I may dare to express my humble feelings—any worldly gratification, for you and my father belong to the same kind, but still, I cannot help feeling immensely happy and proud. After all, you are a little bit my father. Someone said: the rich and the poor are two locked caskets, of which each contains the key to the other (it was Karen Blixen, a Danish writer).

How wonderful!

Sincerely yours,
Karen Challonge

Ms. Challonge is the daughter of the noted French astronomer Daniel Challonge (1895–1977). In the forties, Chandra's theoretical work led to the possibility of the existence of the "negative hydrogen ion" (a hydrogen atom with an additional electron) and its presumed presence on the sun. Daniel Challonge and his collaborators had produced observational evidence for it by studying certain features of the solar spectrum. This work had brought the two scientists together, and on one occasion, at an international meeting in Paris, Chandra was invited for lunch at the Challonge home. Karen was a teenage girl at the time, and that was the only time she had met Chandra face to face. Her letter, and the others as well, received an immediate response from Chandra, often in longhand to those who had written in longhand. "It is a source of gratification that the award has given pleasure to so many," wrote Chandra. "It has for me a sobering effect; one cannot escape the feeling that the occasion has been exaggerated by friendly feelings." To Ms. Challonge he wrote,

November 4, 1983

Dear Karen,
(I hope I may call you that). I was deeply moved by the spontaneity of your letter of October 20; and it brought to me vividly the figure of your noble father. And the pleasure you have expressed in the award and the purity of those expressions make me humble and profoundly grateful at the same time.
I greatly hope that there may be an occasion for us to meet.

Yours ever
S. Chandrasekhar

What was Chandra's private reaction to this public admiration and adulation? Over the years in my conversations with him, I had found Chandra somewhat trenchant and unhappy about the Nobel Prize and the public attention it attracts. On occasions in the past, before he received the Nobel Prize, when he found himself described as a Nobel laureate in bulletins and announcements of national and international conferences, he had been greatly embarrassed. He had reacted rather strongly, taking the organizers to task. Now that he was a recipient, I was quite curious to see how he really felt, since I watched him go through the Stockholm ceremonies in December 1983 in a rather somber mood. He appeared to be overwhelmed by it all, but not overjoyed. He was proper and dignified, but not relaxed as he received the prize from His Majesty the King of Sweden.

I took upon myself the task of exploring in some depth his reactions. The following excerpts are from our conversations in April 1984.

KW: What was your immediate feeling when you heard about it?
SC: I was astonished. It was six o'clock in the morning; I was just out of the shower when I heard the telephone ring. I answered it, expecting it to be a wrong number. The man at the other end asked me whether I had heard the announcement from Stockholm. I said, "No, I haven't." He then told me that I was awarded the Prize. I then said, "I suppose Hawking is the other person who got it." He said no.

No, I didn't expect it. I had even forgotten that it was due to be announced sometime during that week. As it was my birthday, Lalitha and I were thinking of doing something together, going out for dinner or something.

KW: Over the years when we have talked about the Nobel Prize, I have heard you making critical remarks about it. I wonder how you feel now that you are a recipient.

SC: I was really never critical of the people who got it. I was critical only of the atmosphere it creates and critical of the way in which some people seem to go after it; some people eat their hearts out. I recall, for instance, the year in which Heisenberg and Dirac got it. Heisenberg was visiting Cavendish on the occasion of his "Scott Lectures." I was sitting in the last row in the lecture hall. Max Born was next to me. The hall was packed full and, as the great luminaries—Rutherford, Aston, Chadwick, Dirac, Heisenberg and others—walked in, everyone stood up and applauded. Born was in tears; he said, "I should be there, I should be there."

I felt it was distorting to science to a large extent, and I thought I could see it objectively because I never considered myself at any time as a possible candidate. The idea that I might be had never occurred to me.

KW: Was it because astrophysics and astronomy, until recently, were not areas eligible for the Nobel Prize?

SC: Partly that. But mainly because the areas I worked in, areas like hydrodynamics, radiative transfer, post-Newtonian approximations, ellipsoids, are areas that are not in the limelight of science. I did not feel I had a sufficient number of people belonging to the establishment who were familiar with my work. The areas I worked on did not make me visible.

KW: Your research areas then, did you choose them deliberately for that reason—not to be visible?

SC: No. On the other hand, my motives in science were different. I wanted to pursue what I wished to, and in that pursuit the focus was never on getting the prize. Therefore, the fact that it led as a rule into areas that are not in the center of things did not particularly matter. I mean, in my work on radiative transfer, I solved the problem of light scattering that had not been solved in seventy years. Some individuals here and there appreciated it, but it is not something that receives a great deal of publicity. Similarly, the relativistic instability, ellipsoidal figures of equilibrium states—none of this work was the kind that makes newspaper headlines. So the nature of my work, the way I started

doing science, certainly after the forties, never suggested to me even re-
motely that I might be considered for the prize. Take even my recent
work on the black holes. It is outside the mainstream of everything,
including relativity. I worked on it purely as a matter of personal satisfaction. I did not expect, and I do not expect, that this work will be
appreciated in any reasonable future, if at all. So you see, I was adjusted
to the fact that that is the kind of science I was involved in. And this
thing coming up and making a huge perturbation in my life is not something which I have particularly liked.

KW: Do you think it distorts your life, your way of doing science, this
sudden plunge into the public limelight?

SC: Yes, definitely. It is clearly a distortion of values if a Hindu priest
from Pittsburgh calls you and says he wants to perform a *puja* ceremony
for you. I mean, there is apparently a public image of getting this prize
which contributes to this distortion.

KW: What is your attitude to prizes in general? Has there been any
particular one that you have really cherished?

SC: Well, when I was young, during the thirties, election to the Royal
Society meant a great deal because of my early associations and the
expectations of others, my mother for example. You know how it was
regarded in India in those days. Ramanujan was the first Indian, then
J. C. Bose, C. V. Raman. . . . When I was elected, I was very pleased.
Milne was partly responsible for my election; according to him, apparently I had also Eddington's strong support. I would say that up to a
point I was also pleased with the award of the Royal Astronomical
Society Gold Medal in 1953. In these two cases I would be wrong if I did
not say that I looked forward to them in some way. But after that, starting from 1953, there was no scientific recognition which I wanted or
wished or thought about. But when the National Medal of Science was
given to me by President Johnson, I was happy because it was a different
kind of recognition. I was the first astronomer to receive the National
Medal of Science.

KW: Don't you think, though, that things would have been different if
in the thirties the importance of your work had been immediately
recognized?

SC: Perhaps, yes. I might have thought of the Nobel Prize had my
work in the thirties received recognition without the unhappy controversy with Eddington. As it turned out, I had no stimulus, no temptations in that direction. I heard several lectures by Henry Norris Russell
on stellar evolution and stellar structure. I never found any reference to

me by name in any of them. I came to Yerkes in 1936 and wrote the book *Introduction to Stellar Structure* in 1939, but I wasn't given tenure until 1942. Other people like Kuiper and Strömgren, who came to Yerkes at the same time I did, were promoted to tenure three or four years before me.

KW: All this did not bother you then?

SC: No, it didn't bother me then, but on the other hand, the fact that it had happened didn't particularly help me in thinking . . .

KW: About awards, prizes?

SC: Yes, anything like that. I know, in later years, occasionally people have told me . . . for example, when Jim Cronin received the Nobel Prize in 1981, more than one person came to me and said, "You are the person who deserved it most of all." Well, usually I attributed it to kindness and never thought anything more about it. Certainly I did not expect it in 1983. I was completely and totally surprised. [Pause.] I never expected it, and in many ways I would have much preferred not to have received it.

KW: Why do you say that?

SC: Well, as I said before, I find it a distortion of my life. Secondly, at the present time, I find that the public views me in terms of the work done fifty years ago. Friends try to tell me that it is not so, the prize was not intended just for the work I did long ago. But I don't think that is the way it is viewed.

I find it totally distorting. Every article in newspapers, journals like *Physics Today,* or books like *Pioneers in Science,* emphasizes that one thing. I don't know what to make of it. The response I made the other day truly represents my attitude, exactly my attitude, but I believe it was considered out of place for me to have made that statement.

Chandra was referring to the response he made at a dinner to honor 1983 Nobel laureates in science which was held on 2 April 1984 at the Museum of Science and Industry in Chicago. I quote his remarks in part:

. . While an occasion such as this one is personally very gratifying, I must confess to some misgivings as to the appropriateness of selecting for special honor those who have received recognition of a particular kind by their contemporaries. I am perhaps oversensitive to this issue, since I have always remembered what a close friend of earlier years, Professor Edward Arthur Milne, once said. On an occasion, now more than fifty years ago, Milne reminded me that posterity, in time, will give us all our true measure and assign to each of us our due and humble place; and in the end it is the judgment of posterity that really matters. And he further added: he really succeeds who perseveres according to his lights, unaffected by fortune, good or bad. And it is well to remember that there is in general no correlation between the judgment of posterity and the judgment of contemporaries.

I hope you will forgive me if I allow myself a personal reflection. During the seventies, I experienced two major heart episodes. Suppose that one of them had proved fatal, as it well might have. Then there would have been no cause for celebration. But I hope that the judgment by posterity of my efforts in science would not have been diminished on that account. Conversely, I hope that it would not be enhanced on account of a doctor's skills.

Thoughts on His Life

Over the years Chandra and I have talked about scientists' lives in general, his in particular. An exile from his native India, how does he feel about his life in this country? How did he endure a severe heart attack and triple bypass surgery? His scientific career began, like most, with fame, recognition, and glory in mind. It gradually was transformed into a single-minded pursuit of science; aesthetics, gaining a personal perspective, and personal satisfaction became the dominant motives. "Chandra is a hermit, a scientific sage," a friend and an eminent scientist himself once said to me. Has this single-minded pursuit and dedication produced happiness, contentment? What does Chandra think of his life on the whole?

KW: I know this is not an easy question to answer, but let me ask you anyway. On the whole, are you happy with your life in this country?

SC: Well, one has to consider both one's personal contributions to science and one's contributions to the scientific community. There's no doubt that living here made it enormously easy for my own work on the whole. But when I think, for example, that I had fifty Ph.D. students and ran the *Astrophysical Journal* for twenty years—that is as much service to the scientific community as one normally makes or one can make in one's lifetime. But I don't think these kinds of services either make or destroy sciences in this country. On the other hand, suppose I had stayed in India and had not fifty Ph.D.'s but half that number of students, and ran a journal, creating the kind of standards I did with the *Astrophysical Journal.* Suppose that it had not become as eminent as the *Astrophysical Journal.* I think, relative to India, that would have been a far greater contribution. How does one evaluate the relative contributions? But I think I would have been satisfied with half my scientific work if I could have served the Indian scientific community at least to half the extent that I have served the American scientific community. Because in total terms for the future, it would have meant far more for Indian science than in fact it has to American science.

KW: I do not know whether I agree with you completely. There is a big "if" in what you say. The likelihood of your accomplishing what you visualize is very small. It would have been an uphill struggle against all sorts of odds. Politics!

SC: Yes, in fact, the point you raise about politics is the essential one. People say you cannot do science in India because there are no facilities, because there is no atmosphere, etc. In my case at least, none of these things are important. I'm a theoretician. I don't need facilities; I don't need up-to-date observations. I depend very little on external communication. The facilities I need are marginal. But the absence of politics is the key ingredient. All these years I have been here at the University of Chicago, no one has tried to put any spoke in my wheel. So long as what I asked was related to my work, I have got what I wanted. So long as my recommendations were not based on any purely personal motives, they have been fulfilled. People don't speak behind your back, say wicked things about you, or try to undermine you. The key ingredient, the absence of politics, is a very big "if" in the India of the last fifty years.

KW: Yes. I think it would have been extremely difficult for you just to do your science without being drawn into politics. But I do understand how you feel. We all have a double loyalty in some sense.

SC: That's right.

KW: Well, in a more personal way, am I right in saying that your life in this country has been extremely lonely?

SC: That's right. Loneliness has become a way of life.

KW: Normal times, when you are preoccupied with your work and routine business, it's easy to bear. But you have survived a heart attack and had heart bypass surgery. Those must have been tremendously stressful times for Lalitha and you.

SC: They were.

KW: We have talked about Fermi and von Neumann and their terminal illnesses. Do you recall your distressful experience?

SC: Quite vividly. It all started in December 1973. It was on a Sunday, I remember. I felt an enormous tension in my chest. Lalitha called the doctor right away and he sent me to Billings [one of the University of Chicago hospitals in Hyde Park] Emergency. It turned out that it was angina in its preliminary stage. I stayed in the hospital for a week or so, and since nothing very much happened, I was sent home with a warning to be careful about sudden signs. During the beginning months of 1974, I wasn't too happy with my scientific work. The work I was doing with John Friedman had come to an impasse and I was looking for something new. By the spring or early summer of 1974, I started my work on black holes and was looking forward to six months of stay in Munich, Germany, on a Humboldt Fellowship. I was excited about being in the same institute as Heisenberg, whom, as you know, I had met in 1928 in India.

KW: Had you not met him after that?

SC: Yes. I had met him on a few occasions. I had an extensive discussion with him in 1954 in Chicago when he came to visit Fermi. But I was hoping to get to know him better in Munich. I was also planning to take the opportunity to prepare my Ryerson lecture scheduled for the spring of 1975.

KW: This was to be on Shakespeare, Newton, and Beethoven.

SC: Yes. And then it was another Sunday, a Sunday morning in September 1974. I felt enormous pressure in my chest. I asked Lalitha to call the cardiology department. The doctor [Al-Sadir] who had attended me earlier was not in residence at that time. So she called the chairman of the department, Dr. Resnekov. He said that in ten mintues he was leaving for O'Hare, but in any event he thought the matter was important, and he said to Lalitha, "Take your husband to Billings right away. I will call the cardiology department to receive him and take him at once to the intensive care unit." Lalitha immediately called a taxi and we went over to Billings. A doctor and a nurse were waiting for us, and I was taken right away to the intensive care unit. I was told that I had had a very slight heart attack and that I must be kept under observation because something may yet happen. But I didn't feel anything very bad. So I stayed there. But as the evening progressed, I was feeling worse and worse; the doctors were continually present, and at one point they gave me a sedative because at that time I was beginning to feel intense pressure in my chest. And then I remember suddenly one doctor turning to the other and saying, "This is it." Somehow I thought that it meant that I was going to die at that moment. I remember very clearly that when that statement was made I felt an enormous sense of peace. Then suddenly the thought occurred to me, what's going to happen to Lalitha? And at that instant I went off consciousness. I woke up early the next morning, and the doctor told me that it was a heart attack. I asked him how serious it was. He said on a scale of 5, 5 being fatal, mine was a 3-1/2, but if I had not been in the intensive care unit and if the attack had occurred at home, it would have been fatal.

KW: You remember it so vividly.

SC: I remember it. Even now I remember extremely distinctly the way in which I became peaceful. And suddenly this thought about Lalitha occurring to me has remained with me ever since. Anyhow the result of the heart attack was that my visit to Germany was canceled. However, since even before going to the hospital I had sort of planned what my Ryerson lecture was going to be, I had started reading *Othello* in the Yale edition. So, during the illness, I started reading more Shakespeare because the doctor told me that I couldn't do science. On the other hand, Steve Detweiler was at that time on the campus, and we were planning some work together. The doctor gave me special permission to see him one hour a day after the third week of the attack. But no more than that.

KW: How long did you stay in the hospital?

SC: Four weeks. And then the doctor told me that I should be convalescing until December, and he further said that during this period I should not do science in a serious way.

KW: Who was the doctor?

SC: The doctor was [Jafar] Al-Sadir, an Iraqi doctor. He was supposed to be very good. In some ways I consider I was fortunate. I mean starting from October, for three months I was at home all the time and I spent a large part of my time studying the various plays of Shakespeare, his biography, and heard on the tape recordings all his tragedies. We have records of all his plays. I spent three months very largely in devoting myself to Shakespeare.

Nevertheless, I did write two papers with Steve Detweiler on black holes, my first two papers on that subject; in one we clarified the problem of the perturbations of the Schwarzschild black hole, and in the other we discussed the quasi-normal modes. This latter work, incidentally, has turned out to be moderately important because we describe the way in which a black hole which is excited will decay with time. This is how I have described it at later times: If you strike a bell, then the bell rings after it has been struck; the last pure tones of it represent the fundamental modes of the hammer. I mean it's a very pure tone. In the same way a black hole, if it is excited, emits a pure tone in the end with a certain decay time, and the frequency of this wave which is emitted can be computed. This frequency is independent of how it is excited. Many people have studied the collapse of stars and they always find that the last stages of the collapse are described exactly by the quasi-normal modes that Steve and I computed at that time.

So the result of this whole situation was that in December I had prepared my Ryerson lecture more or less, and because of the time and rest I had, I was more confident about the kind of lecture I wanted to give. Almost every Ryerson lecture since has been a technical lecture in which the person expands on his own work. I had chosen to talk on a subject not related to my own professional area and really quite outside the realm of my competence. But the fact that I did have three months to study convinced me that I could embark on this rather, what I thought, dangerous area. It also turned out that, beginning in January 1975, my health was all right; I felt happy about the work I was doing, and I was going to give the Ryerson lecture on a topic with which I felt comfortable; all that helped me.

KW: Was there any hint of [the need for] a bypass surgery during this time?

SC: No. Neither after nor before the heart attack when, just months before, I had a complete check-up and was given a clean bill of health. Nowadays they have better tests. Anyway 1975 was a good year for me, a very successful year because my work on the black holes started going along. I was invited to give the Weyl lectures that fall in Princeton.

KW: That's when you decided to write the book?

SC: Yes. We also went to India for a short visit. 1976 rolled in. My researches on black holes continued to deepen. I separated the Dirac equation. I was also in Cambridge for three months. I mention all this because, as 1976 came to an end, my work on the black holes was reaching a climax. I was near a complete solution of the perturbation of the Kerr metric, but the work was proving to be exceptionally difficult and at the same time my health was beginning to deteriorate. I was experiencing again periodically the pressure in the chest. The doctor said that I ought to have an arteriospectroscopy—I always mix up that word—cardiocatheterization, I think. Well, they wanted to examine my heart again to find out if surgery was needed. But I did not want to have that because it would interrupt my work. If the test showed that I had to have surgery, then it would have been a serious interruption, and the work which I thought was coming to an end would not be finished. So I deliberately did not want to have the catheterization. It was in a way a rather painful period because I felt I was racing with time in order to finish the work. Anyhow, I thought I had completed my work on pertur-

bations by July of that year, and I was to give one of the invited talks at
the General Relativity Conference in Waterloo at the end of the summer. I therefore arranged so that on returning from Waterloo I would
have the test. The test was made, and the doctor said that I had to have
three bypass valves put in. It was an extremely dangerous situation and
I ought to have the heart surgery right away. So we decided—in fact,
Lalitha was very insistent—that the surgery should be performed right
away. The test was made on a Monday; the results were known on
Tuesday; Lalitha called Dr. Kirsner and Martin Schwarzschild for advice, and it was agreed on Wednesday that the surgery would take place
on Thursday morning. Well, the doctor who attended me was [Dr.
Constantine] Anagnostopoulos, an extremely able surgeon. The day before he had told me that the surgery would take a long time and that
when I woke up I would find that I would not be breathing through my
nose, but through tubes inserted through my stomach, and that I would
be completely bandaged, and that my legs would also be bandaged because that is where they would take the veins for grafting, and so on. So
when I woke up the next afternoon at three o'clock, I found myself as I
was told I would be. But then I found that during the entire evening
people were doing many things to me. They were taking X-rays; they
were waking me up frequently. I didn't know what was happening. Of
course I was highly sedated. Suddenly at night I noticed that they had
put an extremely strong searchlight on my face to wake me up. They
forcibly woke me up and said, we have to take you upstairs again because your heart is bleeding more than it should, and blood is accumulating in the lungs so we have to open the heart again. We want you to
sign this paper for the surgery because we can't get your wife at this
time. So they took me upstairs and apparently they had to open the
heart again. I was there all night. The bleeding was stopped by platelets.
This complication made the results of the surgery extremely difficult, because the blood was accumulating all the time and I started hic-
cuping, which is very bad because every time you hiccup you feel as
though your heart is torn apart. It was extremely painful. But of course
there was no danger, according to the doctor, because the surgery was
successful.

Certainly the months of September and October were not pleasant, to
say the least. I used to say that at night it was the long night's journey
into the morning, because I couldn't lie down since I couldn't breathe. I
had to sit up, and I hardly slept because I was in pain all the time. One
of the most painful experiences was when they drew blood out of the

lungs. They did that more than once. On one occasion they had to draw something like a liter and a half of blood. And when the blood is tapped drop by drop, it takes several hours. It's a very painful business.

Well, anyhow I didn't particularly worry about all this because, while it was painful, it was nothing beyond endurance. During my convalescence John Friedman came and made some remarks about my work, which indicated that the work in fact was not complete, that I had to do something more about it. Well, I won't go into the details of this, but the actual solution to the problem that I thought was finished was not in fact finished until January of 1980. It was a long effort and also besides this I did some work on quasi-normal modes and so on. So it was only in February 1980 that I was ready to write the book which I had started in 1975.

KW: I heard your talk in Waterloo that summer and approached you then about writing a series of articles about you. Then we met in December 1977 and began to tape. I'm shocked that I didn't have the slightest hint of what you had undergone just a few months before!

SC: I tend not to worry, not to let health concerns upset too much the daily occupations.

KW: We have talked about Fermi and von Neumann and their illnesses, how they reacted. Can I ask you about your feelings? Were you afraid of death?

SC: No. Not when I had the attack nor at present. Purely for my own sake, I don't have any fear or forebodings about my death. If someone were to tell me I have cancer and will die in three months, I don't think it would make much difference to me. But there are other kinds of problems that worry me. What will happen to Lalitha? I know financially she will be all right. However, wouldn't she be very lonely? I worry about that. I think my own attitude is rather colored probably by the Hindu upbringing. I am not religious in any sense; in fact, I consider myself an atheist. Nonetheless, because the Hindu religion, despite its outward trappings, is an essentially rational way of life, it's easy to live with it. It's so tolerant.

KW: Yes. It's almost impossible for anyone to say that he is not a Hindu because of such and such. Some tenet of Hinduism or the other has a place for his such and such. You cannot escape being a Hindu by saying that you are an atheist. There is ample room for atheism in the Hindu religion.

SC: Yes. No matter what one is, the idea that as one grows older one has to develop a sense of detachment has played a positive role for me. I have tried to achieve detachment—detachment from having students, writing proposals, and so on. Let others worry, run things. But I have a feeling of disappointment because the hope for contentment and a peaceful outlook on life as a result of pursuing a goal has remained unfulfilled.

KW: What?! I don't understand. You mean single-minded pursuit of science, understanding parts of nature, and comprehending nature with such enormous success still leaves you with a feeling of discontentment?

SC: I don't really have a sense of fulfillment. All I have done seems to be not very much.

KW: Don't you think that this is common to everybody?

SC: Well, it may be. But the fact that other people experience it doesn't change the fact that one is experiencing it. It doesn't become less personal on that account.

KW: Do you think Einstein felt that contentment?

SC: I'm not sure. I would love to have known what Einstein's attitude was at the end of his life. My own suspicion is that he was not very happy. I have read a great number of biographies. They never give answers to what scientists like Maxwell, Lord Kelvin, Stokes, or Einstein felt. I wish I knew how they felt. But it does not seem to me that the pursuit of science results in the feeling of contentment or peace after years of pursuit.

KW: Do you mean to say that some sort of faith, simple beliefs are necessary to attain the inner peace and harmony? Pursuit of science is not adequate?

SC: No. I'm not saying that. I'm not sure. There is this marvelous story by Balzac called "The Atheist." It is the story of a doctor, an atheist, who was seen and observed at a church mass on a particular Saint's day twice a year. He was confronted by a friend, "You say you are an atheist. But how does it happen that you attend this same mass, this Saint's day twice a year?"

The doctor explains that he arranges the mass out of respect for a poor man who was a water carrier, who was very religious, and who had helped him when he was young.

The friend then asks him, "But how is it that you are so serious and contemplative during the mass?"

The doctor replies, "Well, the thing I repeat to myself during the mass is, 'I wish I had the faith the water carrier had, but which I do not have.'"

To say that religion could save a person doesn't help me because I don't have the faith. I don't want to put myself in the ranks of all those great names I mentioned, but what is true from my own personal case is that I simply don't have that sense of harmony which I had hoped for when I was young. And I have persevered in science for over fifty years. The time I have devoted to other things is minuscule. But that does not mean I regret the past or I wish I had done things differently. No, I don't have any such feelings. It also does not make me feel that the future is limited.

KW: Aren't such feelings an integral part of this single-minded pursuit of science, or any single-minded pursuit for that matter?

SC: Yes, that's certainly a relevant factor. One realizes how enormous knowledge is and how one's own part is so negligible that one begins to consider that as trivial.

KW: Don't you also think this inner peace and harmony depends to a certain extent on the society we live in, the way science is done? The present-day scientific atmosphere almost nowhere is conducive to a quest for gaining a personal perspective and satisfaction as you aspire to. Science is done in such a competitive atmosphere that no one, no matter how much he or she has contributed, can remain in peace. The desire for recognition and fame continues to dominate.

SC: Yes. That's certainly true. Science at the present time is greatly associated with haste and the desire to be at the top. But my unhappiness or discontent is not due to that, I think. Perhaps it's because of the distortion, in some sense, of my life, of its one-sidedness, of the consequent loneliness, and my inability to escape from it all.

KW: What do you mean?

SC: For example, even at this stage of my life, when I send in a paper for publication, I constantly worry whether it will be accepted. I ask myself, why should I? Why should I? Why shouldn't I be able to spend the rest of my life reading Shakespeare? There are so many things which one would like to do, which one does not do. There are so many things which would be profitable and pleasurable to do which you don't, and one begins to wonder at a later time, was it worth it, all this? I mean, in my own case, I felt when I was young that when one reaches the age of mid-forties or fifties and one is moderately successful, one would have a sense of personal security and assurance combined with some contentment. I certainly haven't found them. I find that very difficult to reconcile with; namely, to pursue certain goals all your life only to become doubtful of those goals at the end. I don't know if this is a common experience for everybody or not, but certainly it's not at all clear to me whether the single-minded pursuit of science at the expense of other, personal aspects of one's life is justifiable. Not so much for oneself, but particularly for those with whom you are associated. The person who suffers the most is, of course, one's wife. Was one justified in imposing that kind of life on that other person?

KW: So, your discontentment is not in the sense that there is still so much to be done?

SC: Not so much. No, I don't feel I have to do as much in science. In fact, if anything, my feeling at the present time is the opposite. I mean I've worked in science for over fifty years now, and I don't see that I can accomplish very much more or that I can add to my scientific contributions by anything that will substantially change my record, unless by accident something or other happens. So if I continue to do science it's largely for my personal pleasure, and also because I do not know what else to do. I mean, you see, I've got so used to a certain way of life, that it's difficult to change. But there are other things in one's life. I always remember the story by Chekhov, "Rothschild's Fiddle." Rothschild is a man who constructs coffins. He had a book where he always wrote his losses. For example, he wrote, "There was a lady from Kharkov. She was sure to die, but she went to Kharkov for higher medical attention, and she died there. What losses!" Because he couldn't make the coffin for her. And then his wife dies. After his wife is buried, he goes to the riverside and sits there and begins to think and suddenly a picture arises in his mind of his once having had a little child and his child had died. And he says, "What losses? Why have I always been rude and cruel to others? Why didn't I think of going on that boat, playing my violin and earning money?" And he says, "What losses?" and so on. You can have your life simply go by. [A fulfilled] life is not necessarily one in which you pursue certain goals, there must be other things.

Notes

The main source for this book has been the Subrahmanyan Chandrasekhar Papers in The University of Chicago Archives. These papers from the years 1928–83 are arranged in six series, of which the following three are important for this book.

Personal correspondence and papers, Boxes 1–9. Folders in this series contain letters to his father and other family members, and also the letters from his father and from his brother Balakrishnan.

Scientific correspondence, Boxes 10–33. Folders in this series contain correspondence between Chandra and scientific colleagues in the United States, Great Britain, Europe, and India.

Astrophysical Journal, Records, Boxes 148–52. This series contains correspondence and other material relating to the *Astrophysical Journal.*

Prologue

1. This is how he is known to his students, colleagues, and associates.

2. Sarma retired as professor of mathematics at Banaras Hindu University and is currently a permanent resident of the United States.

3. Sachs is professor emeritus of the Department of Physics and the Enrico Fermi Institute at the University of Chicago.

4. Sagan is David Duncan Professor of Astronomy and Space Sciences and director of the Laboratory for Planetary Studies at Cornell University.

5. For the published account of this story, see S. Chandrasekhar, *Notes and Records of the Royal Society of London* 30:250.

6. Ashtekar is Distinguished Professor of Physics at Syracuse University. He was a graduate student and research associate at the University of Chicago.

7. S. Chandrasekhar, "The Richtmyer Memorial Lecture—Some Historical Notes," *American Journal of Physics* 37 (1969):577.

8. S. Chandrasekhar, "A Scientific Autobiography, 1943–1983," unpublished ms., S. Chandrasekhar Papers, box 1, folder 1, The University of Chicago Archives.

9. For that purpose, one should consult the six volumes of S. Chandrasekhar, *Selected Papers,* published by the University of Chicago Press.

10. Conversations, Victor F. Weisskopf, professor emeritus of the Department of Physics at the Massachusetts Institute of Technology, his office, Cambridge, Massachusetts, 15 June 1981.

11. Peter L. Kapitsa, "Recollections of Lord Rutherford," *Proceedings of the Royal*

Society A 294 (1966):123–37. Kapitsa, the well-known Russian physicist, spent fifteen years beginning in 1921 at the Cavendish Laboratory, Cambridge, England.

12. After seeing the cartoon in the *New Statesman and Nation,* Chandra requested a copy. Vicky sent him the original, suggesting that he contribute to Oxfam.

13. S. Chandrasekhar, *Eddington, The Most Distinguished Astrophysicist of His Time* (Cambridge: Cambridge University Press, 1983). These lectures and the drawing have also been published in S. Chandrasekhar, *Truth and Beauty* (Chicago: University of Chicago Press, 1987).

14. Jost is a noted Swiss physicist at Eidgenossische Technische Hochschule (E.T.H.), Zurich, Switzerland.

1. The Simple and True

1. Conversations, James W. Cronin, his office, University of Chicago, 10 June 1986.

2. Conversations, Marvin L. Goldberger, his residence, Pasadena, California, 14 June 1980.

3. Conversations, Victor F. Weisskopf, his office, Massachusetts Institute of Technology, 15 June 1981.

4. Conversations, Basilis Xanthopoulos, University of Chicago, 16 January 1986. He is currently professor of physics at the University of Crete and the Research Center of Crete, Iraklion, Greece.

5. Conversations, Kip S. Thorne, his office, California Institute of Technology, 14 June 1980.

6. Conversations, Alan Lightman, his office, Harvard-Smithsonian Center for Astrophysics, 15 May 1980.

7. Until 1983, the "whole class" meant only the two students, Tsung-Dao Lee and Chen-Ning Yang. Since 1983, the term includes the teacher as well. I must note that this story's historical accuracy has been contested. Apparently there were a few others (Donald Osterbrock and Enrico Fermi, for instance) who attended the lectures often but not regularly.

8. Conversations, Yavuz Nutku, Syracuse, New York, 25 March 1978. He is currently at TUBITAK Research Institute for Basic Sciences, Gebze, Kocaeli, Turkey, and Bosphorus University, Bebek, Istanbul, Turkey.

9. Conversations, Carl Sagan, Los Angeles, 20 June 1980.

10. Conversations, Peter Vandervoort, his office, University of Chicago, 15 June 1978.

11. Conversations, Valentine L. Telegdi, at the time of the First International Meeting on the History of Scientific Ideas, Ciutat de Sant Feliu de Guixols, Catalonia, Spain, 23 September 1983.

12. Conversations, T. D. Lee, his office, Columbia University, 23 May 1978.

13. Conversations, Jeremiah Ostriker, his office, Princeton University, 10 March 1986.

14. Jeanette Burnett was the production manager for the *Astrophysical Journal* after Chandra left the editorship. In 1979 she resigned her job for private reasons. Beverly Wheeler went on to be the secretary to the president of the University of Chicago. Jeanne Hopkins was senior manuscript editor for the *Journal* when she

died in 1984. She started working with Chandra in 1967 and, with his encouragement, compiled the *Glossary of Astronomy and Astrophysics* (Chicago: University of Chicago Press, 1976; rev. 2d ed., 1980).

15. Conversations, Jeanne Hopkins, University of Chicago Press, 14 June 1978.

16. Conversations, Donna Elbert, Chicago, 15 January 1986.

17. Conversations, William Press, Harvard Center for Astrophysics, 8 June 1979.

18. Subrahmanyan Chandrasekhar, *On Stars, Their Evolution and Their Stability* (Nobel Foundation, 1984). This Nobel lecture reprint includes a brief autobiographical account.

19. Letter to Chandra from Peter Jakobsen, Astrophysics Division, European Space Research and Technology Center, Noordwijk, The Netherlands, 25 April 1986.

20. Conversations, Lyman Spitzer, his office, Princeton University, 10 March 1986.

21. As it was, Chandra was unsuccessful in finding a job in India which suited his research interests.

2. Choosing the Unconventional

1. C. S. Ayyar, "Family History" (1946), S. Chandrasekhar Papers, box 6, folder 4, The University of Chicago Archives.

2. The Tamil brahmans of South India belong to two major sects—the Vaishnavas, who worship the God Vishnu, and the Shaivites, who worship the God Shiva. The former generally have the common family name Iyengar, and the latter the name Ayyar (also spelled Iyer or Aiyyar).

3. See, for instance, Percival Spear, *A History of India,* vol. 2 (Harmondsworth: Penguin Books Ltd., 1965); and Stanley Wolpert, *A New History of India* (New York: Oxford University Press, 1977).

4. Thomas Babington Macaulay (1800–1859) joined the ruling council as the new law member in 1834 and proved to be a powerful force behind Lord Bentinck's efforts to initiate change.

5. Burke took a leading part in bringing about the impeachment of Warren Hastings, who, after Robert Clive's military conquests, was chiefly responsible for giving a political shape to the conquered territories. Known as the company's *bahadur,* he became the first governor-general of Bengal under the company's rule. But in consolidating and extending the company's rule, he was accused of being a relentless ruler who used despotic tactics. Although he was acquitted of his alleged crimes in the process of impeachment, he forever remained a symbol of tyrannical power in India.

6. *Suttee*—the burning of widows on the funeral pyres of their husbands. *Thuggee*—highway robbery sanctioned by religion, as the robbery was combined with a ritual murder in the service of the Goddess Kali.

7. The so-called permanent land settlement code was introduced by Governor-General Cornwallis in 1793, primarily to ensure a steady, well-regulated flow of income to the company's treasury. Previously, in Moghul times, the cultivators of the land were the owners of the land, but without any official paper title or deed that decreed their private ownership. Likewise those who collected taxes from them for

the rulers held only "interests" in land "assigned" to them for the purposes of tax collection. The new code changed all this. By assigning titles and deeds, it sanctified the private ownership of land mainly by the tax collectors who became *zamindars* (landlords). The latter, under the new code, were under obligation to pay a certain, specified sum of money annually for a certain period of time. When they were unable to pay because of poor rains or flooded crops, their titles and deeds were taken over by bankers and money lenders. This created a new class of landlords, the "absentee" *zamindars,* who naturally became staunch and loyal supporters of the company's rule, since it assured their right to the land by virtue of the paper deed. The displaced landlords, deprived of their traditional prestige, power, and possessions, equally naturally became the opponents of the company's rule.

8. Dalhousie invented the doctrine of "lapse" or "paramountcy" under which a princely state could be annexed if its ruler was judged unfit. Likewise, a state lost its independent status and became part of the British territory if its ruler had no natural heir. The doctrine did not recognize an adopted heir. The annexations of the central Indian state of Jhansi in 1853 and that of Oudh in 1856 were, as it were, the last straws in a series of takeovers of princely states by Dalhousie using this doctrine. Had he allowed, the widowed young queen of Jhansi would have ruled the state with an adopted infant as heir to the throne. As it turned out, four years later, she proved a formidable foe of the British. Her bravery and sacrifice made her a national heroine. *Jhansi ki Rani* (Queen of Jhansi) became a household name. And ignoring a treaty that had been in effect for over half a century, Dalhousie deposed the popular ruler of Oudh under the pretext that he was effete, corrupt, and debauched.

9. On 2 August 1858, the British parliament passed the Government of India Act, through which the British Crown replaced the company. On 1 November 1858 came Queen Victoria's proclamation, expressing her personal concern and responsibility in the new administration. She abolished Dalhousie's "doctrine of lapse" and assured the princes and chiefs that all treaties made with them would be scrupulously honored. India became a full-fledged British colony and Queen Victoria became the Empress of India.

10. Spear, *A History of India,* 2:153.

11. Parvati (1869–1916) was a daughter of Saptarishi Sastriar of Tiruvanakoil, a suburb of Trichinopoly. R. C. was fifteen and Parvati twelve when the marriage ceremony was celebrated. The marriage "consummation," as it is called, took place some two years later.

12. If Raman had not become famous by discovering the Raman effect and winning the Nobel Prize in physics, both the village of Tiruvanakoil and the house where the two brothers were born would have remained anonymous, like the multitudes of other villages and tenements in India. Instead there is a photograph of the house in the permanent exhibit at the Raman Institute in Bangalore. The photograph shows the front of a small house, its thatched roof supported by crooked pillars.

13. Also known as Tiruchinapalli, derived from Trisirapalli, meaning the town of Trisira, the three-headed demon, the brother of the ten-headed demon Ravana.

14. The student was Ramaswami Sivan, who later entered Indian agricultural services. In a similar instance, we are extremely fortunate that Meghnad Saha was not allowed to take the All-India Finance Service Examination, although he was

eminently qualified, because of his associations with political revolutionaries in his youthful years. With the path to a high position in government closed, Saha decided on a career in science. He became one of India's most important astrophysicists, well known for his fundamental contributions to the theory of ionization, which played a singularly important role in our understanding of stars.

15. Vizagapatam or Vishakhapatnam is now the capital of Andhra Pradesh.

16. Annie Besant (1847–1933) was a British woman who became the leader of the Theosophical Society after Mme Blavatsky, the founder of the society which promoted Hindu religious and social ideals. Besant took a very active part in the Indian National Movement for Independence. She was the first woman to be elected president of the Indian National Congress.

17. Again in his "Family History," C. S. Ayyar writes, "One day, when he had been to Kumbakonam, he underwent a trial feat of strength with a Muslim. It was no mere pugilistic display but a trial of strength, or fatigue shall I say. The opponent and himself had to make a hundred movements of the 'Karla' (a long club) at a time, alternately, and whosoever of the two got thoroughly exhausted the earlier, was the loser. Father won the test after three or four hours of constant exercise alternately."

18. At least this was true for the sons. For the daughters it was a different story. Women's education beyond rudimentary reading and writing in Tamil had to await the next generation.

19. Two of R. C.'s other sons died young. The youngest of all his children, C. Ramaswamy (1907–), is a contemporary of Chandra. He is a distinguished meteorologist who is retired and currently lives in Bombay. Of the three daughters, one died when she was barely nine years old. The other two also died at early ages.

20. An All-India Competitive Examination.

21. *Life's Shadows* (a collection of short stories of India in transition; Kamara Guru, 1938), S. Chandrasekhar Papers, box 7, folder 3, The University of Chicago Archives. His other literary publications include *Grammar of South Indian (Karnatic) Music* (Madras: Vidya Shankar, 1976); and "Art and Technique of Violin Play" (in Tamil) and "An Artist's Miscellany" (a book of 24 essays), S. Chandrasekhar Papers, box 6, folders 6, 8–10, The University of Chicago Archives.

22. N. Balakrishna Aiyar was a *tahsiladar* of Pudukottah. A *tahsiladar* was a government official responsible for the administration of a group of villages forming part of a district of a province. *Rao Bahadur* was a title that used to be awarded by the British to an Indian for distinguished service aiding the British cause.

23. The house where they lived was in the vicinity of the Indian Association for the Cultivation of Science building, 210 Bowbazzar Street. S. Ramaseshan, in his C. V. Raman Memorial Lecture in 1978, writes that within six or seven days of reaching Calcutta, Raman noticed the association's sign while on his way to work by tram, and on his way back home he knocked on their door. He found a dusty lecture hall and a large laboratory with a lot of even dustier equipment, mostly of the demonstration type. Amrit Lal Sircar, the secretary of the association and son of the founder, Mahendra Lal Sircar, gladly gave the keys of the building to Raman for his researches. Later this became the famous laboratory where the Raman effect was discovered.

24. Legend attributes the founding of Lahore to Lava, the son of Rama, the hero

of the epic *Ramayana*. But its recorded history goes back only to the seventh century (in the writings of Hwen Tsang, a Chinese pilgrim). Touched by many civilizations during the course of centuries, it had its golden period during Moghul times (1525–1724) when it became the northern capital and the royal residence of Moghul kings. The architectural splendors of those times (e.g., Anarkali tomb, NorJehan's tomb, Wazir Khan's mosque, Shalimar gardens) still exist and attract tourists from all over the world.

25. *Punjab* means the land of five rivers: the Sindh, Ravi, Jhelum, Chenab, and Sutlej. The Sindh, the most famous of them all, provided the basis for the Indus civilization.

3. Determined to Pursue Science

1. The name of the building was Sadananda in the Anarkali area.

2. *Anarkali* means pomegranate blossom. It was the name given to a favorite lady in the harem of Akbar, the Moghul emperor. Prince Salim, the emperor's son, was in love with her. Akbar once saw Anarkali smile at the prince. That was enough of a crime for Akbar to order her to be buried alive. Later, when Prince Salim became the Emperor Jahangir, he built the Anarkali tomb in his beloved's memory, thereby immortalizing her name.

3. Ramaswamy, born in 1907, was the youngest son of R. C. Only three years apart in age, he and Chandra became close friends as they went to high school and college together. Ramaswamy (or Ramudu, as Chandra likes to call him) joined the Indian Meteorological Service after his B.Sc. honors degree in physics, where he became director general. He had the unique distinction of working for and getting a doctorate degree after his retirement. He has written a number of interesting papers on monsoons and climatic conditions in India. As recently as 1982, when he was seventy-five years old, he published a monograph on the subject of monsoons. He and Chandra have kept in close contact. "He is one of the few of the early generation still living," says Chandra.

4. Conversations, Chandra's sisters and nieces, Madras, 2 January 1980.

5. The English textbook that Chandra studied was by J. A. Yates and P. T. Srinivas Iyer, the principal of the college where his grandfather, R. C., had been a professor and vice-principal. By the time Chandra was eight, he had read *Lamb's Tales from Shakespeare,* Chaucer, and the so-called A. L. Bright Story Reader series. He also remembers reading books from a Science Readers series that came from England.

6. G. H. Ranade, in *C. S. Ayyar Commemoration Volume,* p. 17 (two-part volume containing articles by his sons and daughters, relations, and friends for his sixtieth birthday; privately published, 1945).

7. Chandra's younger sisters all have college degrees. Vidya is an accomplished *vina* player and has continued her father's work on the theory of Karnatic music.

8. S. Balakrishnan, in *C. S. Ayyar Commemoration Volume,* p. 26.

9. Conversations, S. Balakrishnan, Bangalore, 19 December 1979.

10. A southern suburb of Madras with the ancient Shaiva shrine of Kapaliswara at its center, it is described as a favorite residential quarter of well-to-do Indians, the "West End" of Hindu Madras.

11. Jagadish Chandra Bose (1858–1937), famous for his meticulous experiments on the properties of short radio waves. He was also a pioneer in the field of biophysics, famous for developing extremely sensitive automatic recorders for measuring plant growth.

12. The two professors who took an interest in Chandra's career were Appa Rao, M.A.L.T., National Professor of Physics, Presidency College, and H. Parameswaran, M.A., Ph.D., D.Sc., F.Ins.P, also a professor of physics. Parameswaran had worked with Lord Rutherford in Cavendish Laboratory and later with Sir William Bragg at the University of London.

13. Conversations, V. S. Jayaraman, Madras, 3 January 1980.

14. The Compton effect demonstrates the quantum nature of light. When light is scattered from electrons, it behaves as though it consists of particles each of energy $h\nu$ and momentum $h\nu/c$, where h is Planck's constant, c is the speed of light, and ν is the frequency of light.

15. Chandra had read Compton's *X-Rays and Electrons* and knew the theory of the Compton effect as well as the difference between the Compton and Raman effects.

16. Satyendra Nath Bose (1894–1974), of Bose-Einstein statistics, had developed in 1924 a new derivation of Planck's law, which made it independent of classical electrodynamics.

17. S. Chandrasekhar, "The Compton Scattering and the New Statistics," *Proceedings of the Royal Society* 125 (1929):231. In this paper Chandra considers Compton scattering of light in degenerate electron gas in a conductor. Using Fermi-Dirac statistics for momentum distribution of the electrons, he derives the spectral intensity distribution and shows that the intensity distribution does not fall off exponentially about the Compton maximum (as one would obtain by assuming the Maxwell distribution) but *parabolically*.

18. The Indian Science Congress Association is patterned after the British and American Associations for the Advancement of Science. Its origin in India was due to the initiative and efforts of two British chemists, J. L. Simonsen and P. S. Mac-Mahan, who came to India in 1910. The first meeting of the association was held in 1924 under the auspices of the Asiatic Society of Bengal in Calcutta. Since then a meeting of the association has been held every year under the leadership of distinguished men of science. For more details, see B. Mukerji and P. K. Bose, eds., *A Short History of the Indian Science Congress Association* (Calcutta: Indian Science Congress Association, 1963).

19. S. Chandrasekhar, "The Ionization Formula and the New Statistics," *Philosophical Magazine* 9 (1930):292; idem, "On the Probability Method in Statistics," *Philosophical Magazine* 9 (1930):621.

20. Meghnad Saha (1894–1956) and S. N. Bose were the first two *theoretical physicists* of India. Saha, an inspiring teacher, produced a generation of Indian physicists who have played an important role in the growth of physics in India.

21. A. S. Eddington, *The Internal Constitution of Stars* (Cambridge: Cambridge University Press, 1926). Chandra had received this book as a prize from Professor Appa Rao for an essay he had written on the quantum theory.

22. He did just well enough to obtain a passing grade. Otherwise, he would have failed the whole high school examination in spite of his brilliant performance in all

other required subjects. The poor performance in history brought his overall average down so that he was neither on the first, nor on the second, but on the third list of students admitted to Presidency College.

23. Two other students in the honors course were C. Ramakrishnan, who later went to England and became an ICS officer, and C. V. Dakhinamurthy, who entered Hyderabad Education Service. Lalitha Doraiswamy was a year junior to Chandra.

24. Besides physics and mathematics, as a third-year student Chandra was attending German classes held at Madras University more than two miles away from home. His uncle Ramudu recalls: "Riding on a bicycle and wearing a *dhoti* in the 'cylindrical' style, he used to rush to the German classes early in the morning. After the class was over, he rushed back home to take his brunch, then back to the university area again on a bicycle. This would have been strenuous for anybody. But he took to this extra cycle journey quite sportingly."

25. For orthodox brahmans, crossing the sea was a sin. Besides that, there was the fear that the young, under the corrupt influences of the West, would start eating meat, indulging in alcoholic drinks, and consorting with western women who did not have the high moral principles of Indian women. Parents were (and the majority still are) afraid that their eligible sons would be "captured" by western women.

4. Discoveries, Personal and Scientific

1. K. R. Ramanathan, a distinguished physicist, an early student of C. V. Raman, who made his career in meteorology. He was the director of the Meteorological Observatory in Bombay.

2. A mathematical model of an isolated mass of gaseous stellar material in equilibrium under its own gravitational forces.

3. That is, the consequences of Einstein's special theory of relativity.

4. S. Chowla, an Indian mathematician known for his contributions to number theory.

5. Known as the "silver-tongued orator of India," V. S. Srinivasa Sastri was a member of the Imperial Legislative Council during the years immediately following the First World War.

6. The so-called Rowlatt Act was intended to make the wartime restrictions on civil rights permanent. In spite of widespread opposition to the law, it was rushed through the Imperial Legislative Council in 1919 and was used mercilessly in curbing civil disobedience throughout India. The notorious General Reginald Dyer acted under this law to carry out the massacre at Jallianwallabagh in Amritsar, Punjab, on 13 April 1919.

7. Chandra, however, discovered later that the essential results he had found had already been published in a Russian journal by J. Frenkel three years before. Consequently, he had to abandon the publication of this paper.

8. Dirac was named to the Lucassian Chair in 1932, the chair once occupied by Isaac Newton.

9. From experimentally determined fundamental constants of nature like electric charge e, Newton's constant G, and Planck's constant h, one can construct some dimensionless numbers. Among such numbers, some are extremely large; for example, the ratio of the electric to the gravitational force in a hydrogen atom is 10^{40}.

Similarly, the number of nucleons in the universe and the age of the universe expressed in appropriate natural units are 10^{80} and 10^{40}, respectively. These numbers are extremely large compared with some other numbers, such as the fine structure constant or the ratio of the mass of the proton to that of the electron. Physicists believe that ultimately an explanation for these numbers will be found. According to Dirac, the only way to explain the large numbers is to relate them to the age of the universe. This is known as the Large Number Hypothesis.

10. The Large Number Hypothesis leads to the conclusion that physical quantities, such as electric charge e and Newton's constant G, previously considered as constant are not really constant, but vary with time. Standard cosmologies cannot accommodate such an idea. Dirac proposed a new cosmology consistent with the Large Number Hypothesis and general theory of relativity.

11. The attendance at these meetings was restricted to members of the society and their guests. A guest was formally introduced by a member simply by writing his guest's name next to his in the registration notebook that kept the attendance record.

12. Louis Harold Gray (1905–65) had a distinguished career in physics as a member of the Cavendish Laboratory. Later he became interested in studying the biological effects of radiation and took a position in Mount Vernon Hospital in London in 1933. He became a pioneer in radiobiology and was elected a Fellow of the Royal Society.

13. A graduate of Girton College, one of the two colleges of Cambridge University for women at that time.

14. The two papers were "The Stellar Coefficients of Absorption and Opacity, Part II," *Proceedings of the Royal Society* 135 (1932): 472–90; and "O_2 Eridani B," *Zeitschrift für Astrophysik* 3 (1931):302–5.

15. Henri-Frederic Amiel (1921–81), a Swiss author.

16. Conversations, Victor F. Weisskopf, professor emeritus of the Department of Physics, MIT, his office, Cambridge, Massachusetts, 15 June 1981.

17. Victor Weisskopf, *Physics in the Twentieth Century: Selected Essays* (Cambridge: MIT Press, 1972), 55.

18. In reply to Chandra's congratulatory note on his winning the Nobel Prize for physics in 1975, Aage Bohr wrote that he still remembered those days of his childhood when Chandra played with him.

5. Fellow of Trinity College

1. Conversations, Victor F. Weisskopf, his office, Cambridge, Massachusetts, 15 June 1981.

2. Vaidyanathaswamy was a reader in the Mathematics Department of Presidency College, Madras; his position was equivalent to that of an associate professor in an American university.

3. Not without some regret for his "extravagance" in spending 18 pounds on the carpets instead of on books.

4. Conversations, Freeman J. Dyson, Institute of Advanced Studies, Princeton, New Jersey, 24 May 1978.

5. The Royal Society had its headquarters in the same house in those years.

6. The exchanges between Jeans and Eddington apparently became so entertaining in the twenties that the mathematician G. H. Hardy became a member of the RAS and attended the meetings just to hear them.

7. One moved gradually forward with age and accomplishment. Chandra says that McCrea now sits in the first row.

8. As the observatory director, Gerasimovič had a house for himself on the observatory grounds with a guest room and space for a live-in maid.

9. In turn, it should perhaps be noted that Chandra must have made a strong impression on the young astronomers and students he came into contact with in Russia. Svetlana Alliluyeva, Joseph Stalin's daughter, in her autobiographical account, *Only One Year* (New York: Harper and Row, 1969), writes about her astronomer friend Lydia Yulyevna and says, "Long ago, in her youth, when Lydia Yulyevna had just started working at the Pulkovo Observatory, she had met the famous Indian Astronomer Chandrasekhar, and afterward had always wanted to have a look at India. In this house [Yulyevna and her husband were both astronomers] India was loved and understood, and maybe that was why I always felt so at home in it" (p. 257).

10. S. Chandrasekhar, "Some Remarks on the State of the Matter in the Interior of Stars," *Zeitschrift für Astrophysik* 5 (1932):321.

6. The Absurd Behavior of Stars

1. Chandra was referring in his equation (1) to the generally accepted formulas for the electron density, the total energy, and the pressure of a completely degenerate electron gas.

2. A. S. Eddington, "Relativistic Degeneracy," *Monthly Notices of the Royal Astronomical Society* 95 (1935):194.

3. Letter from William H. McCrea, November 1979, in response to my request for his recollections and comments.

T. G. Cowling, a physicist of some repute and a friend of Chandra's from the Cambridge days, was also reminded of the controversy after reading an account of Chandra's work in the thirties in a recent popular book. He wrote to Chandra on 5 March 1987 saying, "It was interesting to be reminded about this, your controversy with Eddington. I sat on the fence in this: I had great admiration for Eddington, but found real difficulty in understanding his arguments. I felt that these arguments decreased rather than increased the plausibility of his case."

4. Letter from Rudolf F. Peierls, 5 May 1983, in response to my request for his recollections and comments. His paper: "Note on the Derivation of the Equation of State for a Degenerate Electron Gas," *Monthly Notices of the Royal Astronomical Society* 96 (1936):780.

5. S. Chandrasekhar Papers, box 15, folder 1, The University of Chicago Archives.

6. Eddington kept track of his cycling miles. To reach a figure n, he had to do rides of n miles or more at least n different times. Thus "n is still 75" means he had 75 rides of 75 miles or more.

7. *Observatory* 39 (1916):336, 339.

8. See also Chandra's lectures given in Cambridge to mark the centenary of Eddington in October 1982 and published under the title, *Eddington, The Most Distin-*

guished Astrophysicist of His Time (Cambridge: Cambridge University Press, 1983).

9. The address given in the spring of 1936 was published in the *Smithsonian Report* (1937):131–44.

10. Eddington is referring to Hardy's lectures on Ramanujan, which were also part of the tercentenary celebrations.

11. A remark which held little water. Fowler's paper was quite clear; the problem stated and the result were not at all mysterious.

12. Chandra was back in Cambridge at the time. Eddington had spoken to a large audience and had made a strong impression, eliciting frequently laughter at phrases like "stellar buffoonery," "physicists buzzing about my ears," etc.

13. Eddington, *Science and the Unseen World* (London: Allen and Unwin, 1929), 91.

14. S. Chandrasekhar, "A Limiting Case of Relativistic Equilibrium," in *General Relativity,* ed. L. O'Raifeartaigh, in honor of J. L. Synge (Oxford: Clarendon Press, 1972), 185–99.

15. From Schwarzschild's exact solution to Einstein's equations, one can contemplate a radius for every spherical mass object such that if the object shrinks to that radius, the space around it becomes curved enough by its own gravitational field so that light gets trapped and the object becomes a black hole.

16. The equation of state given by (pressure) = $\frac{1}{3}$ (density)(speed of light)^2.

17. Von Neumann's notes on this work can be found in *Collected Works of John von Neumann* (New York: Pergamon Press, 1961–63) 6:175–76.

18. J. Robert Oppenheimer and George M. Volkoff, "On Massive Neutron Cores," *Physical Review* 55 (15 February 1939):374–81.

19. Henry Norris Russell, "Impossible Planets," *Scientific American* (July 1935): 18–19.

20. R. H. Fowler, *Statistical Mechanics,* 2d ed. (Cambridge: Cambridge University Press, 1936), 652.

7. "I Must Push On in My Directions"

1. His father thought that a fellowship from the United States would be a good thing for Chandra's career.

2. Lalitha's family had the reputation of being quite radical in some respects.

3. This phrase is from Tennyson's poem "Guinevere" in *The Idylls of the King.* Chandra's father said in his letter that Tennyson was his favorite poet when he was young and quotes the following lines from "Guinevere":

> To lead sweet lives in purest charity
> To love one maiden only, cleave to her,
> And worship her by years of noble deeds,
> Until they won her; for indeed I know
> Of no more subtle master under heaven
> Than is the maiden passion for a maid
> Not only to keep down the base in man
> But to teach high thought, and amiable words
> And courtliness and the desire for fame
> And love of truth and all that makes a man.

4. Since theirs would not be a conventional, arranged marriage, they could not expect support in the form of a dowry.

5. Lalitha's father had died when she was ten years old.

6. Doctor of Science degree, considered more advanced than a Ph.D.

7. A friend of Chandra's in Cambridge, H. S. M. Coxeter, the geometer known for his book *Regular Polyhedra,* had introduced Chandra to Paul Donchian of Hartford, Connecticut. Owner of a carpet store, Donchian was also an amateur artist. The illustrations in Coxeter's book are photographs of models by Donchian.

8. It should be noted here that the formal offer of the research associateship came directly from Robert Hutchins, the president of the University of Chicago. Chandra received a cable offering him the position along with a prepaid cable for his answer aboard the ship. This, and also the fact that while in Chicago Struve had arranged a brief meeting between Chandra and Hutchins, were rather mysterious in view of the nature of the appointment, a temporary research associate, which normally should not have required the intervention of the highest authority of the university.

9. The committee, known as the Quinquinium Committee, was appointed by the Viceroy of India.

8. Lalitha

1. Conversations, Lalitha, Syracuse, 10 May 1987.

2. The infinitely flexible and accommodating Hindu religion allowed the essentials of the matrimonial rites and rituals to be compressed within almost any finite interval of time.

3. In some parts of India, the widows were required to carry a bell to warn others of their presence in the street. I should note that I am speaking here of brahman caste and customs. While widow remarriages were (and still are to a great extent) extremely rare and difficult, humiliations of the type described here were not imposed on women in other castes. They led a more normal life.

4. Monica Felton, *A Child Widow's Story* (New York: Harcourt, Brace, and World, Inc., 1967), 68.

5. Built in the early eighteenth century, the giant structure was used for a long time to store ice, hence the name Ice House. Later, toward the beginning of this century, it was converted into a hotel and during the summers wealthy zamindars and rajahs made it their summer palace in which to enjoy the sea breezes.

6. In "Little Sister," an autobiographical essay by Lalitha, which she wrote after her mother's death in 1958 (unpublished).

7. The use of such pet or familiar names is a common occurrence in joint families in India. Kanthamani, Lalitha's oldest sister, and Nithya, Lalitha's youngest aunt, were almost the same age, so they grew up together more like twin sisters than as aunt and niece. Nithya naturally called her sisters Ponnakka (Golden Sister) and Chinnakka (Little Sister). Kanthamani followed suit and called her mother Chinnakka. Of course, the younger sisters of Kanthamani followed her lead.

8. Anatole France, *"Bee," The Princess of the Dwarfs,* trans. Peter Wright (Dents Children's Classics).

9. Conversations, Lalitha, Chicago, 31 October 1987.

9. Scientist in the Midst of Political Turmoil

1. Otto Struve, "The Story of an Observatory," *Popular Astronomy* 55 (1947): 1–29. Otto Struve (1897–1967), born in Kharkov, Russia, came from a family of directors of observatories. His great-grandfather, Wilhelm Struve, was the founder-director of Pulkovo Observatory, and his grandfather retained the directorship in the family. His uncle, Herman Struve, was the director of Berlin-Babelsberg Observatory. Struve's brilliant work as a student in Russia was interrupted by the First World War and the Russian Civil War. Edwin Frost had saved this veteran of war and revolution from near starvation and death by getting him to Yerkes in 1921.

2. The land boom which had brought sudden wealth to southern California came to an end equally suddenly. Spence and other promoters of the project were wiped out, and in 1892 they defaulted on the payments of their contract with Clark.

3. Hale's father's wealth came from his elevator business, which thrived due to the new skyscrapers being built all around Chicago after the fire of 1871 burned down the city.

4. Donald E. Osterbrock, *James E. Keeler: Pioneer American Astrophysicist* (Cambridge: Cambridge University Press, 1984), 185.

5. Williams Bay, Wisconsin, approximately one hundred miles northwest of Chicago, was judged to be the ideal place because it was a safe distance away from the city lights of Chicago and also from the cloudiness due to Lake Michigan.

6. Hale tried in vain to make this observatory in southern California a branch observatory of the University of Chicago. He was also frustrated in his efforts to get an appointment at the university for James Keeler, his mentor and friend who was recognized as the world's best stellar spectroscopist. Harper was unwilling to commit the university to another professorship in astronomy. Further, health reasons of his family, the California climate, and his romantic involvement with Alicia Mosgrove decided in favor of Hale's permanent stay in Pasadena. Hale went on to become one of the great statesmen of science, a promoter and organizer of vast schemes (Osterbrock, *Jaames E. Keeler,* 350). He founded the California Institute of Technology, organized the National Research Council, revived the National Academy of Sciences, and built its present headquarters in Washington, D.C. He also became a major figure in the organization of the International Research Council of Scientific Unions.

7. Struve, "The Story of an Observatory," 13.

8. Ibid., 24.

9. Conversations, Martin Schwarzschild, his office, Princeton University, 24 May 1978.

10. By 1972, when Chandra stopped taking on students, fifty-one had earned their Ph.D. degrees under his supervision.

11. With no official support available to speak of, Chandra depended on the faculty and visitors at Yerkes, occasionally on some speakers from the main campus, but mostly on himself to sustain this activity. Following the Cambridge system, he designated the first colloquium as number 1 and kept on until, in the early sixties, the number 1,000 was reached. Unfortunately the file containing the announcements of these 1,000 colloquia has been destroyed.

12. This work gave rise to one of the most celebrated and most extensively quoted

papers of Chandra: "Stochastic Problems in Physics and Astronomy," *Reviews of Modern Physics* 15 (1943):1–89.

13. S. Chandrasekhar, *Radiative Transfer* (Oxford: Clarendon Press, 1950).

14. S. Chandrasekhar Papers, letter from John von Neumann, box 32, folder 8, The University of Chicago Archives.

15. Conversations, Robert G. Sachs, Chicago, 12 March 1978. Sachs had joined the group in 1942, just after completing his doctoral work at the Johns Hopkins University, Baltimore, Maryland.

16. The only other time Chandra engaged himself in such work was in the early 1960s when he was persuaded to join an elite group of prominent scientists for the Institute of Defense Analyses (Jason Division). But that was without the feeling of moral compulsion or the feeling of service and sacrifice for a great cause. After a few years of being, according to Chandra, mostly "a sleeping member," he ceased his association with the institute.

17. S. Chandrasekhar Papers, letter from Hans Bethe, box 11, folder 13, The University of Chicago Archives.

18. S. Chandrasekhar Papers, letters from Edward A. Milne, box 22, folders 8, 9, 10, The University of Chicago Archives.

19. S. Chandrasekhar Papers, letter from K. S. Krishnan, box 19, folder 16, The University of Chicago Archives.

20. S. Chandrasekhar Papers, letter from Henry Norris Russell, box 27, folder 16, The University of Chicago Archives.

21. Conversations, Peter Vandervoort, University of Chicago, 15 June 1978.

10. The Autocrat of the Editor's Desk

1. S. Chandrasekhar Papers, Report of the Managing Editor, 20 October 1969, box 150, folder 9, The University of Chicago Archives.

2. It is also worth noting that when rapid developments in radio astronomy began to take place in the 1950s, Chandra took it upon himself to encourage radio astronomers to publish in the *ApJ* and its Supplement. In the first issue of the 1959 volume, he said in his editorial,

The accelerated development of radio astronomy in America, and the need for rapid publication in this field, has raised doubts in the minds of some radio astronomers whether the existing astronomical journals in this country are adequate for their needs. For this reason, I should like to state that the *Astrophysical Journal* heartily welcomes authors to submit papers on radio astronomy, and assures them that their papers will be given every consideration. . . . Radio astronomy is a rapidly developing branch of astronomy; and any schism between this newer branch and the older branch of optical astronomy is to be greatly regretted.

3. Conversations, Eugene N. Parker, distinguished service professor, Departments of Astronomy and Astrophysics and Physics, University of Chicago, 12 March 1980.

4. S. Chandrasekhar Papers, letter to Hermann Bondi, box 148, folder 3, The University of Chicago Archives.

5. Transcript of an interview of Chandra by Spencer Weart, who is now the director of the Center for History of Physics, American Institute of Physics, New York. S. Chandrasekhar Papers, box 2, folders 8, 9, The University of Chicago Archives.

6. Angelo Secchi in Italy, William Huggins and J. N. Lockyer in England, P. J. C. Janssen in France, H. C. Vogel in Germany, and a few others did pioneering work in understanding the physical nature of stars.

7. Since 1947, the University of Chicago was absorbing the cost of Yerkes papers in the general budget of the journal when estimating the cost of publication shared by other observatories, individual authors, and smaller institutions. This was a violation of contractual obligation and was pointed out by Otto Struve (4 June 1949, Struve's letter to director of the University of Chicago Press, S. Chandrasekhar Papers, box 150, folder 9, The University of Chicago Archives).

8. Other members of the committee were Dirk Brouwer, N. U. Mayall, and Paul W. Merrill.

9. Letter from Alfred H. Joy to Earnest C. Colwell, July 1950, S. Chandrasekhar Papers, box 150, folder 9, The University of Chicago Archives.

10. In 1941, the AAS had taken over the management of the *Astronomical Journal* founded by B. A. Gould in 1849. Before that it had been managed as a private undertaking by the Gould Foundation directors.

11. Chandra had written in April 1948 to his long-time friend K. S. Krishnan, requesting him to arrange an invitation for him to attend the Indian Science Congress, to which a number of scientists from abroad were invited every year with expenses paid. Chandra had explained in the letter that since both Lalitha's mother and Chandra's father were losing health and advancing in age, Lalitha and he were anxious to visit India. They could not afford the trip on their own (letter to Krishnan, 29 April 1948, S. Chandrasekhar Papers, box 19, folder 16, The University of Chicago Archives). Yet, they had not received a reply from Krishnan one way or the other.

12. Harrison had been prevented by the director of the University of Chicago Press from taking action. Harrison had sent a note to Strömgren citing the objections of the Press, but Strömgren, instead of bringing it to the attention of Chandra, had buried that note in his unattended correspondence during the summer when he was away in Denmark.

13. The members of the first editorial board were Lyman Spitzer (chairman), Gerhard Herzberg, C. D. Shane, Paul Merrill, and Fred Whipple.

14. In order to allay such fears, Chandra took on Kuiper as an associate editor.

15. Chandra recalls going to the president of the university (Lawrence Kimpton at that time) for temporary help to make up the deficit. But the president said that there were no funds; the journal was supposed to be self-sufficient. Chandra then suggested that the necessary amount be transferred from his own salary to the journal account. Thereupon the president found the money from his contingency fund.

16. S. Chandrasekhar Papers, letter to Bondi, box 148, folder 3, The University of Chicago Archives.

17. S. Chandrasekhar Papers, box 148, folders 1–10, The University of Chicago Archives.

18. After this incident, Lick Observatory did not submit any of their papers for publication in *ApJ* for two years.

19. There was also some pressure from some in the astronomical community, notably from Marshall Wrubel (a former student of Chandra), to publish a Letters section.

11. In the Lonely Byways of Science

1. There were a few exceptions: Jeremiah P. Ostriker, Maurice J. Clement, Laurence F. Rossner, and Morris L. Aizenman.

2. Conversations, E. Margaret Burbidge, San Diego, California, 19 June 1980. She was at Yerkes during 1951, 1952-53, and later during 1957-62, when her husband Geoffrey Burbidge was on the faculty at Yerkes.

3. Conversations, Agnes Herzberg, Chicago, 11 March 1981.

4. Conversations, Peter Vandervoort, University of Chicago, 15 June 1978.

5. During his 1951 visit, Homi J. Bhabha had invited Chandra to join him to build the Tata Institute of Fundamental Research in Bombay.

6. In an article entitled, "Subrahmanyan Chandrasekhar, F.R.S.," in *Triveni* 15, no. 2, a quarterly magazine published in Madras, India.

7. According to Howard Boatwright, the former Dean of the Music School, Syracuse University, C. S. Ayyar was the first Indian who had developed such a notation for Indian music. Boatwright was a Fulbright Scholar in 1959-60 in India and had corresponded and met C. S. Ayyar. Conversations, Howard Boatwright, Syracuse, New York, 3 February 1984.

8. Henry G. Gale was the dean of physical sciences, a physicist who had acquired some fame because of his translation of the classic German book on optics by Drude.

9. William H. Bragg (1862-1942), a well-known physicist noted for his work on X-ray diffraction, received the Nobel Prize in physics in 1915 with his son William L. Bragg.

10. Esther Conwell, currently at Xerox Webster Research Center, Webster, New York.

11. Conversations, Edward Levi, Chicago, 7 January 1985.

12. Conversations, John Wilson, Chicago, 29 September 1984.

13. Conversations, Noel Swerdlow, Chicago, 14 June 1986.

14. Conversations, Valentine L. Telegdi, Barcelona, Spain, 23 September 1983.

15. Basilis C. Xanthopoulos is currently professor of physics, University of Crete. Valeria Ferrari, currently at Dipartimento di Fisica, Università di Roma, "La Sapienza."

16. S. Chandrasekhar, "The *Principia*: The Intellectual Achievement That It Is," unpublished ms., 1987.

Epilogue: Conversations with Chandra

1. When monochromatic light is scattered from liquid molecules, one observes new spectral lines with frequencies different from those of the incident light. The differences are characteristic of the internal vibrational and rotational quantum states of the molecules. This effect is known as the Raman effect and has proved to be an enormously useful tool in mapping the internal structure of molecules.

2. Adolf Smekal (1895-1959), a German physicist who had indeed studied and pointed out the theoretical possibility of the Raman effect.

3. While nationalism played a dominant role when it came to facing the British, local affiliations based on religion, caste, and language played a dominant role in

internal matters. Raman was a *madrassi* in Bengal. He had no business dominating an institute in Bengal.

4. The directorship of the Kodaikanal Observatory was offered to Chandra more than once. See chapters 8 and 9.

5. Homi J. Bhabha (1909–66), a contemporary of Chandra's at Cambridge, a scientist who devoted himself to helping develop Indian science. He founded the Tata Institute of Fundamental Research and, as the chairman of the Indian Atomic Energy Commission, he was chiefly responsible for the development of nuclear energy for peaceful uses in India. His tragic death in an Air India plane crash in 1966 ended prematurely an illustrious career.

6. This account is now in print in *Littlewood's Miscellany,* ed. Bela Bollobas (London: Cambridge University Press, 1986).

7. Eugene Wigner, *Symmetries and Reflections* (Bloomington: Indiana University Press, 1967), 261.

8. S. Chandrasekhar, in *The Physicist's Conception of Nature,* ed. J. Mehra (Dordrecht: D. Reidel, 1973), 800–802.

9. Chandra's "Recollections of Mrs. Indira Gandhi," including this incident, are found in the Indira Gandhi memorial volume, published by the Government of India.

10. The two students Chandra is referring to are Meghnad Saha and S. N. Bose.

11. Amagh Nduka, currently vice-chancellor of the Federal University of Technology, Oweri, Nigeria. In 1988, his university conferred an honorary degree on Chandra.

Appendix

Books Written by S. Chandrasekhar

An Introduction to the Study of Stellar Structure. Chicago: University of Chicago Press, 1939. Reprinted, New York: Dover Publications, 1967. Translated into Japanese and Russian.

Principles of Stellar Dynamics. Chicago: University of Chicago Press, 1943. Reprinted, New York: Dover Publications, 1960.

"Stochastic Problems in Physics and Astronomy," *Reviews of Modern Physics* 15 (1943): 1–89. Reprinted in *Selected Papers on Noise and Stochastic Processes,* ed. Nelson Wax. New York: Dover Publications, 1954.

Radiative Transfer. Oxford: Clarendon Press, 1950. Reprinted, New York: Dover Publications, 1960. Translated into Russian.

Hydrodynamic and Hydromagnetic Stability. Oxford: Clarendon Press, 1961. Reprinted, New York: Dover Publications, 1981. Translated into Russian.

Ellipsoidal Figures of Equilibrium. New Haven: Yale University Press, 1968. Reprinted, New York: Dover Publications, 1987. Translated into Russian.

The Mathematical Theory of Black Holes. Oxford: Clarendon Press, 1983. Translated into Russian.

Eddington: The Most Distinguished Astrophysicist of His Time. Cambridge: Cambridge University Press, 1983.

Truth and Beauty: Aesthetics and Motivations in Science. Chicago: University of Chicago Press, 1987.

Selected Papers. 6 volumes. Chicago: University of Chicago Press, 1989–90.

Honors, Medals, and Prizes Awarded to S. Chandrasekhar

1944   Fellow of the Royal Society of London
1947   Adams Prize (Cambridge University)
1952   Bruce Medal (Astronomical Society of the Pacific)
1953   Gold Medal (Royal Astronomical Society)
1955   Member of the National Academy of Sciences
1957   Rumford Medal (American Academy of Arts and Sciences)
1962   Royal Medal (Royal Society of London)
1962   Srinivasa Ramanujan Medal (Indian National Science Academy)
1966   National Medal of Science (United States)
1968   Padma Vibhushan Medal (India)
1971   Henry Draper Medal (National Academy of Sciences)

1973   Smoluchowski Medal (Polish Physical Society)
1974   Dannie Heineman Prize (American Physical Society)
1983   Nobel Prize for Physics (Royal Swedish Academy)
1984   Dr. Tomalla Prize (ETH, Zurich)
1984   Copley Medal (Royal Society of London)
1984   R. D. Birla Memorial Award (Indian Physics Association)
1985   Vainu Bappu Memorial Award (Indian National Science Academy)

Acknowledgments

Writing the biography of a living person is not an easy task, but in all probability, it is more difficult to be the subject of one. I am indeed enormously grateful to Chandra for allowing me to tell the story of his life, for his patience during hours and hours of conversations, and for providing me with all the information I needed regarding his correspondence and documents. While he read both the preliminary and the finished versions of this book and was of immense help in verifying facts, at no stage did he make me feel that I was under any kind of censorship. I enjoyed complete freedom.

I am also grateful to Lalitha for providing me with information regarding her family background and generously allowing me to use some material from her unpublished essays. Likewise, several members of Chandra's family, particularly his brother, Dr. S. Balakrishnan, played a major role in making this book possible.

I was extremely fortunate to have the advice and encouragement of writers Molly Daniels, Raymond Carver, and Tess Gallagher. I owe special thanks to Tess, who read the entire manuscript, and improved it immeasurably. I am indebted to Professor Lynn Margulis for reading parts of the manuscript and for her continual encouragement throughout the course of this work.

This has been a family adventure. My daughters Alaka and Monona edited the earlier versions for style and content and helped me prepare the first draft; the final version passed the careful scrutiny of my daughter Achala. It is difficult to thank my wife Kashi in a few words. She shared my joys and patiently endured the rough periods. Friends Abhay Ashtekar, Marcia King, Dzidra Knecht, Mavis Lozano, Elaine and Keith Lessner, and V. L. N. Sarma were constant sources of encouragement.

It also gives me great pleasure to thank Professors Sheldon L. Glashow and Robert G. Sachs for securing honorary appointments for me as an associate and visiting physicist at Harvard University and the University of Chicago, respectively, so that I could use the splendid library facilities at these institutions. The staff of the Special Collections Department of the Regenstein Library at the University of Chicago, especially Daniel Meyer and Richard Popp, were most helpful in providing me special privileges, a special place to work, as well as research assistance.

The book would not have the character it has were it not for the warm response, the generous sharing of time, and the recollections and insights of Chandra's students, friends, and associates whom I interviewed. Their names are listed below.

I must finally thank Janet Pease for her diligent and enthusiastic help in research, in transcribing and editing the taped interviews, and in preparing several drafts of the manuscript.

329

Interviews

Name	Place Interviewed	Date of Interview
S. Balakrishnan	Bangalore, India	19 December 1979
Howard Boatwright	Syracuse, NY	3 February 1984
E. Margaret Burbidge	San Diego, CA	19 June 1980
Jeanette Burnett	Chicago, IL	15 June 1978
Bimla Buti	Ahmedabad, India	24 November 1984
Alastair G. W. Cameron	Cambridge, MA	23 July 1984
Brandon Carter	Meudon Observatory, Paris, France	17 March 1984
Karen Challonge	Paris, France	14 November 1983
Lalitha Chandrasekhar	Chicago, IL	10 May and 31 October 1987
Chandra's sisters and nieces	Madras, India	2 January 1980
Arthur A. Code	Madison, WI	10 March 1981
Esther Conwell	Rochester, NY	18 May 1978
James W. Cronin	Chicago, IL	10 June 1986
Paul A. M. Dirac	Miami, FL	12 January 1981
Freeman J. Dyson	Princeton, NJ	24 May 1978
Donna Elbert	Chicago, IL	15 January 1986
John Friedman	Chicago, IL	14 June 1978
Robert Geroch	Chicago, IL	15 June 1978
Marvin L. Goldberger	Pasadena, CA	14 June 1980
Mildred Goldberger	Pasadena, CA	14 June 1980
Maurice Goldhaber	Brookhaven, NY	14 May 1979
Agnes Herzberg	Chicago, IL	11 March 1981
Gerhard Herzberg	Ottawa, Canada	19 October 1981
Jeanne Hopkins	Chicago, IL	14 June 1978
V. S. Jayaraman	Madras, India	3 January 1980
Tsing-Dao Lee	New York, NY	23 May 1978
Norman R. Lebovitz	Chicago, IL	1 May 1979
Edward Levi	Chicago, IL	7 January 1985
Alan Lightman	Cambridge, MA	15 May 1980
Robert Marshak	Blacksburg, VA	6 December 1980
William McCrea	Brighton, England	19 July 1983
Leon Mestel	Brighton, England	19 July 1983
Yavuz Nutku	Syracuse, NY	25 March 1978
Donald Osterbrock	Washington, DC	7 December 1985
Jeremiah P. Ostriker	Princeton, NJ	10 March 1986
Eugene Parker	Chicago, IL	12 March 1980
Sotirios Persides	Chicago, IL	16 January 1986
William Press	Cambridge, MA	8 June 1979
William H. Reid	Chicago, IL	14 June 1978
Robert G. Sachs	Chicago, IL	12 March 1978
Carl Sagan	Los Angeles, CA	20 June 1980
Martin Schwarzschild	Princeton, NJ	24 May 1978

Name	Place Interviewed	Date of Interview
David Shoenberg	Cambridge, England	21 August 1979
Lyman Spitzer	Princeton, NJ	10 March 1986
Noel Swerdlow	Chicago, IL	14 June 1986
Valentine L. Telegdi	Barcelona, Spain	23 September 1983
Kip S. Thorne	Pasadena, CA	14 June 1980
Henk van de Hulst	Leyden, Holland	1 December 1983
Mrs. van de Hulst	Leyden, Holland	1 December 1983
Peter Vandervoort	Chicago, IL	15 June 1978
Victor F. Weisskopf	Cambridge, MA	15 June 1981
John Wilson	Chicago, IL	29 September 1984
Basilis Xanthopoulos	Chicago, IL	16 January 1986

Index

Since this is a biography of Subrahmanyan Chandrasekhar, all subject headings should be interpreted as denoting a relationship with Chandrasekhar (e.g., "Attitude toward Science").

Abt, Helmut A., 227
Adams, Walter Sydney, 187
Adrian, Edgar Douglas, 113
Aiyar, Rao Bahadur N. Balakrishna, 46, 313 n.22
Aizenman, Morris L., 324 n.1 to chap. 11
Alliluyeva, Svetlana, 318 n.9
Allison, Sam, 203
Al-Sadir, Jafar (physician), 300, 301
Ambartsumian, Viktor A., 117, 123, 272–74, 275–76
American Astonomical Society, 160–61, 206. See also Astrophysical Journal
Ammal, Ammani, 169
Anagnostopoulos, Constantine, 303
Anderson, H., 239, 242, 269
Artin, Emil, 104, 106
Ashtekar, Abhay, 9–10, 309 n.6
Askey, Richard, 265
Aston, Francis William, 295
"The Atheist" (Balzac), 305
Astronomical Journal, 211, 214
Astronomy and Astrophysics. See Astrophysical Journal, early history of
Astrophysical Journal: International Review of Spectroscopy and Astronomical Physics, 14, 17, 20–21, 121, 201, 205, 206–28, 237–38; associate editor, 209–16; call for Chandrasekhar's impeachment, 219–20; changes under Chandrasekhar, 206–7, 216; cosponsorship of, 210–14, 323 n.12; early history of, 208–9; editor, 216–27; financial management of, 210, 323 nn.7, 15; publication policies, 214–15, 220–21, 225–26, 322 n.2, 323 n.19; relations with authors, 208, 217–20, 221–22, 223, 323 n.18; relations with University of Chicago Press, 223–25; transfer of journal and editorship, 226–28
Atomic Energy Commission, 93, 267
Atomic Structure and Spectral Lines (Sommerfeld), 57, 61, 76, 250
Attitude toward science, 22–23, 27–29, 32, 82, 144, 189–90, 245–46, 247, 248, 295–96. See also Reflections on his life
Awards, honors, and prizes, 5, 12, 14, 28, 68, 103, 115, 145, 203, 286, 296. See also Nobel Prize; Royal Society of England, Royal Astronomical Society of London
Ayyar, C. Subrahmanyan (father), 34, 40, 41, 44–47, 50–52, 54–57, 71, 72, 87–88, 112–13, 148–50, 151–55, 157–59, 164–67, 168, 183–84, 233–35, 257, 258, 319 n.1, 324 n.7
Ayyar, Sitalakshmi (née Divan Bahadur Balakrishna; mother), 46, 47, 52–53, 54, 56–57, 68–71, 90–92

Baade, Walter, 223
Babcock, Harold D., 223
Balakrishnan S. (brother), 47, 50, 53, 58–59, 91, 165, 169, 233
Balaparvathi (Mrs. P. Visvanathan; sister), 47, 54
Baldwin, Stanley, 260
Bannerjee, S. K., 282–83

Bartky, Walter, 200, 216
Beadle, George, 236
Bentinck, William, 35–37
Besant, Annie (née Wood), 43, 313 n.16
Bethe, Hans, 61, 196–97
Bhabha, Homi J., 233, 255–56, 283, 288, 324 n.5, 325 n.5
Biermann, Ludwig, 93
Blackett, Patrick Maynard Stuart, 86, 253, 287
Bloch, Felix, 272
Boatwright, Howard, 324 n.7
Bohnhöffer, Karl, 94
Bohr, Aage, 103, 317 n.18
Bohr, Niels, 98–99, 100–103, 128–31, 132, 133, 147, 156
Bondi, Hermann, 207, 217
Borello, Piero, 3
Born, Max, 92, 94, 99, 161, 250, 295
Bose, D. M., 65
Bose, Jagadish Chandra, 9, 56, 245, 247–49, 296
Bose, Satyendra Nath, 4, 9, 61, 198, 246, 250, 256, 259, 315 nn.16, 20, 325 n.10
Bowen, Carrol G., 216
Bowen, Ira S., 223
Bragg, William H., 236–37, 324 n.9
Bragg, William L., 324 n.9
Brillouin, Leon, 93
Brouwer, Dirk, 323 n.8
Burbidge, E. Margaret, 230, 238, 324 n.2 to chap. 11
Burbidge, Geoffrey, 324 n.2 to chap. 11
Burke, Edmund, 36, 311 n.5
Burnett, Jeanette, 20–21, 224, 227, 310 n.14 to chap. 1
Buti, Bimla, 20, 290

Cabbannes, Henri, 253
Candeth, M. A., 67–68, 106
Cesco, Corlos, 189
Chadwick, James, 295
Challonge, Daniel, 293
Challonge, Karen, 293–94
Chamberlain, Arthur Neville, 260
Chandrasekhar, Lalitha (wife), 9, 10–11, 16, 20, 55, 147, 148–51, 163, 166–72, 175, 176–84, 192–93, 194–95, 199, 212, 226, 229–30, 231–39, 241, 272, 274, 281–86, 289,

299–304, 306–7, 316 n.23, 319 n.2, 320 n.5, 323 n.11; family background, 171–77
Chandrasekhar limit, 4, 10; 29–30, 119, 144–45; initial formulation of, 75–76, 79. See also Milne-Eddington controversy
Chandrasekhar, Parvati (grandmother), 40, 312 n.11
Chandrasekhar, Ramanathan (1837–1906; great-grandfather), 34
Chandrasekhar, Ramanathan (1866–1910; grandfather), 34, 39–44, 313 n.17
Chettiar, Alagappa, 264
A Child Widow's Story (Felton), 174 n
Chowla, S., 78, 80, 86, 158, 161–62, 316 n.4
Citizenship, 231–32, 281–85; reaction to, in India, 232–35, 285–86
Clark, Alvin G., 143, 185–86, 321 n.2
Clement, Maurice J., 324 n.1 to chap. 11
Clive, Robert, 311 n.5
Code, Arthur Dodd, 189, 230
Collected Works of Enrico Fermi (Fermi and Chandrasekhar), 265
Colwell, Ernest C., 210–13, 323 n.9
Compton, Arthur Holly, 60, 61, 66, 76, 315 nn.14, 17
"The Compton Scattering and the New Statistics" (Chandrasekhar), 62–63
Conwell, Esther, 189, 238, 324 n.10
Cornwallis, Charles, 311 n.7
Cowling, T. G., 318 n.3
Coxeter, H. S. M., 320 n.7
Cripps, Sir Stafford, 191
Cronin, Annette, 241
Cronin, James, 14, 33, 241, 280, 297, 310 n.1
Cultural background, 36, 38, 47, 48, 51, 53, 71, 177–78, 231, 254–55, 304, 311 n.6, 320 nn.2, 3, 320 n.7 to chap. 8, 324 n.3 to epilogue; caste system: Brahmans, 34–35, 39, 43, 149, 171, 311 nn.2, 6, 316 n.25, 320 n.3; regarding marriage, 43, 54, 111, 149–51, 166, 170–71, 320 n.2, 320 n.4 to chap. 7; regarding women, 49, 52, 53–54, 172, 173–74, 313 n.18, 320 n.3. See also India, British rule in; India, independence move-

ment; Life in Great Britain, cultural adjustment

Dakhinamurthy, C. V., 316 n.23
Dalhousie, James Ramsay, 35, 37, 312 n.8
Darwin, C. G., 99, 113
Davenport, Harold, 113, 114, 198, 199, 231
Day, B. B., 257
DeBroglie, Louis, 138
Debye, Peter, 61
Delbrück, Max, 100, 105
Desai, Morarji, 286
Detweiler, Steve, 301–2
Dingle, Herbert, 161
Dirac, Paul Adrien Maurice, 8, 62, 68, 81, 82–83, 83–84, 86, 96, 98, 99, 101–2, 106, 111, 112, 114, 128, 131–32, 133, 137, 247, 248, 255, 295, 316 nn.8, 9, 317 n.10
Discrimination encountered, 9, 53–54, 65–66, 74, 112, 165, 194–95, 204, 231, 235–37, 285, 296–97. See also Citizenship; India, independence movement; Political context
A Doll's House (Ibsen), translation of, 15, 52, 159
Donchian, Paul S., 160, 320 n.7
Doraiswamy, Captain (father-in-law), 55, 175–76
Doraiswamy, Lalitha. See Chandra-sekhar, Lalitha
Doraiswamy, Radha (sister-in-law), 176
Doraiswamy, Savitri (mother-in-law), 55, 176–77
Dreyer, J. L. E., 115
Dumond, Jesse W., 63
Dyer, Reginald, 316 n.6
Dyson, Sir Frank, 116
Dyson, Freeman, 114, 317 n.4

Eckart, Carl, 121, 271
Eckart, Klara. See Neumann, Klara von
Eddington, Sir Arthur Stanley, 8, 12, 31, 76, 79, 82, 83, 84, 86, 92, 97, 108, 111, 112, 113, 115, 119–45, 148, 152, 165, 188, 199, 247, 248, 296, 318 n.6 to chap. 5, 318 n.6 to chap. 6, 319 n.12
Einstein, Albert, 12, 14, 30–31, 143, 192, 259, 305
Elbert, Donna, 21–22, 242

On Ellipsoidal Figures of Equilibrium (Chandrasekhar), 26
Elvey, C. T., 187
Ewald, Paul, 61

Family background, 34–35, 47–63; interest in education, 35, 39–40, 41–43, 44–45, 52, 56–57, 171, 314 n.7; interest in literature and the arts, 15, 44, 45, 51–53, 55, 113, 151, 235, 324 n.7. See also Chandra-sekhar, Lalitha: family background; Cultural background
"Family History" (C. S. Ayyar), 40
Feather, Norman, 98
Felton, Monica, 174
Fermi, Enrico, 8, 17, 19–20, 62, 196, 199, 203, 208, 234, 237, 239, 241, 265, 268, 269–71, 272, 283, 310 n.7
Fermi, Laura, 20, 234, 270
Ferrari, Valeria, 241, 324 n.15
Fondness for children, 230–31
Fowler, Ralph Howard, 29, 62–63, 70, 76, 77–81, 82, 83, 85, 86, 96, 98, 99, 107–8, 111, 114, 115, 120, 121, 125–26, 127, 133, 137, 141, 144–45, 149, 158, 165, 247, 258, 259
Fowler, William A., 13, 93, 291
Franck, James, 200
Frenkel, J., 316 n.7
Freund, Peter, 280
Freundlich, Erwin Finlay, 94, 122
Freye, Frieda. See Gray, Frieda
Friedman, John, 300
Fröhlich, Herbert, 61
Frost, Edwin Brant, 187, 209, 321 n.1
Fyson, P. F., 66, 67–68, 257, 258

Gale, Henry G., 236, 324 n.8
Gandhi, Indira, 286–89
Gandhi, Mohandas Karamchand, 5, 9, 45, 175, 191–92, 245, 259
Ganesan, A. S., 54
Garwin, Richard, 241
Gauss, Carl Friedrich, 99
Gell-Mann, Murray, 241
Gerasimovič, B. P., 116–17, 318 n.9
Ghosh, J. C., 65
Ginzburg, V. I., 275, 278, 279
Goldberger, Marvin, 14–15, 21–22, 23, 203

Goldberger, Murph, 292
Gould, Benjamin Apthorp, 323 n.10
Granlund, Paul, 262, 265
Gray, Frieda Marjorie Picot (née Freye),
    86–87, 113, 199, 317 n.13
Gray, Louis Harold, 86–87, 199, 317 n.12
Grunau, Eric, 94
Gurney, Ronald, 194

Hale, George Ellery, 185–87, 208–9,
    321 nn.3, 6
Halifax, Lord. See Wood, Edward Fred-
    erick Lindley
Hardy, Godfrey Harold, 113, 141, 147,
    158, 162, 257, 258, 259–60, 261–62,
    264, 318 n.6, 319 n.10
Harper, William Rainey, 186–87
Harrison, R. Wendell, 213, 323 n.12
Hartek, Paul, 94
Hastings, Warren, 311 n.5
Have, Margere, 100
Hawking, Stephen W., 294
Heart attack and surgery, 9, 299
Heisenberg, Werner, 8, 61, 62, 64, 93,
    98, 99, 100, 295
Heitler, Walter, 61
Henrich, Louis R., 189
Henry Fellowship, 147–48
Herzberg, Agnes, 230
Herzberg, Gerhard, 255, 323 n.13
Hilbert, David, 27
Hilbert (Reid), 27
Hopkins, Frederick, 113
Hopkins, Jeanne, 20–21, 224, 310 n.14
    to chap. 1
Hubble, Edwin, 194, 223
Huggins, William, 323 n.6
Hutchins, Robert, 185, 187–88, 199–
    201, 231, 235–36, 239, 284, 320 n.8

Ibsen, Henrik Johan, 15, 52, 159
Immigration and Naturalization Act
    (McCarran Act). See Citizenship
India: British civil service system in,
    38, 45, 49, 51, 56; British educational
    system in, 34, 36–39, 48–49, 50,
    255–59; British rule in, 35–38,
    311 nn.5, 7, 312 nn.8, 9, 313 n.22,
    316 n.6; conflicts within the scientific
    community, 154–55, 161–62, 166,
    248–52, 254, 255–56, 289–91,
    320 n.9 to chap. 7; history of science

in, 246–48; independence move-
    ment, 5, 41–42, 65–66, 86, 114, 191,
    245, 246–47, 258, 283; women's
    rights movement in, 174–75, 320 n.5
    to chap. 8. See also Cultural back-
    ground, regarding women; Gandhi,
    Mohandas Karamchand; Nehru, Ja-
    waharlal; Raman effect
Indian Association for the Cultivation
    of Science, 60–61, 249
Indian Institute of Science, 166, 178–
    79, 249, 250
Indian Journal of Physics, 63
Indian National Congress Movement.
    See India, independence movement
Indian Science Congress Association,
    63, 64–65, 315 n.18
Influence of Chandrasekhar, 24–25,
    64–65, 318 n.9
The Internal Constitution of Stars (Ed-
    dington), 64, 76, 128
International Astronomical Union
    meeting, 133–34, 135–38, 143,
    152–53
An Introduction to the Study of Stellar
    Structure (Chandrasekhar), 24, 32,
    189, 297
Israel, Werner, 292
Ivanov, V. V., 274, 276–77, 278, 281
Iyer, Chithy, 172–73
Iyer, P. T. Srinivas, 314 n.5
Iyer, Subbalakshmi, 172–75
Iyer, Subramania, 172
Iyer, Visalaksi, 172

Jakobsen, Peter, 25, 311 n.19
Janssen, Pierre-Jules-Cesar, 323 n.6
Jayaraman, V. S., 57–58
Jeans, Sir James Hopwood, 115–16, 133,
    141–42, 144, 318 n.6 to chap. 5
Jinnah, Muhammad Ali, 192
Johnson, Lyndon Baines, 296
Joliot-Curie, Irène, 138
Jost, Res, 12, 310 n.14
Joy, Alfred H., 210–14, 323 n.9

Kadanoff, Leo, 280
Kahn, Steve, 25
Kancheev, A. A., 118
Kapitsa, Peter, 11, 86, 114, 278–79,
    309 n.11
Keeler, James E., 209, 321 n.6

Kelvin, Lord, *See* Thomson, William
Kemp, James, 292
Kent, Robert H., 193–94
Kimpton, Lawrence, 202, 242, 323 n.15
Klein, Oskar, 99, 100
Kopferman, Hans, 100, 105
Kozyrev, Nikolai A., 117
Krat, V. A., 277
Krishnan, K. S., 60, 61–62, 64, 169, 198, 249–52, 254, 274, 282, 323 n.11
Krogdahl, Margaret Kiess, 189
Krogdahl, Wasley S., 189
Kuiper, Gerard, 137, 160, 162, 182, 184, 188, 189, 204, 234, 297, 323 n.14
Kuiper, Paul, 230

Landau, Lev D., 117, 121–22, 278–79
Laporte, Otto, 61
Lawrence, Ernest Orlando, 161
Lawson, Andrew, 203
Lebovitz, Norman R., 26
Ledoux, Paul, 189
Lee, Tsung-Dao, 17, 20, 230, 310 nn.7, 12 to chap. 1
Levi, Edward, 228, 239
Levi-Civita, Tullio, 143
Libby, Willard, 241
Life in Great Britain (Cambridge, 1930–36), 8, 9, 72–168; correspondence with family, 88–92; cultural adjustment, 73, 75, 87–88, 92, 95, 112; decision to go to the United States, 165; difficulties entering Cambridge, 70, 77–79; difficulties entering the United States, 180–81; Fellowship at Trinity College, 108–110; friends, 86–87, 114; invitations from the United States, 153, 157, 162–63, 165, 320 n.8; lectureship at Harvard, 159–61, 163; loneliness, 81–82, 86–87, 91–92, 95, 100–101, 114; Sheep Shanks Exhibition, 84; thesis presentation, 107–8; trip to Belgium, 103–4; trip to Denmark, 98–103, 104, 122, 150–51; trip to Germany, 92–94; trip to Russia, 116–19; trip to Yerkes Observatory, 164. *See also* Dirac, Paul Adrien Maurice; Eddington, Sir Arthur Stanley; Fowler, Ralph Howard; Henry Fellowship; International Astronomical Union Meeting; Milne-Eddington controversy

Life in India (1910–30); 9–71; beginning his scientific career, 63; departure for England, 70–71; disillusionment and withdrawal, 53–54; education, 26–27, 43, 44, 48–50, 53, 54, 55–57, 60–62, 64–68, 314 n.5, 316 n.24; extracurricular activities, 57–58; Government of India Scholarship, 67–68, 70, 99, 106–7, 258, 282. *See also* Cultural background; Family background; India; Residences, Chandra Vilas
Life in the United States (1936–   ), 5, 7, 9, 181–243; cultural adjustment, 182–83; curriculum in astronomy, 188–89; curriculum and department change, 201–3, 229; disillusionment with colleagues, 203–5; job offer from Princeton, 200–201; participation in university affairs, 199–200, 239–41; as professor emeritus, 241–43; reaction to, in India, 5; trip to Princeton, 192–93; trip to Russia, 272–81; war-related work, 5, 17, 193–94, 196–97, 266, 322 n.16. *See also Astrophysical Journal;* Citizenship; Political context; Pressure to return to India
Life's Shadows (C. S. Ayyar), 51
Lifshitz, E. M., 278
Lightman, Alan, 24, 25, 122, 310 n.6
Lindblad, Bertil, 153
Linlithgow, Lord (Victor Alexander John Hope), 250
Literature and the arts, interest in, 15–16, 25, 33, 87, 93, 94, 104, 118, 156–57, 159, 300, 301, 305, 306, 307
Littlewood, J. E., 78, 81, 113, 158, 162, 258, 259, 263
Lockyer, Sir Joseph Norman, 323 n.6
Lukash, V., 275
Lynch, Christina, 174–75

Macaulay, Thomas Babington, 35–37, 311 n.4
McCrae, William, 115, 124, 134, 153, 318 n.7 to chap. 5
MacMahan, P. S., 315 n.18
Mahabarata (Indian epic), 18, 38, 176
Manhattan Project. *See* Life in the United States, war-related work
Mantois, M., 186, 321 n.2

Markarian, B. E., 276
Mathematical Theory of Relativity (Eddington), 30–31
Maxwell, James Clerk, 11, 141, 305
Mayall, N. U., 323 n.8
Mayer, Joe, 241
Mayer, Maria, 241, 271
Merrill, Paul W., 214–15, 217, 323 nn.8, 13
Millikan, Robert, 254
Milne, Edward Arthur, 12, 81, 84–85, 86, 92, 94, 96–97, 98, 102, 105, 106, 108, 109–10, 111, 112, 114, 115, 119–27, 132–33, 140–41, 144, 148, 150, 158, 165, 188, 197–99, 231, 259, 296, 297–98
Milne-Eddington controversy, 10, 29, 30–32, 119–38; astronomers' views on, 132, 134, 144–45; effects of, 131–32, 138, 145, 146, 150; physicists' views on, 128–32, 135, 144
Minkowski, Rudolph, 223
Mirozoyan, L., 275
Mnatsakanian, N. A., 276, 278
"Model Stellar Photospheres" (Chandrasekhar), 97
Modern Algebra (Van der Waerden), 26–27, 93
Modern Analysis (Whittaker and Watson), 93
Modesty in science, 3–4, 21, 74–75, 188, 272
Møller, Christian, 133
Molotov, Vyacheslav Mikhaylovich, 278
Monthly Notices of the Royal Astronomical Society (MNRAS), 94, 97, 121, 135
Morgan, Emily, 230
Morgan, William W., 187, 188, 204, 210–11, 215–16
Moseley, Henry Gwyn Jeffreys, 140
Mott, Neville F., 62–63, 98
Mukherjee, Sir Ashutosh, 248
Mukherjee, J. N., 65
Mulders, Jerry, 160
Mulliken, Robert S., 285
Münch, Guido, 25, 189, 292
Myer, Joseph, 194

Naidu, Sarojini, 9
Nath, Nagendra, 161–62

Nduka, Amagh, 291, 325 n.11
Nehru, Jawaharlal, 5, 9, 45, 66, 191, 245, 251–52, 255, 258, 259–60, 264, 283, 285, 289
Nelson, Gaylord, 284
Neugebauer, Otto, 287
Neuman, John von, 8, 17, 143–44, 192–94, 196–97, 265–68, 269–71
Neumann, Klara von (later Klara Eckart), 267, 269, 271
Nobel Prize, 8, 13–14, 27, 33, 145, 291–93; reaction to, 294–96, 297–98
Nordheim, L., 93
Novikov, Igor, 275, 278
Nutku, Yavuz, 17, 18, 310 n.8

Observatory, 124–25, 132
Oppenheimer, J. Robert, 144, 196–97, 266, 267
Oriental Exclusion Act. *See* Citizenship
Osterbrock, Donald E., 189, 227, 310 n.7
Ostriker, Jeremiah, 20, 33, 189, 310 n.13 to chap. 1, 324 n.1 to chap. 11

Pai, Keshava, 68
Parameswaran, H., 63, 66, 67–68, 70, 106, 148, 169, 178, 249, 257, 315 n.12
Parker, Eugene N., 207, 322 n.3
Patel, Sardar Vallabhabhai, 9
Pauli, Wolfgang, 61, 62, 98, 99, 128, 129, 131, 132, 192, 256
Payne, W. W., 209
Pease, Francis Galdheim, 187
Peierls, Rudolf, 61, 135
Pekeris, Chaim, 256
Penrose, Roger, 290–91
Penzias, A. A., 292
Phillips, Rev. T. E. R., 115
Philosophical Magazine, 63, 67
Physical Review, 225
Physics Today, 297
Pioneers of Science, 297
Placzek, George, 15, 100, 103
Planck, Max, 29
Plaskett, Harry H., 115, 127, 152, 153, 165
Poincaré, Jules-Henri, 28
Political context, 9, 65–66, 114, 190–92, 232–35, 258, 259–60, 273, 283–84, 285; effect on science,

277–79, 299. *See also* Gandhi,
Mohandas Karamchand; India, inde-
pendence movement; Life in the
United States, trip to Russia; Life in
the United States, war-related work;
Nehru, Jawaharlal
Political philosophy, 279–81
"Position as a Differential Operator in
Quantum Mechanics" (Chandra-
sekhar), 80
Pradhan, Trilochan, 290
Press, William, 23, 24, 311n.17
Pressure to return to India, 104, 106–7,
110–12, 153–56, 157–59, 161–62,
163, 164–65, 183–84, 282–83,
285–86, 288, 311n.21, 325n.4. *See
also* Citizenship
Principia (Newton), 242–43
Principles of Quantum Mechanics (Di-
rac), 77
Principles of Stellar Dynamics
(Chandrasekhar), 24, 32
Pringsheim, Peter, 63
Proceedings of the Royal Society, 62–63,
64, 67, 101–2, 116
Proxmire, William, 284
Purcell, E. M., 292

Radhakrishnan, Sarvepalli, 55, 181, 289
Radiative Transfer (Chandrasekhar), 24,
26, 32, 190
Ragjopal, C. T., 264–65
Rajalakshmi (Mrs. A. S. Ganesan; sister), 47, 54, 168
Ramakrishnan, C. A., 74, 88, 316n.23
Raman, Sir Chandrasekhara Venkata
(uncle), 4, 9, 41, 42, 43, 44–46,
49–50, 51, 56, 59, 60–61, 63, 70, 73,
107, 154–55, 161–62, 166, 171, 177,
179, 183–84, 198, 236–37, 245, 246,
247, 248–54, 257, 258–59, 263,
282–83, 296, 312n.12, 324n.3 to
epilogue
Raman effect, 60–61, 63, 246, 248–52,
313n.23, 324n.1 to epilogue
Raman, Lokasundari Ammal (aunt), 46,
70
Ramanathan, K. R., 72, 153–54, 252,
316n.1
Ramanathan, S. (brother), 47, 54, 91, 169
Ramanujan Institute of Mathematics,
264–65

Ramanujan, Srinivasa Aaiyangar, 9, 12,
56, 69–70, 73, 109, 110, 245,
246–47, 258, 260–65, 296
*Ramanujan, Twelve Lectures on Subjects
Suggested by his Life and Work*
(Hardy), 261–62
Ramaseshan, S., 253, 289–90, 313n.23
Ramaswamy, C. (uncle), 48, 313n.19,
314n.3
Ramayana (Indian epic), 38, 53, 176
Ranade, G. H., 51–52
A Random Walk in Science (Weber), 24
Rao, Appa, 70, 148, 169, 315nn.12, 21
Rao, K. Ananda, 257
Rayleigh, Lord. *See* Strutt, John
William
Reflections on his life, 298–307; dis-
content with the life of science,
305–7
Reputation, 4–6, 11, 12, 14–15, 28–29,
33, 169, 189, 239
Research and study, periods of, 23–24
Residences: Chandra Vilas, 54–55,
58–59, 169, 314n.10; Cambridge, 80,
113; Hyde Park, 16; Williams Bay,
Wisconsin, 181–82
Review of Modern Physics, 266–67
Ritchey, George Willis, 187
Rosby, Gustaf, 199–200
Rosenfeld, Leon, 100–103, 116, 122,
128–31, 132, 147, 150–51, 199
Rosseland, Svein, 134, 254
Rossner, Laurence F., 324n.1 to chap. 11
"Rothschild's Fiddle" (Chekov), 307
Royal Astronomical Society of London,
30, 85, 97, 102, 108, 124–26, 318n.7
to chap. 5; elected Fellow, 115–16
Royal Institute of Science (Bombay), 72
Royal Society of England, 11, 31, 33, 96;
election to, 68, 197–99, 247, 282
Rummer, Georg, 93
Russell, Henry Norris, 108, 116,
133–34, 137–38, 144, 153, 161, 192,
200–201, 296–97
Rutherford, Lord Ernest, 8, 31, 86, 113,
247, 249, 255, 258, 295

Sachs, Ray, 290–91
Sachs, Robert G., 5, 7, 9, 193–95,
309n.3, 322n.15
Sacks, Jean, 206, 228
Sagan, Carl, 6, 19, 309n.4

Saha, Meghnad N., 4, 9, 64–65, 166, 198, 245, 246, 247, 248–49, 250, 258, 291, 312n.14, 315n.20, 325n.10
Sahade, Jorge, 189
Sakharov, Andrey, 273, 279
Saldhana, Father, 74–75
Salpeter, Edwin E., 220
Sarabhai, Vikram, 286
Sarada (Mrs. R. Dorai Rajan; sister), 47, 72, 91, 111, 149–50, 169
Sarma, V. L. N., 4, 5, 309n.2
Sastri, V. S. Srinivasa, 79, 316n.5
Sastriar, Parvati, 40, 312n.11
Savitri (Mrs. R. R. Sarma; sister), 47, 91, 169, 231
Schilt, Jan, 237
Schmidt, Maarten, 224–25
Schnirrelmann, L., 93
Scholarly productivity: lectures and colloquia presented, 6–7, 7–8, 10, 12, 13, 17, 63, 72, 85, 94, 102, 103, 108, 115, 117, 124, 148, 160, 161, 164, 169, 182, 226, 229, 242, 263, 272, 275, 285, 286–87, 300, 301, 302, 303, 321n.11; as teacher and lecturer, 6, 14, 16–20, 188–89, 202, 222–23, 229, 237–38; writings, 20, 23–24, 62–64, 75–76, 80, 84–86, 94, 96–97, 101–2, 103–5, 121, 143, 147, 189, 197, 301, 321n.11; writings, style of, 25–26. See also Astrophysical Journal
Schönberg, Mario, 189
Schrödinger, Erwin, 62
Schwarzschild, Karl, 140
Schwarzschild, Martin, 188–89, 194, 195–96, 212–13, 214, 217, 223, 227, 234, 303
"A Scientific Autobiography" (Chandrasekhar), 10–11
Secchi, Angelo, 323n.6
Segrè, Emilio, 270
Shane, C. D., 323n.13
Shapley, Harlow, 116, 148, 153, 155–56, 157, 159–60, 161, 162, 214
Shastri, Lal Bahadur, 287
Shoenberg, David, 114, 180, 199, 231
Sidereal Messenger. See Astrophysical Journal, early history of
Silbertstein, Ludwig, 31
Simonsen, J. L., 315n.18
Sircar, Amrit Lal, 313n.23

Sircar, Mahendra Lal, 313n.23
Sitalaxmi (Mrs. A. S. Sivaramakrishnan; aunt), 111
Sitter, Willem de, 30
Sivan, Ramaswami, 312n.14
Smekal, Adolf, 249, 324n.2 to epilogue
Smith, Earlam, 67–68, 257
Smyth, Henry Dewolf, 192
Sobolev, Viktor V., 274, 276–77, 281
Sommerfeld, Arnold, 61–62, 66, 76, 79, 249
Space, Time and Gravitation (Eddington), 30–31
Spence, Edward F., 185–86, 321n.2
Spitzer, Lyman, 25, 210, 212–13, 214, 223, 292, 311n.20, 323n.13
"On the Statistics of Similar Particles" (Chandrasekhar), 101–2
Steenholt, Gunnar, 123
"Stellar Absorption Coefficients" (Chandrasekhar), 94
"Stellar Coefficients of Absorption, Part II" (Chandrasekhar), 96
Stevenson, Adlai Ewing, 232, 283–84, 285
Stokes, Sir George Gabriel, 305
Stone, Marshall, 200
Strömgren, Bengt, 100, 162–63, 182, 184, 188, 201–3, 204, 215–16, 297, 323n.12
Strömgren, Sigrid, 202
Strutt, John William (Lord Rayleigh), 28–29, 259
Struve, Otto, 145, 162–63, 164–65, 181, 182, 187–89, 199–200, 204, 209, 210, 235–36, 320n.8, 321n.1, 323n.7
Sundari (Mrs. C. Ramaswamy; sister), 47, 91, 169, 231
Sutherland, Gordon, 156
Swerdlow, Noel, 239–41
Sykes, John, 24

Tagore, Rabindranath, 9, 245
Taube, Henry A., 13
Telegdi, Valentine, 19–20, 241
Teller, Edward, 93, 196, 241, 267
Teukolsky, Saul, 24
Thacker, M. S., 256, 285–86
On the Theory of Relativity (Eddington), 30–31
"On the Theory of Stars" (Landau), 121

Thomas, L. H., 194
Thomson, George Paget, 68
Thomson, Joseph John, 29, 86, 109, 110, 113
Thomson, William, 305
Thorne, Kip, 16, 23, 310 n.5
Time Travel and Papa Joe's Pipe (Lightman), 122
Towards Freedom (Nehru), 114
Trehan, Surindar K., 290
Troubles—doubts and uncertainties: about leaving for England, 68–71, 73, 90–92; at Cambridge, 79, 95–96, 97–98, 105–7, 126–38; about marriage, 150–52, 163, 168, 169–70; about leaving for the United States, 166
Troubles—public humiliations, 30, 32, 126, 202–3. *See also* Milne-Eddington controversy
Truman, Harry S., 284

Unsöld, Albrecht, 61
Urey, Harold, 199, 239, 241

Vaidyanathaswamy, 78, 106, 317 n.2
Vishwanathan, S. (brother), 47, 50, 91, 112, 165
Van Biesbroeck, George, 182, 187
Van der Waerden, Bartel L., 26–27
van Maanen, Adriaan, 153
Vandervoort, Peter, 19, 202–3, 238, 310 n.10
Victoria (queen of Great Britain), 312 n.9
Vidya (Mrs. V. S. Shankar; sister), 47, 55, 58–59, 91, 169, 231, 235, 314 n.7
Vishveshwara, C. V., 290
Vogel, H. C., 323 n.6
Volkoff, George M., 144, 266
Vorontsov-Veliaminov, B. A., 118

Waddell, John, III, 221–22

Waerden, Bartel L. Van der, 26–27
Wares, Gordon W., 189
Watson, George N., 93
Weart, Spencer, 322 n.5
Weil, André, 264
Weisskopf, Victor, 11, 15, 25, 100, 103, 105–6, 196, 309 n.10
Wentzel, Gregor, 61, 203, 241
Weyl, Hermann, 27, 192
Wheeler, Beverely, 20–21, 310 n.14 to chap. 1
Whipple, Fred, 160, 181, 323 n.13
Whittaker, Edmund T., 93
Wigner, Eugene, 196, 266, 267, 268, 269, 279–80
Williams, E. J., 100
Williams, J. H., 93
Williamson, Ralph E., 189
Wilson, Charles Thomson Rees, 149
Wilson, John T., 17, 239–40
Wilson, Robert, 93
Wood, Annie. *See* Besant, Annie
Wood, Edward Frederick Lindley (Lord Halifax), 191–92
Works influenced by, 26–27, 57, 62, 64, 76, 77, 93
Wrubel, Marshall, 323 n.19

X-Rays and Electrons (Compton), 76, 250
Xanthopoulos, Basilis C., 16, 241, 310 n.4

Yang, Chen-Ning, 17, 310 n.7
Yates, J. A., 314 n.5
Yerkes, Charles T., 186–87
Yerkes Observatory, history of, 185–88, 321 n.5
Yulyevna, Lydia, 318 n.9

Zeitschrift für Astrophysik, 122
Zel'dovich, Ya. B., 274, 275, 278
Zweifel, F., 275